Methodological Developments in Biochemistry

4. Subcellular Studies

Edited by Eric Reid

Wolfson Bioanalytical Centre
University of Surrey

Longman

Longman
1724 - 1974

LONGMAN GROUP LIMITED

London

Associated companies, branches and representatives
throughout the world

First published 1974

ISBN 0 582 46026 3
Library of Congress Catalog Card No. 73–85209
Printed in England

by Lowe and Brydone (Printers) Limited

Editor's Preface

A recurring theme throughout this book is the assignment of intracellular bioconstituents to particular subcellular elements, and the choice of the requisite 'marker constituents' with due regard to present-day views on cell dynamics. This theme was vigorously debated by the array of subcellular experts who attended the Subcellular Methodology Symposium (Guildford, 12-14 July 1973) that has led to this book. Other main themes, evident from the Contents list, include the isolation of nucleoprotein particles and of membranous elements and membrane-associated proteins. The opening Articles deal with novel immunological and affinity approaches; but the key to most of the studies is centrifugal separation, the methodology of which is described and discussed together with interpretations and possible troubles.

The present book is, then, a wide-ranging assembly of working procedures, coupled with factual and conceptual observations. It should appeal to research workers, lecturers, and final-year students (particularly in connection with projects) in the subcellular field. Like its predecessors, which it complements, this Volume is sufficiently inexpensive to find a home on the desks of individual workers who, like the Editor, are despondent about the escalating cost of advanced books in the biochemical field. There has been magnificent cooperation from a range of distinguished contributors, who evidently felt the enterprise to justify their spending precious time in putting pen to paper. This was reflected in compliments received from participants in the Symposium and the associated Workshop Courses, the 1973 'festival of centrifugation' being the most ambitious since such ventures were started by the Wolfson Bioanalytical Centre in 1969. The dedication and enthusiasm of the Centre's research staff also attracted favourable comment. Perhaps the unusual blend of subcellular biochemistry and contract research in the Centre confers hybrid vigour, which the Editor hopes has benefitted this book.

Where Figures are based on material that has appeared elsewhere, this is acknowledged in the text. Particular thanks in this connection are expressed to Academic Press, New York (Arts. 16 & 34), American Society of Biological Chemists (Art. 23), American Society of Plant Physiologists (Art. 34), Biochemical Journal (Arts. 23 & 33), Elsevier (Arts. 27 & 34), Federation of European Biochemical Societies (Arts. 11 & 12), Rockefeller University Press (Art. 23), and Springer-Verlag (Arts. 1 & 8).

Wolfson Bioanalytical Centre
University of Surrey
Guildford GU2 5XH, U.K.

9 May 1974.

CONTENTS

NOTE: The articles are listed below in groups sufficiently similar to those in the Contents lists of Vols. 1 and 3 to facilitate comparison.
An asterisk (*) against an article draws attention to an especially relevant item in the final NOTES AND COMMENTS section, Art. 39 (where the grouping is similar).

Membrane-associated proteins

Asterisks (*) on preceding pages draw attention to item(s) in Art. 39

LIST OF AUTHORS

Full addresses are given in the headings to the cited Articles
(look up CONTENTS to find page no.). 'Art. 39' consists of 'Notes'.

R. Bachofen - Art. 3
Univ. of Zurich, Switzerland

E.J. Barrett - Art. 25
Univ. Coll., Cork, Ireland

Y. Ben-Shaul - Art. 35
Tel-Aviv Univ., Israel

A. Bergstrand - Art. 22
Karolinska Inst., Stockholm, Sweden

G.D. Birnie - Arts. 11 & 12
Beatson Cancer Res. Inst.,
Glasgow, U.K.

N. Black - Art. 9
Univ. of Surrey, Guildford, U.K.

M. Boothby - Art. 24
Washington Univ., St. Louis, U.S.A.

J.-P. Buchet - Art. 18
Univ. Cath. de Louvain, Leuven,
Belgium

T.G. Cartledge - Art. 32
Univ. of Bath, U.K.

G.B. Cline - Arts. 4 & 28
Univ. of Alabama, Birmingham, U.S.A.

E. Cohen - Art. 35
Tel-Aviv Univ., Israel

Rosemary A. Cooper - Art. 39
Llandaff Coll. of Technology,
Cardiff, U.K.

R.A. Cox - Art. 13
Natl. Inst. for Med. Res.,
Mill Hill, London, U.K.

N. Crawford - Art. 30
Univ. of Birmingham, U.K.

M.J. Crumpton - Art. 39
Natl. Inst. for Med. Res., London, U.K.

Martha M. Dagg - Art. 4
Univ. of Alabama, Birmingham, U.S.A.

M. Dobrota - Arts. 19 & 39
Univ. of Surrey, Guildford, U.K.

P.F. Duggan - Art. 25
Univ. Coll., Cork, Ireland

J.L.E. Ericsson - Art. 22
Karolinska Inst., Stockholm, Sweden

L. Funding - Arts. 6 & 7
Univ. of Aarhus, Denmark

N.B. Furlong - Art. 2
Univ. of Hohenheim, Stuttgart, Germany

K. Gammon - Art. 24
Washington Univ., St. Louis, U.S.A.

R.A. Gams - Art. 28
Univ. of Alabama, Birmingham, U.S.A.

C.C. Gilhuus-Moe - Art. 10
Nyegaard & Co., Oslo, Norway

H. Glaumann - Art. 22
Karolinska Inst., Stockholm, Sweden

J.D. Glickson - Art. 28
Univ. of Alabama, Birmingham, U.S.A.

J. Glomset - Art. 39
Washington Univ., Seattle, U.S.A.

Elizabeth Godwin - Art. 13
Natl. Inst. for Med. Res.,
Mill Hill, London, U.K.

T. Hallinan - Art. 39 *(two Notes)*
Royal Free Hosp. School of Med.,
London, U.K.

B. Halliwell - Art. 34
Univ. of Oxford, U.K.

J.R. Harris - Art. 38
Univ. of St. Andrews, U.K.

G.C. Hartman - Art. 9
Univ. of Surrey, Guildford, U.K.

M.J. Hayman - Art. 39
Natl. Inst. for Med. Res.,
Mill Hill, London, U.K.

D.R. Headon - Art. 25
Univ. Coll., Cork, Ireland

H.F. Helander - Art. 29
Univ. of Alabama/Veterans Admin.
Hosp., Birmingham, Alabama, U.S.A.

Anna Hell - Arts. 11 & 12
Beatson Cancer Res. Inst.,
Glasgow, U.K.

R. Henning - Art. 20
Universitat Köln, Germany

H.J. Hilderson - Art. 15
RUCA - Univ. of Antwerp, Belgium

R.H. Hinton - Arts. 9 & 10
Univ. of Surrey, Guildford, U.K.

Piroska Huvos - Art. 13
Natl. Inst. for Med. Res.,
Mill Hill, London, U.K.

P.J. Jacques - Arts. 16 & 39
Univ. Cath. de Louvain, Leuven, Belgium

H. Jaus - Art. 2
Univ. of Hohenheim, Stuttgart, Germany

K. Kaminski - Art. 39
Univ. de Liège, Belgium

H. Keating - Art. 25
Univ. Coll., Cork, Ireland

T.W. Keenan - Art. 21
Purdue Univ., Indiana, U.S.A.

S.A. Kempson - Art. 27
Queen Elizabeth Coll., London, U.K.

A.M. Kidwai - Art. 26
Univ. of Alberta, Edmonton, Canada

Marlene Koplitz - Art. 39
Washington Univ., Seattle, U.S.A.

C.R. Krishna Murti - Art. 39
Central Drug Res. Inst., Lucknow India

R. Lauwerys - Art. 18
Univ. Cath. de Louvain, Brussels, Belgium

D. Lloyd - Arts. 32 & 33
Univ. Coll., Cardiff, U.K.

D. Lowe - Art. 39
Royal Free Hosp. School of Med.,
London, U.K.

Elizabeth MacPhail - Art. 11
Beatson Cancer Res. Inst.,
Glasgow, U.K.

A.H. Maddy - Art. 37
Univ. of Edinburgh, U.K.

J.A.T.P. Meuwissen - Arts. 7 & 39
Rega Inst., Leuven, Belgium

D.J. Morré - Art. 21
Purdue Univ., Indiana, U.S.A.

Barbara M. Mullock - Art. 10
Univ. of Surrey, Guildford, U.K.

J.W. Owens - Art. 24
Washington Univ., St. Louis, U.S.A.

Helena Persson - Art. 22
Karolinska Inst., Stockholm, Sweden

A.R. Poole -Arts. 1 & 39
Strangeways Lab., Cambridge, U.K.

R.K. Poole - Art. 33
Univ. Coll., Cardiff, U.K.

A.H. Price - Art. 29
Univ. of Alabama/Veterans Admin. Hosp.,
Birmingham, Alabama, U.S.A.

R.G. Price - Art. 27
Queen Elizabeth Coll., London, U.K.

T.D. Prospero - Art. 39
Univ. of Surrey, Guildford, U.K.

D. Rickwood - Art. 12
Beatson Cancer Res. Inst., Glasgow, U.K.

M. Riddle - Art. 39
Washington Univ., Seattle, U.S.A.

H. Roels - Art. 18
Univ. Cath. de Louvain, Leuven, Belgium

W. Romen - Art. 2
Univ. of Hohenheim, Stuttgart, Germany

Betty Rorive - Art. 39
Univ. de Liège, Belgium

R.B. Ryel - Arts. 4 & 28
Univ. of Alabama, Birmingham, U.S.A.

G. Sachs - Art. 29
Univ. of Alabama, Birmingham, U.S.A.

B. Schlatterer - Art. 2
Univ. of Hohenheim, Stuttgart, Germany

Y. Shain - Art. 35
Tel-Aviv Univ., Israel

P. Sheeler - Art. 5
Univ. of California, Northridge, U.S.A.

G. Siebert - Art. 2
Univ. of Hohenheim, Stuttgart, Germany

R. Sinclair - Art. 9
Univ. of Surrey, Guildford, U.K.

E.A. Smuckler - Art. 39
Washington Univ., Seattle, U.S.A.

J.G. Spenney - Art. 29
Univ. of Alabama/Veterans Admin. Hosp., Birmingham, Alabama, U.S.A.

P. Stahl - Art. 24
Washington Univ. School of Med., St. Louis, U.S.A.

J. Steensgaard - Arts. 6 & 7
Univ. of Aarhus, Denmark

J.L. Stirling - Art. 27
Queen Elizabeth Coll., London, U.K.

A. Strych - Art. 29
Univ. of Alabama/Veterans Admin. Hosp., Birmingham, Alabama, U.S.A.

D.G. Taylor - Art. 30
Univ. of Birmingham, U.K.

Denyse Thinès-Sempoux - Art. 17
Univ. Cath. de Louvain, Leuven, Belgium

O. Touster - Art. 23
Vanderbilt Univ., Nashville, Tenn., U.S.A.

P. Tulkens - Art. 31
Univ. Cath. de Louvain, Leuven, Belgium

E.L. Vigil - Art. 21
Purdue Univ., Indiana, U.S.A.

R. Wattiaux - Art. 8
Facultés Univ. Notre-Dame de la Paix, Namur, Belgium

R. Williamson - Art. 14
Beatson Cancer Res. Inst., Glasgow, U.K.

W.N. Yunghans - Art. 21
Purdue Univ., Indiana, U.S.A.

H. Zola - Art. 36
Wellcome Res. Labs., Beckenham, Kent, U.K.

1   IMMUNOLOGICAL METHODS FOR THE STUDY OF THE CELLULAR LOCALIZATIONS OF PROTEINS AND THEIR SECRETION†

A.R. Poole
Tissue Physiology Department
Strangeways Research Laboratory, Wort's Causeway
Cambridge CB1 4RN, U.K.

Immunological methods are described to localize proteins both inside and out-side cells. Intracellular localizations involve the use of fixed tissues and of Fab' antibody sub-units*, labelled with peroxidase or fluorescein isothio-cyanate. Extracellular proteins can be captured in living tissues by preci-pitation with specific antibody to form an immunoprecipitate which can sub-sequently be detected in fixed tissues with labelled Fab' by a published method. The preparation, characterization and use of these immunochemical re-agents are described, and examples are given of their application.

The advantages of an immunological approach rest mainly on the characteristic ability of antibodies, or their Fab' sub-units, to react specifically with the protein (antigen) against which they were raised. Hence, by allowing labelled antibodies to bind specifically to antigens in a tissue, it is pos-sible to locate these antigens inside and outside cells. In this article the techniques employed in this work will be outlined, and discussed in some de-tail where this is deemed necessary.

GENERAL PRINCIPLES OF INTRACELLULAR AND EXTRACELLULAR LOCALIZATION TECHNIQUES

The stages involved in the localization of proteins in a tissue are outlined in Fig. 1a and b. Intracellular antigens (Fig. 1a) are localized within tis-sues which have already been fixed in such a way as to maintain antigenicity of the protein and retain permeability of cellular membranes to Fab'. IgG (mol. wt. $\sim$ 160,000) cannot be successfully used for intracellular localiza-tion studies with adequately fixed cell membranes since it is too large a molecule to readily penetrate a fixed membrane [1]. Moreover, it also binds non-specifically to tissues via the Fc piece. This can mask any specific binding which may occur.

    The localization of extracellular proteins (Fig. 1b) relies on the initial extracellular capture of the secreted protein under investigation

---

† *For homogenate and other applications, see NOTE & COMMENTS at end of book.*
*   Fab' *contains one of the antigen binding sites of IgG and has a mol. wt. of* $\sim$ *45,000.*

(a)

Fig. 1. Diagrammatic
representation of
localizations of
tissue antigen.
    a. Intracellular.
    b. Extracellular.

Tissue fixed. Cell membranes outlined in black.

Tissue antigen(●)fixed in cell

Attachment of sheep Fab'(l)directed against
antigen. IgG(λ)cannot easily penetrate
fixed cell membrane & also binds non-specifically

Attachment of pig Fab'(l)directed against sheep Fab'(l),
which has been labelled(○)with fluorescein isothiocyanate or peroxidase

(b)

Steps 1 & 2 in culture, step 3 in fixed tissue

Tissue antigen (●) secreted from cell in culture
in presence of sheep antibody IgG(λ)directed
against antigen

Capture of antigen by sheep IgG
with formation of immunoprecipitates

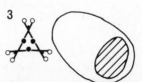

Attachment of pig Fab'(γ)directed against sheep IgG
(or Fab'),which has been labelled (○) with fluorescein
isothiocyanate or peroxidase

by specific antibody in a living tissue. This step, which corresponds to step 1 in intracellular studies, is usually performed in culture employing either antiserum (incubated at 56° for 30 min to inactivate complement and hence prevent cell lysis) or IgG. Tissues are usually cultured for a period of time sufficient to capture enough antigen to form detectable immunoprecipitates which can subsequently be demonstrated in fixed tissues. This period can range from 1 h to 48 h.

In step 2 in both studies the bound IgG or Fab' is localized by reacting the tissue with Fab', directed against the first step antibody or subunit, which is labelled with a fluorochrome (for light microscopy) or with peroxidase (for light and electron microscopy). The fluorochrome can be visualized by fluorescence microscopy. Peroxidase may be reacted, as described below, to produce a final brown/black reaction product which is electron-dense.

In all of this work, control sera are employed from either non-immunized animals of the same species or animals containing antisera which are raised against another antigen which is immunologically quite distinct from that being studied.

The stepwise preparation and characterization of reagents for immunocytochemical studies will now be discussed, followed by a description of their use in localizing proteins.

MATERIALS AND LOCALIZATION TECHNIQUES

Preparation, characterization and use of immunocytochemical reagents

Purification of protein under investigation

The methods for the purification of the protein (immunogen) are varied and reference should be made to the literature. It is essential that the immunogen be pure, otherwise a non-specific antiserum will be obtained which has limited use.

Immunization and raising a specific antiserum

Animals (e.g. rabbits, guinea pigs, sheep or pigs) may be injected intramuscularly (hind and forelegs) on days 0 and 15 with an aqueous emulsion of a 1:1 mixture of the immunogen and Freund's complete adjuvant (Difco or Wellcome). The resultant antiserum should contain antibodies by day 25. Antisera of relatively high titre can be obtained after the injection of 2-5 mg of protein, and titres can be raised by further injections if necessary. If the antiserum is non-specific, a specific antiserum may be raised in another animal of the same species by re-injecting precipitin lines composed of the antigen and antibody [2, 3].

Characterization of antiserum

The precipitating capacity of an antiserum and its specificity should be
determined qualitatively by gel diffusion analyses (double immunodiffusion
and/or immunoelectrophoresis) [4, 5] against purified antigen, against highly
concentrated fractions isolated during antigen purification, and against
highly concentrated 0.1% Triton X-100 treated tissue homogenates (including
that of the tissue from which the antigen was purified) or against serum if
the antigen is present in serum. A specific antiserum should always show
only one precipitin line, when viewed with dark-ground illumination and after
protein staining.

Immunodiffusion and immunoelectrophoresis plates can be prepared using
Kodak slide cover glasses (82 x 82 cm). Plates are pre-coated with a layer
of agarose, to ensure adherence of final agarose gels during washing and
staining procedures, by dipping plates cleaned in hot detergent into a
coating solution of 2 g agarose and 0.1 g thiomersal in 1 litre of distilled
water. They are then allowed to drain and dry in a suitable rack.

Gels are prepared as follows. To 100 ml buffer add 1.0 g agarose and
heat with continuous stirring until clear. Add 0.5 ml of 10% sodium azide
when the solution has cooled to 55°, and mix. Pour 11 ml of gel at this
temperature onto a pre-coated plate on a levelling table. After the gel has
set it can be stored at 4° in a humidified environment until used. For
double immunodiffusion work a 20 mM phosphate buffer at pH 7.2 is used with
0.15 M sodium chloride (PBS)*. A 45 mM barbitone buffer pH 8.6 can be used
with 4 mM calcium lactate for immunoelectrophoresis.

Gels left for 24 h and 48 h can be examined first with dark-ground
illumination, but should always be stained for protein to ensure detection
of all precipitin lines. After washing with PBS* for 48 h, distilled water
for 24 h, and then ethanol for 4 h, plates can be dried on the bench overnight with a Whatman 54 filter paper overlay. They are then stained with
Coomassie brilliant blue (0.1 mg/ml, in a mixture of 35 ml 98% formic acid,
20 g sodium formate, 630 ml water and 330 ml ethanol) for 24 h, and washed
in this medium to remove excess stain before rinsing in distilled water and
drying at room temperature.

The titre of antisera can conveniently be quantitatively assessed
using radial immunodiffusion analyses with phosphate gels, prepared as described above and used according to the principles described by Mancini et al.
[6] and Vaerman et al. [7]. The antigen is introduced into the gel and antiserum is placed in 5 μl wells in increasing dilution. As antibodies diffuse
out into the gel, discs of precipitation are formed. After 48 h their sizes
are recorded and plotted as $diam^2$ against antiserum concentration. Alternatively, 'rocket' immunoelectrophoresis [8] can be used to assess the titre of
antisera, whereby the antibodies are electrophoresed into a gel containing

* The phosphate-buffered saline is abbreviated PBS

the antigen to form a rocket-shaped precipitate of height proportional to the antibody content.

Isolation of IgG

This is usually done by initially precipitating the serum three times with 33% ammonium sulphate. The final precipitate is then dialyzed, and IgG is purified by ion-exchange chromatography using a method such as that described by Aalund *et al.* [9]. The purity of the IgG should be checked using immuno-diffusion and immunoelectrophoresis with antisera to IgG and to whole serum proteins.

Preparation of Fab'

This can be prepared from purified immunoglobulin or ammonium sulphate preci-pitates by enzymic digestion with pepsin followed by reduction to univalent Fab' [10]. Hence IgG at 25 mg/ml $(A_{280}^{1 mg/ml} = 1.4)$ is digested in 0.2 M acetate buffer at pH 4.5 with pepsin (2% of IgG w/w, Worthington) for 24 h at 40°. The incubation mixture contains 0.1% azide. After incubation the pH is adjusted to 8.6 with NaOH to inactivate pepsin. The solution is centrifuged and the supernatant is dialyzed against PBS. The $(Fab')_2$ content can be arbi-tarily determined by using an extinction coefficient (A) for 1 mg/ml = 1.5, or with radial immunodiffusion analyses employing an antiserum to $(Fab')_2$ of the species used.

$(Fab')_2$ can be purified from the pepsin digest of purified IgG by chromatography on Sephadex G-200. This enables the separation of undigested IgG (usually a small 1st peak), $(Fab')_2$ (2nd peak) and Fc digestion products (3rd peak with a long 'tail'). $(Fab')_2$ is reduced to Fab' with 10 mM cys-teine for 30 min. Fab' is characterized by the fact that it will react with antigen but cannot precipitate it because it is monovalent. Hence one can block immunoprecipitation of antigen with native IgG antibody. This test is performed in phosphate-buffered agarose gel soaked before use for 2 h in 1 mM dithiothreitol in PBS.

Labelling of (Fab')₂ with fluorescein

$(Fab')_2$ preparations are labelled using the method of The and Feltkamp [11]. To 1 mg protein is added 10 µl of fluorescein isothiocyanate isomer 1 (1 mg/ml) (BDH) in 0.15 M Na_2HPO_4. The pH is adjusted to 9.5 with 0.1 M Na_3PO_4 and the mixture left for 1 h at room temperature. This labelled $(Fab')_2$ is then imme-diately separated from unreacted fluorochrome by chromatography on Sephadex G-25 using PBS. Azide and thiol compounds inhibit this labelling reaction and should be removed by dialysis beforehand.

The protein content of the final product and its fluorescein content can be recorded spectrophotometrically at 280 and 493 nm respectively with reference to the nomograph by Goldman [12], since the extinction characteris-tics of IgG and $(Fab')_2$ are very similar [13]. Molar ratios may be determined assuming a molecular weight of $\sim$ 90,000 for (Fab')$_2$ and of 389 for fluorescein.

Labelling of (Fab')$_2$ with peroxidase

Fab' preparations should be reacted with peroxidase using the technique described by Avrameas & Ternynck [14] whereby peroxidase activated with glutaraldehyde is bound to (Fab')$_2$. The product contains conjugates of one molecule of (Fab')$_2$ bound to two molecules of peroxidase. These are sufficiently small after reduction (mol wt. $\sim$ 90,000) to penetrate membranes fixed with formaldehyde for 1 h.

In addition to gel chromatography [14], the binding of peroxidase to Fab' can be verified by permitting (Fab')$_2$ labelled with peroxidase to interact with antigen in gel immunodiffusion. Peroxidase should then be demonstrable in the precipitin line when the gel is reacted for peroxidase according to Graham and Karnovsky [15]. The peroxidase activity of the conjugate should be measured, suitably by the method described by Fahimi & Herzog [16]. The Fab' protein content of the conjugate can be measured by radial immunodiffusion.

Immunocytochemical techniques for tissues

Preparation of tissue for light microscopy

Tissues should be immersed in 7% gelatin in 0.9% NaCl at 30° in a plastic tube and then frozen immediately in liquid nitrogen for 90 sec by total immersion. Tubes may then be capped and stored at -20° before use. The plastic may be removed mechanically exposing the tissue embedded in frozen gelatin. The whole is then mounted on a microtome chuck with distilled water and frozen-sectioned in a cryostat at 4-8 μm. Frozen sections are applied to clean microscope slides and allowed to air-dry for about 30 sec. They are then fixed immediately in 4% formaldehyde in PBS freshly prepared from paraformaldehyde [15]. This is done by heating a suspension of 4 g paraformaldehyde in 50 ml PBS to 75° and then adding a few drops of 2 N NaOH from a Pasteur pipette until the solution clears. It is then cooled to room temperature and the pH is readjusted to pH 7.6 with HCl (from about pH 8.6). The final volume is made up to 100 ml with PBS. The fixative should be used immediately.

Sections or cell monolayers are fixed from 5 min up to 30 min at room temperature and washed for 30-60 min in PBS containing 2 mM cysteine.

Preparation of tissue for electron-microscopic studies

Tissues should be initially fixed in small pieces about 2 mm thick either in 4% formaldehyde in 0.1 M cacodylate buffer pH 7.4 or in a mixture of formaldehyde and glutaraldehyde [15] for a period of 1 h at 4°. After rinsing in 0.1 M cacodylate buffer for 2 h at 4° it can be sectioned either unfrozen at

10-40 μm [17] or after freezing, as previously described [18]. Sections are usually collected in 0.1 M cacodylate buffer containing 2 mM cysteine.

Treatment of tissues with Fab' reagents

Sections soaked in PBS or 0.1 M cacodylate buffer containing 2 mM cysteine are reacted with Fab' reagents which have been first converted to and maintained in a monomeric state by exposing them to 10 mM cysteine for 30 min, before use in 2 mM cysteine. This solution is also used for washing sections after treatment with Fab' reagents.

When localizing an intracellular antigen, sections of tissues are initially treated with unlabelled immune Fab' directed against the antigen (or non-immune Fab' as a control) for a period of 30-60 min for sections up to 10 μm or for 4 h for 40 μm sections. They are then washed for a similar period and finally reacted with Fab' (labelled with fluorescein isothiocyanate or peroxidase) which is directed against the Fab' employed in the first step.

In the case of extracellular localizations (Fig. 1b) step 1, having been performed in culture, is omitted.

Tissues are finally washed and may then be examined for fluorescence or the peroxidase may be localized as described below.

Localization of fluorochrome-labelled Fab' in a tissue

The stained and washed specimen can sometimes be counterstained with the dye eriochrome black (Difco; 1/10 - 1/500 dilution with PBS) which stains cell cytoplasms so that they fluoresce red. This can help to clarify green extracellular fluorescence due to fluorescein. After staining, tissues are immediately rinsed for 1 min in distilled water and are mounted in a mixture of 1 vol Tris buffer, 0.2 M, pH 8.6, and 9 vol glycerol. They are then immediately examined, otherwise fluorescence fades considerably within 24 h if the initial fluorescence is weak.

Dark-ground fluorescence microscopy is employed using blue light (420-490 nm) to excite the fluorescein. Light of wavelength above 525 nm is examined for green fluorescein fluorescence. Filter combinations can be obtained from Reichert or Polaron (London). My preference is for a Reichert Zetopan microscope with Polaron filters (Olsen) and fitted with high power and wide-angle low-power dark-ground condensers and a xenon lamp.

Fluorescence can be photographically recorded using, for black and white photography, Tri-X film uprated to about 1200 ASA during development (12 min) with Acuspeed developer (Patterson). Colour photographs are prepared with Anscochrome 500 film (colour transparencies), or with Kodak Ektachrome (high speed, daylight) up-rated to 500 ASA during development.

2a

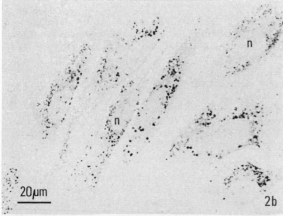

2b

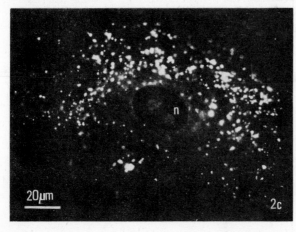

2c

Fig. 2. Intracellular localization of the lysosomal protease cathepsin D in cultured rabbit cells.
n *denotes* nucleus.
(A.R. Poole, unpublished results.)

a & b. Localization by immunoperoxidase staining in rabbit fibroblasts. Cells were initially fixed for 5 min in 4% formaldehyde in PBS. After washing for 30 min with PBS containing 2 mM cysteine they were treated for 30 min with:
a, Fab' isolated from a pool of non-immune sheep sera, or
b, Fab' isolated from a sheep antiserum to rabbit cathepsin D.
They were then washed with PBS and cysteine for 30 min and treated with Fab' prepared from a pig antiserum to sheep Fab' and labelled with peroxidase. After a final wash, the cells were reacted for peroxidase and examined by bright-field microscopy. *Brown/black particulate staining was observed only in lysosome-like structures in the cytoplasm of cells treated with Fab' isolated from the antiserum to rabbit cathepsin D. Control cells (Fig. 2a) were unstained.*
c. Localization by immunofluorescence in a synovial cell. Cells were fixed and treated with Fab' isolated from a sheep antiserum to rabbit cathepsin D as des-

[continued opposite

The demonstration of peroxidase-labelled Fab'

Peroxidase can be identified by virtue of its ability to catalyse the oxidation of diaminobenzidine in the presence of hydrogen peroxide. The product is very osmiophilic and the resultant product after osmication is thus electron-dense and of a brown/black colour. The method of reaction for peroxidase has been described by Graham & Karnovsky [15] and the use of this technique has been described by Kraehenbuhl *et al.* [17]. After osmication sections mounted on microscope slides can be dehydrated and permanently embedded, or, as an alternative,sections in suspension can be dehydrated and embedded in araldite by standard techniques. Sections 1-2 μm thick can be cut and examined with the light microscope, reserving ultrathin sections for electron microscopy.

EXAMPLES OF APPLICATIONS

These methods may be illustrated by reference to work in this laboratory on the intracellular localization of the lysosomal protease cathepsin D [19] (Fig. 2). The immunocytochemical method of localizing the enzyme was the only cytochemical approach possible. The ultrastructural localization of proteins using peroxidase-labelled antibodies is well illustrated by the work of Mendell and co-workers [20].

Studies of the secretion and extracellular localizations of cathepsin D using the techniques developed in this laboratory, as briefly described here, are exemplified by Fig. 3 and other work of the author and his colleagues [21, 22].

Acknowledgements

This work was supported by grants from the Nuffield Foundation, the Arthritis and Rheumatism Council and the Medical Research Council. Dr. Zena Werb kindly supplied the cultured rabbit cells described in Fig. 2.

References

1. Poole, A.R., in *Electron Microscopy and Cytochemistry* [Proc. 2nd Int. Symp., Drienerlo, The Netherlands (1973)] (Wisse, E., The Daems, W., Molenaar, I. & van Duijn, P., eds.), North Holland, Amsterdam (1974), pp. 159-162.

Legend to Fig. 2, *continued from opposite*
cribed above. They were then reacted with Fab', prepared from a pig antiserum to sheep Fab' and labelled with fluorescein isothiocyanate. After a final wash they were examined for fluorescence with dark-ground microscopy. *Green fluorescence was observed in similar sites to those seen in Fig. 2b. Control cells treated in step 1 with non-immune sheep Fab' exhibited no fluorescence.*

Fig. 3.
Legend opposite.

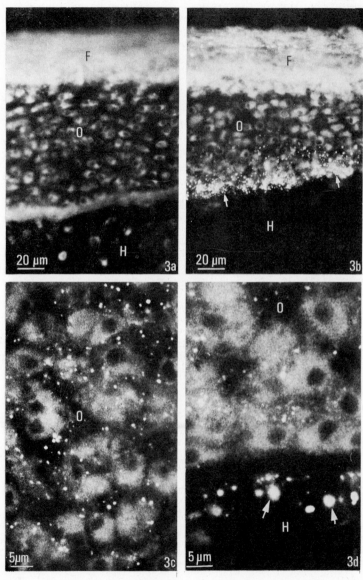

Legend to Fig. 3 *(opposite)*.
Extracellular localization of cathepsin D in the ossifying diaphysis of an embryonic chick limb bone cultured as described below, frozen, sectioned, fixed and stained with rabbit anti (sheep IgG) IgG labelled with fluorescein isothiocyanate, counterstained with eriochrome black (EB) and viewed with dark-ground fluorescence microscopy.

a. Tangential section of mid-diaphysis of a femur cultured for 4 days in 5% calf serum (CS) in medium BGJ_5, then for 2 days in 10% non-immune serum and finally grown for 2 days in 5% CS in BGJ_5. *All cell cytoplasms were strongly stained with EB in both the fibroblastic* (F) *and osteoblastic* (O) *layers of the periosteum; cells of the hypertrophic cartilage* (H) *were weakly stained. Mineralized tissue* (arrows) *between zones* (O) *and* (H) *was sometimes detected by its staining with* EB.

b, c, d. Tangential section (b) and longitudinal sections (c, d) of mid-diaphyses of femora cultured for 4 days in 5% CS in BGJ_5, then for 2 days in 10% sheep antiserum to chicken cathepsin D and finally in 5% calf serum in BGJ_5. *Staining with EB was similar to that observed in Fig. 2a but, although in Fig. 3b some mineralized tissue present between zones* O *and* H *was stained red with EB, it cannot be clearly distinguished in this black and white micrograph. Intense diffuse and particulate green extracellular staining for IgG shown here was only observed in rudiments cultured with antiserum to chicken cathepsin D and was present in the osteoblastic zone (Figs. 3b, c and d) and at the inner margin of this zone with the hypertrophic cartilage (Figs. 3b and d). Figs. 3a and b were taken with Tri-X film and Figs. 3c and d with Anscochrome 500 film.*

From ref. [21], by kind permission of the Editor of *Calcified Tissue Research* and of the publisher (Springer-Verlag).

---

2. Lachman, P.J., in *Standardization in Immunofluorescence* (Holborow, E.J., ed.), Blackwell, Oxford (1970), pp. 235-239.

3. Weston, P.D. & Poole, A.R., in *Lysosomes in Biology and Pathology,* Vol.3 (Dingle, J.T., ed.), North-Holland/Elsevier, Amsterdam (1973) pp. 426-464.

4. Williams, C.A., in *Methods in Immunology and Immunochemistry,* Vol. 3 (Williams, C.A. & Chase, M.W., eds.), Academic Press, New York (1971) pp. 234-294.

5. Ouchterlony, O., in *Handbook of Experimental Immunology* (Weir, D.M., ed.) Blackwell, Oxford (1967) pp. 655-706.

6. Mancini, G., Carbonara, A.O. & Heremans, J.J., *Immunochemistry 2* (1965) 235-254.

7. Vaerman, J.P., Lebacq-Verheyden, A.M., Scolari, L. & Heremans, J.J., *Immunochemistry 6* (1969) 279-285.

8.   Weeke, B., in *A Manual of Quantitative Immunoelectrophoresis. Methods and Applications* (Axelsen, N.H., Kroll, J. & Weeke, B. eds.), Universitetsforloget, Oslo (1973) pp. 37-46.

9.   Aalund, O., Osebold, J.W. & Murphy, F.A., *Arch. Biochem. Biophys. 109* (1965) 142-149.

10. Nisonoff, A., Wissler, F.C., Lipman, L.N. & Woernley, D.C., *Arch. Biochem. Biophys. 89* (1969) 230-244.

11. The, T.H. & Feltkamp, T.E.W., *Immunology 18* (1970) 875-881.

12. Goldman, M., *Fluorescent Antibody Methods,* Academic Press, London (1968).

13. Little, J.R. & Donahue, H., in *Methods in Immunology and Immunochemistry,* Vol. 2, (Williams, C.A. & Chase, M.W., eds.) Academic Press, New York (1968) pp. 343-364.

14. Avrameas, S. & Ternynck, T., *Immunochemistry 8* (1971) 1175-1179.

15. Graham, R.C. & Karnovsky, M.J., *J. Histochem. Cytochem. 14* (1966) 291-302.

16. Fahimi, H.D. & Herzog, V., *J. Histochem. Cytochem. 21* (1973) 499-503.

17. Kraehenbuhl, J.P., de Grandi, P.B. & Campiche, M.A., *J. Cell Biol. 50* (1971) 432-445.

18. Kuhlmann, W.D. & Miller, H.R.P., *J. Ultrastr. Res. 35* (1971) 370-385.

19. Dingle, J.T., Poole, A.R., Lazarus, G.L. & Barrett, A.J., *J. Exp. Med. 137* (1973) 1124-1141.

20. Mendell, J.F., Whitaker, J.N. & Engel, W.K., *J. Immunol. 111* (1973) 847-856.

21. Poole, A.R., Hembry, R.M. & Dingle, J.T., *Calc. Tissue Res. 12* (1973) 313-321.

22. Poole, A.R., Hembry, R.M. & Dingle, J.T., *J. Cell Science 13* (1974) *in press.*

2 'NUCLEAR' COLUMNS IN BIOTECHNOLOGY; ANHYDROUS ISOLATION TECHNIQUES

Günther Siebert, N. Burr Furlong*, Werner Romen[†],
Bert Schlatterer and Heinrich Jaus
Department of Biological Chemistry and Nutrition
University of Hohenheim
D-7 Stuttgart 70, Germany

Progress often depends on the availability of new methods. It is the purpose of this paper to describe two techniques which expand the possibilities of studies with isolated nuclei. The first part is concerned with streamlining the methodology of multiple and routine assays of nuclear enzymatic activities, exemplified with the well-known nuclear marker enzyme NAD pyrophosphorylase; in addition, the biosynthetic capacities of nuclei permit the use of this method for preparative purposes.

The second part concerns one of the two principal ways for the isolation of nuclei, namely the so-called non-aqueous or anhydrous procedure first described by Behrens[1]. A modern version, downscaled for work with cultured cells, has been developed. Some additional evidence is presented for the absence of damaging effects of organic solvents used for such anhydrous procedures.

NUCLEAR COLUMNS

Sauermann [2] first described a technique of fixing isolated nuclei onto finely ground pieces of cellulose nitrate; such 'mounted' nuclei are then used for preparing a nuclear column. The general procedure is outlined in Fig. 1.

Instead of isolating nuclei in high-density sucrose [3] from fresh liver, one may lyophilize liver tissue in the same manner as regularly done for non-aqueous procedures [4]. Such liver powders may then be stored in the deep-freeze for many months, and provide not only a convenient starting material for preparing nuclei but also serve as a normal-liver reference material where, for example, alterations under experimental or pathological conditions are to be studied [5]. A modification of the original Blobel-Potter procedure [3] is, however, necessary, viz.

* *On leave of absence from University of Texas M.D. Anderson Tumor Hospital and Institute, Houston, Tex., U.S.A.*

† *Department of Pathology, University of Würzburg, Germany*

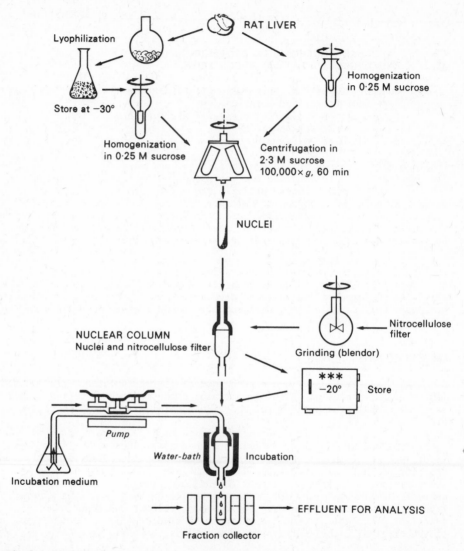

Fig.1. A flow-diagram of the preparation and use of nuclear columns. *For more details see refs. [2, 5].*

the use of 2.1 M instead of 2.3 M sucrose for high-speed centrifugations; the nuclei derived from lyophilized liver do not sediment well in the higher-strength sucrose solution.

In the presence of 0.25 M sucrose, freshly prepared nuclear columns may be stored for a few months in the deep-freeze. For use in experiments, the thawed columns are washed free from sucrose with the desired buffer.

When nuclear columns are perfused with substrate solutions ('incubation media'), a three-way valve easily permits a change of the perfusion media in respect of any parameter which one wishes to check, such as [ ], pH, or inhibitors.

The effluent from nuclear columns is analyzed for product formation, a step in the whole procedure which is capable of being performed by any automated procedure. If, however, biosynthetic capacities of nuclei contained in nuclear columns are utilized for preparative purposes, such as forming labelled or analogue-containing NAD, the fraction collector in Fig. 1 may be replaced by an ion-exchange column absorbing the desired compound.

Examples of experiments with nuclear columns are given in Figs. 2 - 4. In Fig. 2, the rate of perfusion of the nuclear column with substrate solutions has been varied within the range where the enzyme is kept saturated with substrate; NAD production is found to be constant and independent of a 45-fold variation of perfusion velocity. Not shown are experiments which include non-saturating substrate concentrations, where the rate of NAD production declines with decreasing perfusion velocities.

Nuclear columns keep their activity unchanged over 22 h when a small correction for altered hydrodynamic properties is made (Fig. 3).

An example of a different nuclear biochemical activity is given in Fig.4, where the cleavage of NAD is followed in a nuclear column as well as in conventional incubation systems. Hitherto it has been a general experience that the rate of NAD cleavage decreases rapidly in regular enzyme assays [6], thus making it difficult to arrive at zero-time initial velocities. In nuclear columns, however, the splitting of NAD proceeds with constant velocity, except when nicotinamide is introduced as an inhibitor of the following enzymatic reactions - NAD glycohydrolase [7], dinucleotide pyrophosphatase [8] and poly-ADPR synthetase [9]. Both the onset of inhibition, and the release therefrom, by omitting nicotinamide, occur with some delay which may indicate the time necessary for equilibration of solute concentrations within the nuclear column.

In experiments like that shown in Fig.4, all attempts to detect

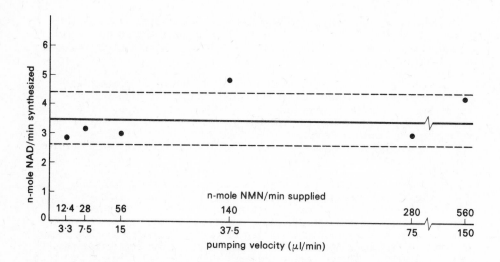

Fig. 2. Activity of the nuclear marker enzyme NAD pyrophosphorylase
assayed in a nuclear column. The flow rate of the substrate solution
is varied as indicated in the abscissa. *Composition of the substrate
solution:* 0.05 M glycyl-glycine buffer pH 7.4, 0.015 MgCl$_2$,
1 × 10^{-2} M ATP, 6 × 10^{-3} M nicotinamide mononucleotide (NMN), 0.2 M
nicotinamide. Aliquots of the effluent fractions are assayed for NAD
formed according to [12]. Each measured point is the average of 3-4
determinations of NAD. The horizontal line represents the mean value
of 3.5 nmoles NAD synthesized per min ± mean error (dashed lines).

poly-ADPR in effluent fractions were unsuccessful [10]. This result is
taken as evidence that - whereas ADPR production may easily be demonstrated [11] - any formation of poly-ADPR from NAD would have to occur
in a simultaneous, compulsory manner with its transfer onto nuclear
acceptor proteins, without poly-ADPR being released into free solution
to any marked extent.

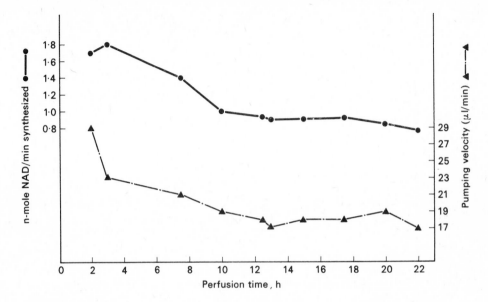

Fig. 3. Long-term use of a nuclear column of 300 µl content for the preparation of NAD. The perfused substrate solution is composed as in Fig. 2 except that non-saturating concentrations of NMN (1.5×10^{-3}M) are used. *The yield of NAD in this experiment is 1.5 mg. In experiments not shown here, yields of NAD up to 70% of the supplied NMN have been obtained.*

Nuclear columns obviously provide an experimental system which, on one hand, facilitates biochemical work with isolated nuclei and permits, on the other hand, the use of structurally bound nuclear enzymes as potential adjuncts to biotechnology, this having gained much in other fields by the development of carrier-bound enzymes.

PROPERTIES OF ANHYDROUS TISSUE POWDERS

In the non-aqueous procedure for the isolation of nuclei [4], livers frozen *in situ* are lyophilized. A sample of this parenchymal powder serves as the control for any effects of organic solvents (petroleum ether for grinding, cyclohexane-carbon tetrachloride mixtures for centrifugation) which may arise from their contact with the liver powder. Whereas the preservation of the fine structure of isolated nuclei has been demonstrated earlier [14], Fig. 5 gives an example of endoplasmic

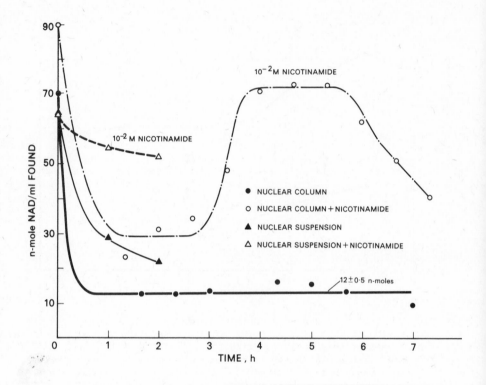

Fig. 4.   Cleavage of NAD by rat liver nuclei incubated as a suspension, in comparison with a nuclear column prepared from the same batch of nuclei.

Nuclear suspension: rat liver nuclei are prepared according to the method of Blobel & Potter [3].   The nuclear pellet - containing approximately 2 mg of protein, determined by the Lowry method [13] is suspended in 5 ml of incubation medium consisting of 0.1 M Tris/HCl buffer pH 8.2, 4×10^{-3} M 2-mercaptoethanol, 6×10^{-3} M KCl, 10^{-2} M $MgCl_2.6H_2O$ and 0.65×10^{-4} M NAD.   The suspension is incubated at 25° and aliquots removed after 60 and 120 min.   These samples are cooled down immediately to 0° and centrifuged at high speed.   The clear supernatant is analysed for NAD with alcohol dehydrogenase according to [12].
Nuclear suspension + nicotinamide: This experiment is run in exactly as described above, with 10^{-2} M nicotinamide in the medium.

Fig.4. *(continued from opposite).*
Nuclear column: this is prepared according to Sauermann [1]. It is incubated at 25° and perfused with the incubation medium described above for nuclear suspension, containing 0.7×10^{-4} M NAD. NAD-splitting is followed by analyzing the clear effluent for NAD content with the method described above. Flow rate = 28 µl/min.
The rate of NAD-splitting is constant at 82% ± 4% cleavage after the first 30 min to the end of the 7 h incubation period.
Nuclear column + nicotinamide: the column is initially perfused during 2 h 50 min with incubation medium containing NAD (0.9×10^{-4} M) as above. Then the column is switched on to a medium containing in addition 10^{-2} M nicotinamide. After a period of 2 h 40 min the column is switched back to the original medium and run for another period of 1 h 20 min. Flow rate = 28 µl/min.

---

reticulum and glycogen particles in rat-liver parenchymal powder after 2 weeks of exposure to the solvents used for centrifugation followed by 3.5 months of exposure to petroleum ether at +2°. It may be safely stated that ultrastructural morphology is preserved well enough as to allow easy identification of almost every structural element of parenchymal liver cells. It seems noteworthy that in the almost complete absence of lipids - due to solvent treatment - osmium tetroxide still 'stains' cellular elements quite definitely.

From the same sample of liver parenchymal powder as shown in Fig. 5, nuclei have been isolated and assayed for the preservation of two soluble glycolytic enzymes, lactate dehydrogenase and aldolase. Table 1 demonstrates the full preservation of the enzymatic activities after several months of solvent exposure.

After very severe treatment of isolated nuclei by sonication, these two enzymes are inactivated to a large extent; yet the nuclear marker enzyme NAD pyrophosphorylase still survives completely (Table 2). Sonication for 35 min in the presence of glass beads is withstood by NAD pyrophosphorylase, probably because it is a structurally bound enzyme whereas the two soluble enzymes (extractable into physiological saline) studied are destroyed proportionately more than the total soluble protein.

Another example of unexpectedly high resistance is provided when the heat stability of enzymes is studied in anhydrous, solvent-exposed liver powder. As shown in Fig. 6, with application of heat for 2 h the observed half-inactivation temperatures lie between 100° and 110°. These data are interesting because rehydration of treated liver powders for assays of enzymatic activities apparently occurs, up to 70°, with no loss of activity at all, thus suggesting that reformation of the 'optimal' conformation of these enzyme molecules proceeds with an astonishing degree of accuracy.

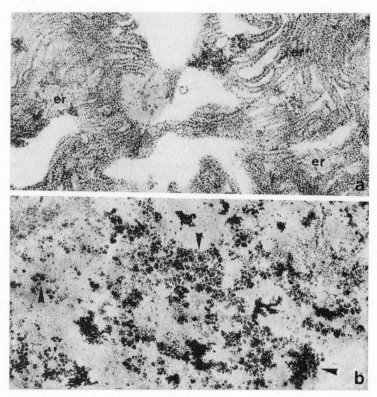

Fig. 5. An electron micrograph of rat liver parenchymal powder [4] exposed to cyclohexane-carbon tetrachloride mixtures for 2 weeks and then to petroleum ether for 3.5 months in the cold-room. Fixation by OsO_4, contrast by lead hydroxide.

5a: *rough endoplasmic reticulum (e.r.).*

5b: *glycogen particles (arrows), mostly arranged in a rosette-like fashion.*

　　　　Taking together the results of Tables 1 & 2 and of Fig. 6, it may be concluded that the absence of water most probably explains this very strong thermal stability of proteins. If solvent-treated parenchymal powder from rat liver is checked for residual water (110° at 16 mm Hg for 2 h; determination of weight loss), about 5% 'water' is found with 'anhydrous' tissue powders. It is not known if there are any other volatile compounds besides water in such powders, but Fig. 6 demonstrates 40-50% retention of enzyme activities under these 'drying' conditions. Once a tissue has a sufficiently low residual content of

Table 1. Preservation of enzyme activities in rat-liver cell fractions after solvent exposure.

Enzyme activities are given as mU/mg protein. For lactate dehydrogenase and aldolase, 10 mg portions of dry cell fractions are extracted with 1 ml 0.14 M NaCl, centrifuged at high speed, and the clear supernatant is suitably diluted for spectrophotometric assays; lactate dehydrogenase according to [15], aldolase according to [16]. Protein is determined by the method of Lowry [13]. Assays for NAD pyrophosphorylase are performed as indicated in Fig. 2.

	Lactate dehydrogenase	Aldolase	NAD pyro-phosphorylase
Parenchymal powder, 2 weeks' exposure	2700	23	0.2
Nuclei, 2 weeks' exposure	3100	24	1.1
Nuclei, 15 weeks' exposure	3100	23	1.1

Table 2. Preservation of NAD pyrophosphorylase in anhydrous nuclei after sonication under extreme conditions in petroleum ether.

Dry nuclei (100 mg) from rat liver are suspended in 16 ml petroleum ether, 3.2 ml glass beads of 0.5 mm diameter are added, and the suspension is cooled to -60°. Sonication is performed with the Branson Sonifier Model J 17 V at 100 W, 20 kc/sec at 10 min intervals, interrupted by 12 min cooling periods, for a total sonication time of 35 min [17]. For the assay methods see Legend to Table 1.

	Before sonication	After sonication
NAD pyrophosphorylase (U/g dry weight)	2.6	2.9
Lactate dehydrogenase (U/g dry weight)	510	20
Aldolase (U/g dry weight)	7	0.7
Soluble protein (% of dry weight)	22	6

water one obviously would not expect additional damaging effects of the organic solvents used in non-aqueous isolation procedures. Some encouragement for making more frequent use of anhydrous techniques may well be derived from these data.

A micromethod for the anhydrous isolation of nuclei from L cells

L cells (L-929, Flow Laboratories Inc., Research grade) are grown as a monolayer in Basal Medium Eagle supplemented with 10% foetal calf serum, 500 µg/ml penicillin and 100 µg/ml steptomycin. Confluent cells are rinsed quickly with physiological saline and harvested by scraping them off the glass wall. The cells are sedimented in a refrigerated centrifuge, and then the pellet is frozen and lyophilized.

The powdered cells are suspended in about 50 parts of petroleum ether and ground by several passages in a continuous all-glass homogenizer [18], obtained from Bühler Co., Tübingen. Smears of the homogenate are stained with toluidine blue [19] and the extent of grinding determined. Usually 8-12 passages suffice for freeing the majority of nuclei from adhering cytoplasm; further homogenization results in a fragmentation of nuclei. Consistently, cytoplasmic particles that are ground off are much smaller than the nuclei. The final homogenate is centrifuged in the cold at low speed and the sediment taken up in 10 ml of petroleum ether (about 100-150 mg dry cells).

Zonal centrifugation is performed for the separation of the different particles in the homogenate. The B-XV rotor is used in a Heraeus Christ Omega II 70 000 centrifuge. A gradient linear with respect to the radius is calculated and built up by means of a LKB Ultrograd gradient mixer synchronized with a peristaltic pump (Bühler Co., Tübingen). The gradient is introduced into the rotor while spinning at 2300 rev/min. In order to slow down the sedimentation rate of the particles in the central part of the rotor, first a viscous zone made from mineral oil and carbon tetrachloride is introduced by means of a bypass of the tube coming from the mixer and the valve of the Ultrograd. This layer is followed by another zone of homogeneous density, consisting of a mixture of cyclohexane and carbon tetrachloride of density 1.290 (Fig. 7). The linear gradient is then formed by appropriate mixing of two mixtures of these two solvents, having densities of 1.290 and 1.450, respectively.

A portion (10 ml) of the homogenate is injected on the top of the viscous zone, followed by an overlay of 15 ml of petroleum ether. The zonal separation occurs at the same speed for 10 min, and the gradient is then displaced with carbon tetrachloride; fractions of 50 ml are collected.

Aliquots of the fractions are monitored at 492 nm (nephelometric determination) and the differences in transmission between the pure solvent mixture and the respective suspensions are recorded. Densities of the fractions are measured by a hygrometer (Fig. 7). The nature of the particles is identified by microscopical observation. On the basis of nephelometric and microscopical data, fractions containing nuclei and cytoplasm respectively are pooled, centrifuged after adding petroleum ether to lower the density, and washed free from mineral oil by 3-4 centrifugations in petroleum ether. The cellular fractions are dried in a vacuum desiccator charged with $CaCl_2$ and pieces of solid paraffin.

As reported earlier [19, 20], calculated density gradients of organic solvents do not differ significantly from observed densities (Fig. 7). Of the cellular material introduced in the zonal rotor, 65% is recovered in the final dry powders; it consists of about 70% cytoplasm

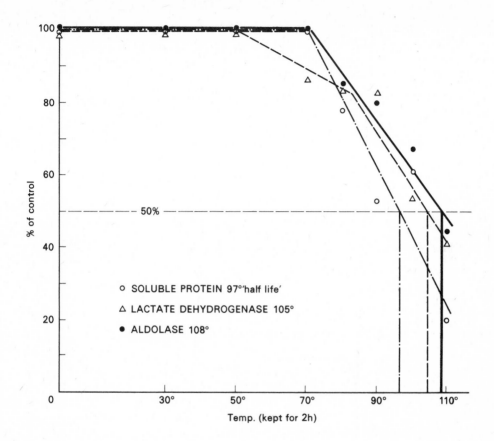

Fig. 6.   Heat stability of rat-liver enzymes in the anhydrous state.
◦—◦ soluble protein;   △—△ lactate dehydrogenase;   •—• aldolase.

Aliquots (300 mg) of solvent-treated parenchymal powders are heated to
the desired temperature at 16 mm Hg pressure in the presence of P_2O_5
for 2 h.   After cooling in a desiccator, the weight loss is determined,
and the tissue powder extracted and assayed with the methods described
in the legend to Table 1.
The dashed horizontal line represents 50% 'survival' of the measured
parameters; dashed vertical lines indicate 'half-life' temperatures for
these parameters.

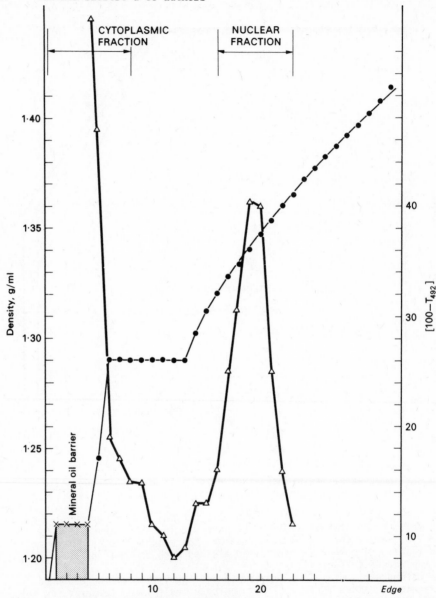

Fig. 7. *Legend opposite.*

(fractions 1-8 of Fig. 7) and 20% nuclei (fractions 16-24 of Fig.7). In the homogenate of dried L cells, an average RNA/DNA ratio of 1.4 is found; the nuclear fraction prepared as above demonstrates a value of 0.4. Since the L cells contain about 34% of their total volume as nuclear volume [21], a change of RNA/DNA rations by a factor of 2.9 would seem to indicate a reasonable purification of the nuclear fraction. The yield of nuclei is in the range of 40% of the theoretical value (12 mg dry weight from 90 mg cells). Homogenization and zonal centrifugation are easily performed within one working day.

An almost complete separation of nuclei from the homogenate is feasible by the procedure described. However, further improvements will have to concentrate on the efficiency and accuracy of the homogenization step.

ACTIVITIES OF DNA POLYMERASE IN ANHYDROUS CELL FRACTIONS FROM RAT LIVER AND L CELLS

Another potential nuclear marker enzyme, DNA polymerase, has been examined in different cell fractions with respect to some of its properties.

DNA polymerase activities in L-cell preparations.

Dried powders, obtained from various stages during the preparation of L-cell nuclei described above, are weighed and suspended in an amount of STM buffer (0.22 M sucrose, 20 mM Tris-Cl of pH 7.5, 2 mM $MgCl_2$) to give approximately 5 mg of the powder per ml. In addition, 6.4×10^5 fresh L cells are homogenized in 1 ml of the same buffer and subjected sequentially to 1,000 and to 10,000 × g for 20 min to produce residues enriched in nuclei and in mitochondria respectively. A motor-driven Teflon pestle is used for homogenization, and cell breakage is monitored by microscope; the homogenate has been judged to have less than 10% whole cells at the start of the differential centrifugation. Residues are suspended in STM buffer.

Legend to Fig. 7 *(opposite)*

Zonal separation of subcellular particles from L cells in a gradient system of organic solvents.

Separation conditions: 2300 rev/min, 10 min, 2°, Heraeus Christ Omega II 7000 centrifuge, B-XV (Ti) rotor. Loading and unloading time each 30 min.
Abscissa: Fraction numbers (50 ml collections); *right ordinate;* transmission at 492 nm, corrected for solvents as blanks; *left ordinate:* observed density of the recovered gradient fractions; the *shaded area* at bottom left indicates the viscosity barrier constructed from mineral oil in carbon tetrachloride.
o—o distribution of subcellular particles monitored at 492 nm.
△—△ observed density gradient after unloading of the rotor.

Table 3. Distribution of DNA polymerase activity in L cells

	Aqueous preparation	Non-aqueous preparation
Specific activity*		
homogenate	1.7	3.7
nuclei	0.8	3.1
cytoplasm	3.1	0.3
Total units†		
nuclei	28	108
cytoplasm	201	20
Per cent activity in nucleus	12%	84%

*
 nmoles [^{32}P-]dATP incorporated per h and per mg protein.
†*Arbitrary units: specific activity × relative protein.*

These fractions, both aqueous and non-aqueous, are analyzed for DNA polymerase [22] and protein [13]. The specific activities of DNA polymerase are determined, and data on cell and nuclear dimensions [21] are used to estimate the distribution of DNA polymerase in the nuclear compared with cytoplasmic portions of L cells. These data are presented in Table 3.

It is obvious from these data that the use of non-aqueous techniques to prepare nuclei gives evidence for an apparent localization of DNA polymerase, in the fractions obtained, profoundly different from that indicated by analyses of standard aqueous fractions. These data confirm and extend the conclusions reported by Keir *et al.*[23] who have found a higher proportion of DNA polymerase in non-aqueously prepared nuclei from regenerating rat liver. Their observations are complicated by the presence of potent nucleases in the liver fractions. Although the kinetics of the incorporation reactions reported here for L cells are not strictly linear over 60 min of incubation, the extent of the deviation would suggest that variations in nuclease activities in these preparations could not have influenced the pattern of the results obtained to a significant extent.

A second major inference can be drawn from the data of Table 3. We find that the specific activity of DNA polymerase detectable in the non-aqueous fractions at the whole homogenate stage actually exceeds that of the 'aqueous' counterpart. This observation suggests that there could not have been any major selective destruction of cytoplasmic polymerase, thus essentially eliminating the possibility that the distribution of enzyme activities is the result of a differential inactivation by the non-aqueous manipulations or solvents.

Table 4. DNA polymerase activities in resuspended non-aqueous liver fractions.

	Specific activity*		Units[†]		% Soluble
	Residue	*Soluble*	*Residue*	*Soluble*	
Parenchymal powder	0.186	0.175	0.219	0.597	73%
Same after homogenization and solvent exposure	0.043	0.192	0.069	0.474	87%
Cytoplasm	0.034	0.153	0.065	0.382	85%
Nuclei	0.120	0.636	0.155	1.456	90%

*n moles [^{32}P-] dATP incorporated per h and per mg protein.
[†]*Arbitrary units: specific activity × relative protein.*

The simplest explanation for this observation is that the bulk of DNA polymerases is readily solubilized in the usual aqueous media and appears in the post-microsomal supernatant, yet insofar as its *in vivo* localization is reflected in the fractions prepared in the absence of water, most of the detectable polymerase activity may actually be associated with the nucleus.

DNA polymerase activities in non-aqueously prepared liver nuclei

Since liver tissue may be obtained in bulk amounts, experiments have been performed with non-aqueously prepared extracts of this tissue. If DNA polymerase can be easily leached from nuclei in the presence of water, we should expect the treatment of anhydrous nuclei with aqueous buffers to result in the transfer of DNA polymerase to a soluble form. Table 4 lists data from which we can calculate the percentage of readily soluble DNA polymerase in four fractions obtained in the preparation of non-aqueous nuclei. Dry powders are suspended in STM buffer at 2° and the solutions gently homogenized by hand. After centrifugation at 10,000 × g for 20 min, the supernates are decanted and the residues resuspended in the original volume of STM buffer. All fractions are analyzed for protein and DNA polymerase.

Since there is measurable activity associated with the residue fraction, it then becomes of interest to determine whether adjustment of ionic strength would be capable of making more of the polymerase soluble. Consequently, the residue is suspended sequentially in buffers containing 0.2 M phosphate of pH 7.3 and the same augmented with 1M NaCl. The residue is suspended in STM buffer and all fractions are dialyzed for 3 h against 2 mM phosphate of pH 7.3 which is 0.1 M in both EDTA and mercaptoethanol. Values for DNA polymerase activity and protein are given in Table 5. Total units are obtained by multiplying total protein and enzyme specific activity.

Table 5. Extractability of DNA polymerase activity from non-aqueously prepared rat-liver nuclei by increasing ionic strength.

Treatment	Total Units
Suspension of nuclear powder	33
Supernatant after low salt exposure	15
Soluble in 0.2 M phosphate	8
Soluble in 1.0 M NaCl	14
Final residue	<0.1

All of the detectable polymerase is released by 1 M salt treatment, and a significant fraction (about 40%) is firmly bound to an extent that resists dissociation even by 0.2 M phosphate alone. It was observed that dialysis to lower salt produced a gelatinous residue in some instances, probably due to the reformation of nucleoprotein, but this suspension was still an effective source of DNA polymerase activity.

In connection with the work on poly-ADPR biosynthesis described above, we found it convenient to test the effect of active poly-ADPR synthesis on DNA polymerase activity. Liver nuclei, prepared in high-density sucrose media as described above [3], are suspended in solutions of NAD to allow of synthesis of poly-ADPR, or are prevented from doing so by the inclusion of 20 mM nicotinamide. DNA polymerase activities are determined on these solutions before and after incubation with NAD. The data obtained indicate that no major alteration in DNA polymerases occurs as a result of changes in poly-ADPR, nor is there a significant alteration of the association of DNA polymerase with the nuclear residue.

In the light of the experience reported above, the great majority of DNA polymerase appears to be of a highly preferential nuclear localization, but its solubility properties prevent this enzyme activity from being a useful nuclear marker.

Acknowledgements

The authors are grateful for discussions and suggestions to Dr. L.S. Dietrich, J. Nittinger and R.A. Shakoori. The research was collaborative, supported in part by grant G-120 from the Robert A. Welch Foundation, Houston, Texas, by a Travel Fellowship (IARC/T834) from the International Agency for Research on Cancer of the World Health Organization, as well as by the Deutsche Forschungsgemeinschaft, Fonds der Chermie, and Universitäts-bund Hohenheim.

References

1. Behrens, M., *Hoppe-Seyler's Z. Physiol. Chem. 209* (1932) 59-74.

2. Sauermann, G., *Biochem. Biophys. Res. Commun. 39* (1970) 738-743.

3. Blobel, G. & Potter, V.R., *Science (Wash. D.C.) 154* (1966) 1662-1665.

4. Siebert, G., *Methods in Cancer Research* (Busch, M., ed.), *Vol. 3*, pp. 287-301; and *Vol. 2*, pp. 47-59 Academic Press, New York (1967).

5. Jaus, H., Siebert, G. & Sauermann, G., *Hoppe-Seyler's Z., Physiol. Chem. 354* (1973) 1124-1128.

6. Nishizuka, Y., Ueda, K., Nakazawa, K. & Hayaishi, O., *J. Biol. Chem. 242* (1967) 3164-3171.

7. Zatman, I., Kaplan, N.O. & Colowick,S.P., *J. Biol. Chem. 200.* (1953) 197-212.

8. Kesselring, K. & Siebert, G., *Hoppe-Seyler's Z., Physiol. Chem. 337* (1964) 79-92.

9. Doly, I., Meilhac, M., Chambon, P. & Mandel, P., *Hoppe-Seyler's Z., Physiol. Chem. 353* (1972) 843-844.

10. Jaus, H., *unpublished results.*

11. Dietrich, L.S. & Siebert, G., *Hoppe-Seyler's, Z., Physiol. Chem. 354* (1973) 1133-1140.

12. Klingenberg, M., in: *Methoden enzymat. Analyse* (Bergmeyer, H.U., ed.), 2nd edn., *Vol. 2* (1970) pp. 1975-1990.

13. Lowry, O.H., Rosebrough, N.J., Farr, A.L. & Randall, R.J., *J. Biol. Chem. 193* (1951) 265-275.

14. Siebert, G., Humphrey, G.B., Themann, H. & Kersten, W., *Hoppe-Seyler's Z. Physiol. Chem. 340* (1965) 51-72.

15. Delbrück, A., Zebe, E. & Bücher, T., *Biochem. Z. 331* (1959) 273-296.

16. Bergmeyer, H.U., Gawehn, K. & Grabl, M. in *Methoden enzymat.Analyse.* (Bergmeyer, H.U., ed.) 2nd edn. *Vol. 1* (1970) pp. 391-392.

17. Shakoori, R.A., *Thesis* University of Hohenheim (1972) p.40.

18. Schenk, H.E.A., *Hoppe-Seyler's Z. Physiol. Chem. 352* (1971).

19. Shakoori, A.R., Romen, W., Oelschläger, W., Schlatterer, B. & Siebert, G., *Hoppe-Seyler's, Z. Physiol. Chem. 353* (1972) 1735-1748.

20. Siebert, G., Schlatterer, B. & Shakoori, A.R., in *Methodological Developments in Biochemistry* (E. Reid, ed.) *Vol. 3*, Longman, London, (1973) pp. 157-158.

21. Nittinger, J. & Siebert, G., *unpublished observations.*

22. Furlong, N.B., *Methods in Cancer Research,* (Busch, H., ed.) *Vol. 3*, Academic Press, New York, (1967) pp. 27-45.

23. Keir, H., Smellie, R.M.S. & Siebert, G., *Nature (Lond.) 196* (1962) 752-754.

3   A SIMPLE AND VERSATILE APPARATUS FOR FORMING REPRODUCIBLE GRADIENTS WITH MORE THAN ONE VARIABLE

Reinhard Bachofen
Institute of General Botany
University of Zürich
Zürich, Switzerland

With the apparatus described, gradients of any desired shape can be formed, allowing the introduction of further variables other than density into the gradient, e.g. special reactants, pH, etc. Two examples of separations obtained with membrane components of R. rubrum *are described.*

Gradients of linear or exponential shape can be easily formed for centrifugations in zonal rotors by using only simple laboratory glassware. On the other hand there are several sophisticated gradient formers on the market which allow to program and reproduce gradients of any desired shape. With these simple devices as well as with expensive gradient formers only one parameter, usually the density of the solution, is varied and can be altered by the program. In some biochemical investigations it may be of special interest to change the value of a solute parameter, e.g. ionic strength, besides and independent of the density. In our early experiments this was achieved by filling the rotor successively with two different exponential gradients, separated by a special reaction band [1, 2]. Much more versatility was obtained by using a multiport valve.

THE APPARATUS AND ITS OPERATION

The whole set-up consisted of the following components (Figs. 1 & 2):

- 24 solvent reservoirs: ordinary plastic bottles of 500 ml volume were used, filled with solutions of the desired composition and density. The latter was calculated by sub-dividing the continuous gradient into a step gradient of 20 to 22 portions.

- MP = multiport valve (MV 710-1524, WFN Labortechnik, Geldernstr. 113, 5 Köln 60): 20 to 22 ports of the 24-port valve were usually used for the gradient, the rest for concentrated sucrose solution for the cushion and the displacement solution and distilled water for rinsing the tubing and the cuvettes of the auxiliary equipment. The multiport valve is driven by compressed air.

- PC = program control (WFN, SG-001): it includes a timer giving pulses at 1 sec intervals. Any time between 1 and 99,999 sec may be pre-set as interval between a port change of the multiport valve. On the other hand the

Fig. 1. The set-up for gradient formation with the multiport valve.
See text for components as denoted by letters.

program unit can be used with an external signal source; in our case a micro-switch mounted on the peristaltic pump gives 3 pulses per revolution. In our system a pulse from the pump is equal to 0.47 ml solution. Any amount of gradient solution can be calibrated within ± 0.5 ml in terms of pulses from the pump. This way the program is independent of the speed of the pump, and the latter can be slowed down or stopped during the program without changing it.

If steps of varying volume are desired, an accessory programer with punched tape (WFN S4-004) is available.

- PU = pump: a six-channel variable peristaltic pump (Serva, Heidelberg) with added microswitch was used. The pump rate was calibrated before each run.

- TC = temperature control: the gradient solution was pumped through a stainless steel coil immersed in a water bath (Haake FK 10, temperature

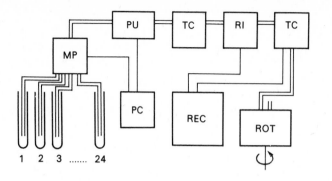

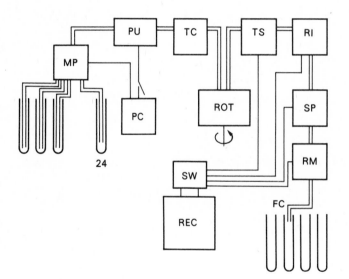

FC fraction
    collector
MP multiport valve
PC program control
PU peristaltic pump
REC recorder
RI refractometer
RM radioactivity
    monitor
ROT zonal rotor
    Ti-14
SP spectrophotometer
SW channel switch
TC temperature
    control
TS temperature
    sensor

Fig. 2. Schematic representation of the set-up for zonal centrifugation.

range from -10° to +100°), which served also for the cooling circuit of the upper bearing.

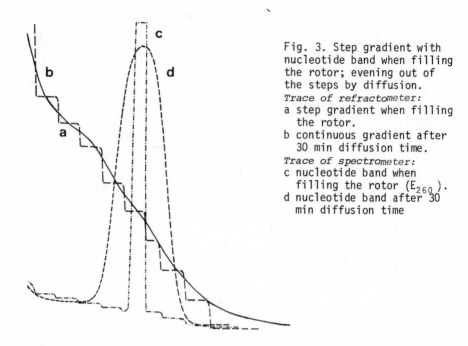

Fig. 3. Step gradient with
nucleotide band when filling
the rotor; evening out of
the steps by diffusion.
Trace of refractometer:
a step gradient when filling
   the rotor.
b continuous gradient after
   30 min diffusion time.
Trace of spectrometer:
c nucleotide band when
   filling the rotor (E_{260}).
d nucleotide band after 30
   min diffusion time

- RI = differential refractometer (Winopal, Isernhagen): this was used to control the shape of the sucrose gradients.

- TS = temperature sensor (Yellow Springs, Ohio): this served for controlling the temperature of the gradient solution before entering the rotor.

- REC = 2-channel potentiometric recorder (Philips 8010): this was used for recording the signals of the refractometer and the telethermometer.

After separation the contents of the rotor were pumped through the thermometer TS, the refractometer RI, followed in line by a spectrophotometer SP (Hitachi 101) and a radioactivity detector (Picker or Berthold LB 242). The 4 signals were alternately fed into the recorder (SW = switch). The solution was finally collected in a fraction collector. If not more than 24 fractions are desired, the multiport valve described at the beginning can be used as a fraction collector.

Fig.3 shows the gradient in steps while filling the rotor with the multiport valve, and the evening out by the rapidly occurring diffusion to give a continuous gradient (measured on pumping the gradient out of the rotor immediately after filling). A sharp band of special composition (e.g.

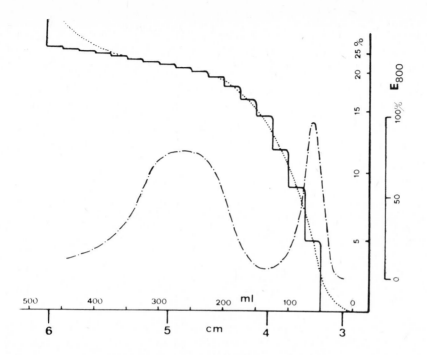

Fig. 4. Separation of the active reaction centres from the bulk of chromatophores from *Rhodospirillum rubrum* G-9.

full line: step gradient when filling the rotor

dotted line: gradient after centrifugation

interrupted line: E_{800} (absorbance)

---

nucleotides) spreads rapidly to a Gaussian shape peak.

A gradient-forming apparatus similar to that described here was
recently demonstrated for column chromatography [3].

EXAMPLES OF RESULTS

(a) Separation of photosynthetic reaction centres of photosynthetic bacteria

Chromatophores obtained from the blue-green mutant G-9 of the photosynthetic
bacterium *Rhodospirillum rubrum* were treated with the detergent LDAO (lauryl-
dimethylaminoxide   [4] for 30 min at 10° and then layered onto an iso-

volumetric gradient ranging from 25 to 26% sucrose. The gradient was calcu-
lated from data of Pollack and Price [5]. The centrifugation was done in a
Ti-14 rotor for 2 h 45 min at 45,000 rev/min ($\omega^2.t = 2.1 \times 10^{11}$ $rad^2 \times sec^{-1}$).
At the end of the run the following separation was achieved (Fig. 4):

(1)   a sharp band at the beginning of the gradient containing the reaction
centre (detected by a large difference in absorbance when measuring at 870 nm
an oxidized sample against a reduced one [4];
(2)   a broad band of chromatophore fragments of varying size due to variable
solubilization of the chromatophores by the detergent: since no difference
between oxidized and reduced samples in absorbancy at 870 nm was observed, it
can be assumed that this material was depleted of reaction centres;
(3)   a small band of non-solubilized larger chromatophores in the cushion.
On running the sample in a gradient from 2 to 20%, the reaction centre particles were further separated from the detergent incubation mixture.

(b) Introduction of a reaction band into the gradient, giving evidence for
conformational changes of membrane components of *Rhodospirillum rubrum*

Chromatophores were energized by incubation with P_i in the light or PP_i in
the dark.   The energized membranes were separated from the incubation medium
and washed free of adsorbed P_i in a buffered gradient in a zonal centrifugation.   The isopycnically sedimented chromatophores were used for a
detailed analysis of the membrane-bound P, of the nature and kind of binding
of the adenine nucleotides, and of the esterification capacity of the membranes, i.e. the capacity to form soluble ATP when incubated with ADP and
Mg^{2+}.   Table 1 shows some results obtained from 2 different centrifugations,
with and without a reaction band in the middle of the gradient [6].   Short
contact of the energized chromatophores with ADP during centrifugation
resulted in:

(1)   a much higher P_i content of the membranes, probably due to a slower
release of the P_i from the membrane after contact with ADP, a phenomenon
which cannot be explained at the moment and is under further investigation;
(2)   a 65% increase of acid-soluble nucleotides;
(3)   a 40% decrease of the nucleotides which can be extracted only with
detergents (SDS);
(4)   a drastic change in the ratio of ADP to ATP in the total nucleotides;
(5)   an 80% decrease in the esterification capacity.

Our results may be best explained assuming conformational changes of the
coupling factor, the site of ATP-formation on the photosynthetic membrane.
Hydrophobically bound ATP (only soluble in SDS) seems to be formed during
energetization and may be the phosphorylated intermediate responsible for
the esterification capacity.   After contact with ADP the latter is phosphorylated with free P_i to ATP, which appears in an acid-soluble form.   At the
same time the original hydrophobically bound ATP is hydrolyzed.   The hypothesis based on these results is summarized schematically in Fig. 5 [6].

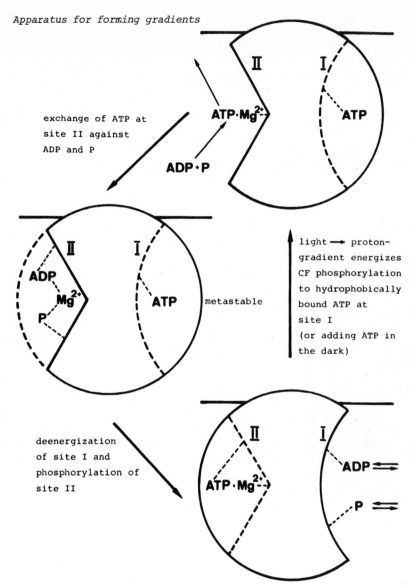

Fig. 5. Phosphorylation hypothesis with conformational changes of the coupling factor [6].

Table 1. Properties of centrifuged chromatophores after incubation in the light in the presence of $^{32}P_i$ (expressed as n-moles P_i/mg bacteriochlorophyll [6].

Reaction band	Analysis of the chromatophores			
	total P	*acid-soluble nucleotides*	*SDS-soluble nucleotides*	*Esterification capacity*
− ADP	62.0	1.5	ADP 2.4	
			ATP 15.3	1.6
+ ADP	1540.3	2.5	ADP 5.9	
			ATP 5.2	0.3

In general, introducing a second variable other than density opens the possibility to perform biochemical reactions within the rotor, which may be important when working with unstable intermediates.

Acknowledgments

I thank Drs. J. Dahl, H. Lutz and U. Pfluger for their part of the experimental work and the Swiss National Science Foundation for generous financial support (grant 3.626.71).

References

1. Bachofen, R. in *Proc. 2nd Int. Congr. Photosynthesis* (Forti, G., Avron M. & Melandri, A., eds.) The Hague (1972) p.1151-1157.

2. Bachofen, R. & Lutz, H.U.,in *Methodological Developments in Biochemistry* (Reid, ed.) *Vol. 3,* Longman, London (1973), p. 233-240.

3. Scott, R.P.W. & Kucera, P.,*J. Chromatog. Science 11* (1973) 83-87.

4. Wang, R.T. & Clayton, R.K., *Photochem. Photobiol. 17* (1973) 57-61.

5. Pollack, M.S. & Price, C.A., *Anal. Biochem. 42* (1971) 38-47.

6. Lutz, H.U., Dahl, J.S. & Bachofen, R., *manuscript in preparation.*

4 THEORY, DESIGN AND CROSS CONTAMINATION OF DISCONTINUOUS DENSITY GRADIENTS

George B. Cline, Martha K. Dagg & Richard B. Ryel
Departments of Biology and Oncology and Hematology
University of Alabama in Birmingham
Birmingham, Alabama 35294, U.S.A.

Discontinuous density gradients have obvious advantages when trying to sharpen zones of material which have banded isopycnically. These gradients also have obvious advantages when forming gradients which contain several supplementary materials such as enzymes, precipitants or solvents. This type of gradient has been criticized for rate separations because of possible trapping of material on the 'steps' as the particles sediment through the gradient. Cross-contamination can occur when steps are made indiscriminately without respect for sedimentation rates of the various particles. This study used two different sizes of latex beads in the B-XIV zonal rotor to assess the low level of cross-contamination with discontinuous density gradients. The particles in the zones were counted by low-magnification electron microscopy. The theory and design of discontinuous density gradients is discussed.

Separations on a rate basis through density gradients have been adequately described by theory [1-5]. Ways to increase both resolution and gradient capacity are constantly being explored. One method of increasing resolution has been to use 'inner gradients' (a series of gradients within a gradient) and attempt to stop the separation when a zone of particles has piled up against one of the steeper inner gradients. While this method has been used to good advantage in a number of separations, one recurring criticism has been that such inner gradients might artificially trap particles which would otherwise sediment through that portion of the gradient. To determine the extent of this trapping we performed some simple experiments with two distinctly different sizes of latex beads. The large bead (1.011 µm) quickly sedimented to a quasi-equilibrium position while the smaller bead (0.109 µm) sedimented more slowly through the gradient. We were interested only in the degree of cross-contamination in the two different zones of latex beads in discontinuous gradients; thus no comparison with continuous gradients was made.

MATERIALS AND METHODS

Latex beads were a gift from Dr. Norman G. Anderson, Molecular Anatomy Program, Oak Ridge National Laboratory. Stock solutions were made as 1:20 dilutions and equal volumes were mixed as the starting sample for most separations. The large bead was 1.011 µm (with S.D. ± 0.005); the small bead was 0.109 µm (S.D. ± 0.0027).

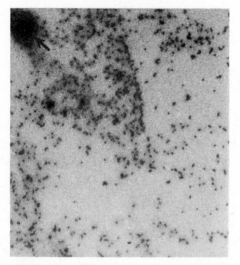

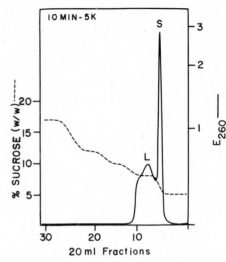

Fig. 1. Electron micrograph of a mixture of 1.001 µm
and 0.109 µm latex beads used as starting material
for the separations. *Arrow points to one large bead
among many small beads.*

Fig. 2 *(above)*
& Figs. 3 & 4
(below, left & right)
— *Legends opposite.*

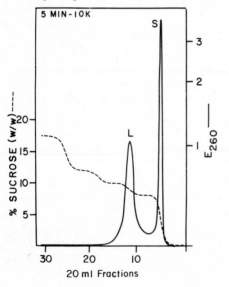

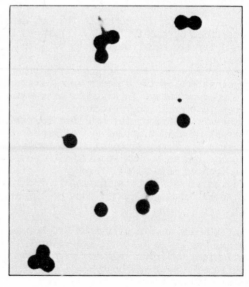

Sucrose gradients were made from stock 66% (w/w) solution. Most gradients, except where noted, were made alike and were pumped into the rotor sequentially with a peristaltic pump. The gradient was made of 125 ml of 10%, 125 ml of 12% and a cushion of 17% sucrose. The sample volume was 12 ml and the sample was adjusted to 7% sucrose. The overlay was 100 ml of 5% sucrose except where noted. Gradients were monitored at 260 nm during unloading.

Centrifugation times and speeds were varied to place zones of beads at different positions in the rotor. Times ranged from 10 min at 5,000 rev/min to 155 min at 28,000 rev/min for the Spinco aluminium B-XIV rotor. Equivalent $\omega^2 t$ values were calculated from the speed-time factors for computational purposes.

Recovered gradient fractions were dialyzed against water to remove the sucrose and then were evaporated partially to concentrate the beads. The concentrated samples and starting samples were placed on Formvar-carbon-coated 200 mesh grids for examination in a Phillips EM 75A electron microscope.

RESULTS

An electron micrograph of the starting latex-bead mixture is shown in Fig.1. Although equal volumes of the starting bead stocks were mixed, there was a predominance of small particles. An *arrow* points to one large particle amongst approximately 700 small particles.

Fig. 2 shows an incipient separation where the large beads have just started to emerge from the starting zone and are seen in the middle of the first step. The profile of the large bead zone is somewhat bimodal while the small bead zone shape is still essentially gaussian. Fig. 3 shows the position of the beads under slightly longer $\omega^2 t$ conditions. The small bead is still in the starting zone while most of the large beads have sedimented through the first step and have piled up on the second inner gradient. A small number of the large beads have already penetrated the second step and are sedimenting to the second inner gradient at around fraction 18. Fig. 4 is an electron micrograph of material from the large bead zone. Very few small beads were found in this zone. (The *arrow* points to one small bead amongst 13 large beads; many fields contained no small particles at all.)

Legends to Figures opposite
Fig. 2 *(top right)*. Incipient separation of the large from the small beads on a discontinuous density gradient. L, large-bead zone; S, small-bead zone. Centrifugation for 10 min at 5,000 rev/min. (K denotes 1,000 × *g*.)
Fig. 3 *(bottom left)*. A 5 min, 10,000 rev/min, separation of large from small latex beads. *Large-bead zone is centred on second inner gradient.*
Fig. 4 *(bottom right)*. Electron micrograph of beads from L zone from Fig. 3. *One small bead evident in this field — low level of small-bead contamination.*

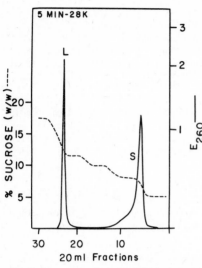

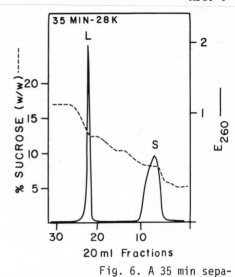

Fig. 5.   A 5 min, 28,000 rev/min, separation of large from small beads. *Some small beads are sedimenting through first step around Fraction 10.*

Fig. 6. A 35 min separation at 28,000 rev/min. *Large beads banded quasi-isopycnically to left; small beads sedimenting through 1st step to right.*

Fig. 7 *(below)*. A 30 min longer separation than Fig.6. *Small beads still sedimenting into 2nd inner gradient.*

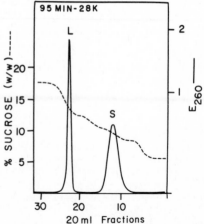

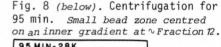

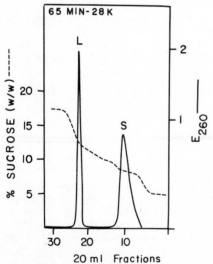

Fig. 8 *(below)*. Centrifugation for 95 min. *Small bead zone centred on an inner gradient at ∿Fraction 12.*

Fig. 5 shows the results of centrifuging for 5 min at 28,000 rev/min. The large beads have sedimented to a quasi-isopycnic position at 12.8% sucrose while the small beads have broken through the first step. Some of the small beads appear to be forming a second zone centred in fraction 10.

Fig. 6 shows the results of a 35 min separation at 28,000 rev/min. The large bead zone is centred at 13% sucrose while the small beads are sedimenting through the first step. The second inner gradient centred at about fraction 12 has slowed up the rapid sedimentation of the faster moving small beads. The zone is skewed toward the second inner gradient.

Fig. 7 is a 65 min separation which further shows the sedimenting properties of the small latex beads. Most of the beads have piled up on the second inner gradient while the trailing beads are sedimenting down the first step. Fig. 8 shows the position of the smaller beads 30 min later (95 min). The beads are nearly centred on the second inner gradient and thus show a nearly gaussian distribution. Fig. 9 shows an electron micrograph of the material found in the small bead zone. No large beads were found while scanning several prepared grids. The picture shows a high concentration of the small beads.

Fig. 10 shows the position of the bead zones one hour later (155 min). The large beads have sedimented slightly further than before and are centred in 13.5% sucrose while the small beads have sedimented against what is left of the third inner gradient. The zone is skewed toward the lower sedimenting side. Some of the smaller particles appear to have traversed the third inner gradient and are beginning to band isopycnically with the large beads.

To demonstrate the effects of having very large steps for particles to sediment through, we set up one rotor with a step of 450 ml of 8% sucrose and used only large beads. Since these had sedimented nicely to fraction 12 in 5 min at 10,000 rev/min (Fig.3) we centrifuged the large beads 2 min longer without the second and third inner gradients (Fig.11). This short increase in time in combination with no further gradients resulted in about 25% of the large beads quickly sedimenting to a quasi-isopycnic level while most of the beads were distributed throughout the step. There appears to be some increase in concentration of beads when the sucrose concentration starts to increase in fraction 24.

DISCUSSION

Few problems have been encountered when using discontinuous density gradients for isopycnic separations. When this type of gradient is used for rate separations, problems of cross-contamination may occur if the sizes of the particles do not differ greatly. Similar cross-contamination between zones is apparent also with continuous gradients. In these studies, we did not find more than an occasional contaminating particle when separating a mix-

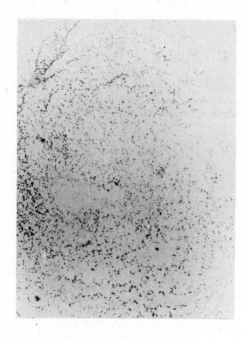

ture of 1.00 μm and 0.109 μm latex beads on a rate basis in discontinuous sucrose-density gradients.

When using discontinuous gradients for rate separations, it is important to follow a set of conditions or steps. The first step, if the particles to be separated are undescribed biophysically, is to make a separation on a linear gradient (isokinetic is probably best) to find the sedimentation

---

Fig. 9 *(left)*. Electron micrograph of beads from Fraction 12 in Fig. 8.

Fig. 10 *(below, left)*. Centrifugation for 155 min at 28,000 rev/min. *Small beads beginning to sediment into large-bead zone.*

Fig. 11 *((below, right)*. The results of centrifuging large beads only through a long step of 8% sucrose. *Some beads are banding isopycnically in Fraction 29 whilst the majority are distributed throughout the 450 ml step.*

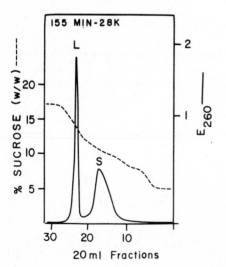

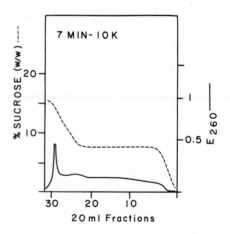

characteristics of the particles. This is shown diagramatically in Fig. 12A. If two zones of particles are partially separated, the separation can be improved by adding a step and creating an inner gradient. To determine the solution needed to make the step, find the density centre of each of the zones (i.e. 24% and 16% sucrose in Fig. 12A). The average corresponds to 20%, the density of the sucrose step to be added for maximal separation of the zones. Fig. 12B shows typical results. Zone narrowing is also achieved because of the steepness of the inner gradients.

Zones go through a variety of shapes when sedimenting through a discontinuous gradient. Fig. 13 shows a composite of shapes noted for the sedimentation of latex beads. Zone A represents the usual starting zone and the nearly gaussian distribution of particles before sedimentation occurs. As sedimentation occurs through the first inner gradient (the one supporting the starting zone), the particles at the greatest radius *(note arrow)* sediment down the first step faster than the majority of particles. The zone is thus skewed toward the periphery of the rotor and appears as zone B. When the fastest sedimenting particles encounter the second inner gradient, their rate of sedimentation is slowed while the rest of the beads catch up. Zone C shows such a skewed situation with the *arrow* showing the beads just entering the zone. Although not shown in this composite figure, when such skewed zones sediment into and are centred on an inner gradient, they appear gaussian (see Fig. 8). When the particles finally sediment to their isopycnic density, they again form a nearly gaussian zone if centrifuged long enough.

When using discontinuous gradients to separate particulates whose sedimentation characteristics are unknown, it is important to note that cross-contamination of zones can occur and one type of particle may actually form two zones. Major reasons for such results include the overloading of the inner gradients near the starting zone and arbitrary centrifugation times. Also important is the length of each step in discontinuous gradients. Steps near the starting zone should not occupy more than about 3 to 5 mm radius while 7 mm is probably an optimal step width further out in the rotor. Although no such data appear in this study, it seems from other studies with biological particles that the length of the step is more important than its height [6].

Discontinuous gradients are the gradients of choice for highest resolution separations, either rate or isopycnic. This type of gradient is simple to make and can be varied easily. Additions of such materials as a second solute, salts, enzymes, precipitants (organic solvents, ammonium sulphate, polyethylene glycol, or heavy metals for gradient resolubilization) and extractants are easily made.

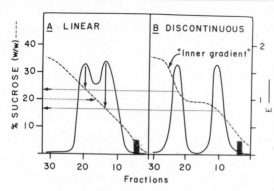

Fig. 12. A composite of two theoretical separations.
A, separation of two types of particles on a linear gradient along with a method of calculating the density of a step to be added in centre of gradient for increased separation.
B shows the results after addition of a step.

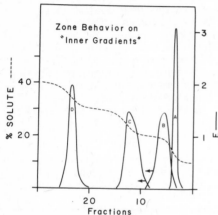

Fig. 13. A composite of zone shapes noted in the sedimentation of 0.109 μm beads through a discontinuous sucrose density gradient.
A, starting zone.
B, shape seen when the beads sediment through the first inner gradient and down the first step.
Zone C, skewed distribution when the beads enounter the 2nd inner gradient. *(Arrows show the fastest sedimenting portion of Zone B and the slowest portion of Zone C.)*
Zone D, distribution of beads when the beads band isopycnically at 13.7% sucrose.

References

1. de Duve, C., Berthet, J. & Beaufay, H., *Prog. Biophysics Biophys. Chem. 9* (1959) 325-369.

2. Brakke, M.K., *Adv. Virus Res. 7* (1960) 193-224.

3. Anderson, N.G.,   *Nat. Cancer Inst. Monogr. No. 21* (1966) 9-40.

4. Berman, A.S., *ibid.No. 21* (1966) 41-76.

5. Schumaker, V.N.,   *Adv. Biol. Med. Physics 2* (1967) 246-339.

6. Cline, G.B. & Ryel, R.B.,   *Methods in Enzymology* (Academic Press, New York) *22* (1971) 168-204.

5   REORIENTING DENSITY GRADIENTS AND THE SZ-14 ROTOR

Phillip Sheeler
Department of Biology
California State University
Northridge, California 91324, U.S.A.

Gradient reorientation is an alternative approach to the dynamic unloading of conventional zonal rotors using rotating-seal assemblies. Reorienting gradient rotors are permitted to slowly decelerate to rest before the density gradient is removed. During the final phase of deceleration, the radial density gradient is reoriented into the vertical position, while the entrained cylindrical particle zones are transformed into a series of horizontal layers. Although the SZ-14 rotor employs no rotating face seals, it may be quickly loaded with gradient and sample either dynamically or statically. This is especially advantageous for rate-zonal separations of large particles such as whole cells and the larger subcellular components. The rotor can also be used to process particle suspensions on a continuous-flow basis.

'Reorienting gradient' centrifugation can be carried out either in centrifuge tubes or in the bowl of a zonal rotor. In the former case, *fixed-angle* rotors are used and the density gradient caused to reorient between the vertical and radial positions within the centrifuge tube during rotor acceleration and deceleration. This is in contrast to density gradient centrifugation in *swinging-bucket* rotors where the orientations of both the centrifuge tube and the gradient change simultaneously. Whereas conventional zonal rotors (i.e. A-12, B-14 and B-29) are unloaded *dynamically,* re-orienting gradient zonal rotors are designed to provide static unloading of the density gradient.

REORIENTING GRADIENT CENTRIFUGATION USING FIXED-ANGLE ROTORS

Here the gradient is prepared in a centrifuge tube and the sample carefully layered onto the surface (Fig. 1A). At this stage, isodense regions of the gradient are arranged as a series of circular, horizontal layers; these layers become elliptical when the tube is placed in its inclined position within the rotor (Fig. 1B). During rotor acceleration, the isodense layers become small sections of *paraboloids of revolution* whose foci lie on the axis of rotation (Fig. 1C). The paraboloids become increasingly steep, eventually approaching 'verticality' (Fig. 1D). At this stage, isodense regions of the gradient are distributed through the centrifuge tube as small segments of concentric vertical cylinders, all points in each segment being equidistant from the axis of rotation. Generally, gradient reorientation from the vertical into the radial position is completed by the time the rotor attains 1,000 to 2,000 rev/min.

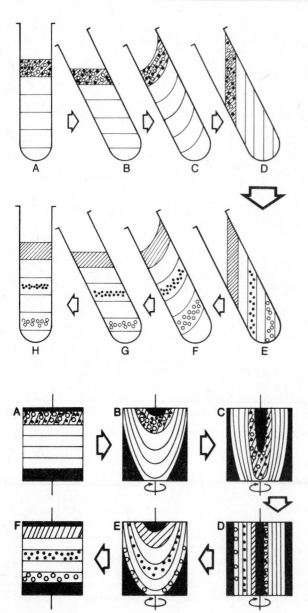

Fig.1. Essential features of reorienting gradient centrifugation using a fixed-angle rotor. Parallel lines represent isodense layers with a *continuous* density gradient. *See text for details.*

Fig. 2. *(below,left).* Stages in the use of a zonal rotor for reorienting gradient centrifugation. *See text for details.*

Rotor acceleration is accompanied by radial sedimentation of the sample particles (Fig. 1E). Particles encountering the marginal tube wall are conducted down the tube under the combined influences of centrifugal force and convection. Generally, reorienting gradient centrifugation in fixed-angle rotors is used for particle separations based on *isopycnic* banding point differences - the technique rarely being used for *rate* separations (i.e. separations based on differences in particle sedimentation coefficients). Consequently, particles sediment through the gradient until they reach an isodense region, at which point they form a band (Fig. 1E).

Deceleration of the rotor is accompanied by reorientation of the density gradient, and the separated particle bands from the

Fig. 3. Basic parts of th
the SZ-14 reorienting
gradient zonal rotor: *A*,
rotor bowl; *B*, core/septa
piece; *C,* lid; *D,* axial
chamber of distributor;
E, annular chamber of
distributor; *F*, distri-
butor connector.

Fig. 4.
Core/septa piece with
attached distributor:
A, septum; *B*, distri-
butor connector inserted
into axial chamber of
distributor; *D*, openings
of core lines; *E*, position
of septa line opening.

radial into the vertical
position (Figs. 1F and
1G). The centrifuge tube
is then removed from the
rotor (Fig. 1H) and the
bands sequentially coll-
ected by displacing the
gradient via the bottom
or the top. This techni-
que has successfully separated a variety of sub-
cellular particles and
macromolecules [1-3].

REORIENTING GRADIENT ZONAL CENTRIFUGATION

In zonal rotors [4] the gradient can be envisaged as a family of concentric iso-dense cylinders, the densest at the rotor wall and the lightest immediately about the core. Gradient, sample and separated particle zones are maintained in the radial position at all times. Reorienting gradient zonal rotors are similar in shape and design to conventional zonals but do not require rotat-ing-seal assemblies and may be loaded and unloaded at rest. In using a zonal rotor for reorienting gradient centrifugation, the empty rotor is loaded with vertical gradient and sample (Fig. 2A). Gradual rotor acceler-ation causes the sample zone and each horizontal isodense plane to bow down-ward at the axis of rotation and upward at the periphery to form a paraboloid of revolution (Fig. 2B). With continued acceleration, the paraboloids become increasingly steep (Fig. 2C), eventually approaching verticality (Fig. 2D, right). The original vertical gradient is thereby completely reoriented with-in the rotor chamber to form a conventional radial density gradient.

At operating speed, particles sediment radially to form a series of separate cylindrical zones (Fig. 2D, *left*). During deceleration, the gradi-ent and entrained particle zones reorient from the radial to the vertical position (Figs. 2E and 2F). Particles comprising the original sample are now distributed through the gradient as separate layers (Fig. 2F) which may be collected by withdrawing the gradient through the top or the bottom. This principle and the first experimental 'reograd' zonal rotor were des-cribed by Anderson *et al.* in 1964 [5]. Experimental rotors have since been described by the author and co-workers [2, 6, 7] and led to the production of the Sorvall SZ-14 reograd rotor [8, 9].

THE SZ-14 REORIENTING GRADIENT ZONAL ROTOR (Figs. 3 & 4), AND ITS OPERATION*

The rotor bowl contains an annular chamber surrounding an axial shaft, both being tapered near the bottom to form an annular V-shaped groove. It is from the vertex of this groove that the density gradient is collected statically after centrifugation. A core/septa piece slides over the shaft and divides the chamber into 6 sector-shaped compartments. It contains two sets of channels: 6 *septa lines*, which descend at an angle through each septum and open at the vertex of the V-groove and *core lines* which descend a short distance through the core, turn radially, and open on the core surface ad-jacent (clockwise) to each septum.

The *distributor*, which also contains two sets of channels, is bolted to the top of the shaft. One set of channels passes radially from the *axial* distributor chamber to connect with the septa lines; the other set passes radially from the *annular* distributor chamber to connect with the core lines. The hollow *distributor connector* is removably inserted into the axial distri-butor chamber. The *lid* completes the assembly.

This description has been abridged. - Editor.

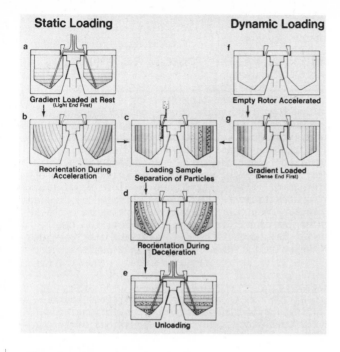

Static Loading   **Dynamic Loading**

a — Gradient Loaded at Rest (Light End First)

b — Reorientation During Acceleration

c — Loading Sample Separation of Particles

d — Reorientation During Deceleration

e — Unloading

f — Empty Rotor Accelerated

g — Gradient Loaded (Dense End First)

Fig. 5. Operation of the SZ-14 zonal rotor. *See text for details.*

The SZ-14 holds about 1,350 ml of combined gradient and sample and is used with Sorvall centrifuges at speeds up to 19,500 rev/min (40,500 g). A 'rate-controller' provides the smooth and gradual rotor acceleration and deceleration required between 0 and 1,000 rev/min to prevent gradient disturbances and mixing during reorientation.

For static loading (Fig. 5), the gradient pumped (light end first) into the distributor connector. The stream passes through the septa lines to the vertex of the V-groove and is displaced upward by the denser fluid flowing immediately behind, giving a vertical density gradient (Fig. 5a). After removing the distributor connector and tubing, the rotor is slowly accelerated to 1,000 rev/min to reorient the gradient (Figs. 5b and 5c).

For dynamic loading, the empty rotor is accelerated to about 2,000 rev/min and the gradient pumped (dense end first) into the annular distributor chamber. The gradient is directed into the distributor through a small nozzle, and swept centrifugally through the core lines and radially on the septa surfaces to fill the rotor beginning at the wall (Figs. 5f and 5g). A radial gradient is thus formed without a reorientation step.

With the rotor spinning at about 2,000 rev/min, the sample is introduced into the annular distributor chamber, and carried centrifugally through the core lines and onto the exposed centripetal surface of the gradient (Fig. 5c, *left*). The rotor is then accelerated (or decelerated) to the desired operating speed. Once particle separation has been achieved (Fig. 5c, *right*), the rotor is again decelerated. The gradient and particle zones reorient

Fig. 5d), distributing the separated particles as horizontal layers (Fig. 5e). The gradient is pumped out and the particles collected in order of decreasing density or sedimentation rate.

SPECIAL FEATURES OF THE REORIENTING GRADIENT TECHNIQUE, AND APPLICATIONS

Compared with dynamically unloaded zonal rotors, rotor design, operation and maintenance are greatly simplified by eliminating rotating and static-seal assemblies. Since in the absence of face-seals cross-leakage between the wall and core lines cannot occur, gradients formed using viscous materials (e.g. sucrose, Ficoll, dextran) can be quickly pumped into (or out of) the SZ-14. The sample too can be rapidly deposited on the gradient surface. While this is of little significance for *isopycnic* separations, it materially influences *rate-zonal* separations where the starting zone thickness affects the ultimate resolution attainable. (The distance two particle populations having different sedimentation coefficients must travel through the gradient in order to be resolved increases in proportion to the starting zone thickness). Since sedimentation begins immediately upon entry into the rotor, the longer the time between the introduction of the first and last portions of the sample, the thicker the starting zone. Since no face-seals are present, a 30-50 ml sample can easily be introduced in seconds, producing a narrow starting zone.

Unless they have already reached their isopycnic positions, particles continue to sediment during *dynamic* unloading of a conventional zonal rotor. This can significantly complicate rate-zonal separations of larger particles such as whole cells, cell nuclei or mitochondria which undergo appreciable sedimentation even at moderate centrifugal forces. Negligible particle sedimentation occurs during the static unloading of the reorienting gradient zonal rotor.

Using the SZ-14 at low speeds, Wells and James [10] separated large quantities of log-phase fission yeast cells into age-related cell size classes, enabling biochemical events occurring throughout the cell cycle to be followed. A similar 'culture fractionation' in the SZ-14 has been carried out with budding yeasts [11]. The SZ-14 has been used for liver homogenate fractionation [7], the isolation of plasma membranes [12], the isolation of helminth mitochondria by isopycnic banding [13], and rate-zonal sub-fractionation of enzymatically diverse liver mitochondria [14].

CONTINUOUS FLOW CENTRIFUGATION

By replacing the zonal distributor with the *continuous-flow distributor* and *inlet/outlet assembly* (Fig. 6), the SZ-14 can be used for continuous flow centrifugation. Two bearings in the upper section of the inlet/outlet assembly and the distributor rotate with the spinning rotor. Again, no face-seals are used: the particle suspension is introduced centrifugally, while the supernatant is expelled using a lift device [15]. Continuous-flow centri-

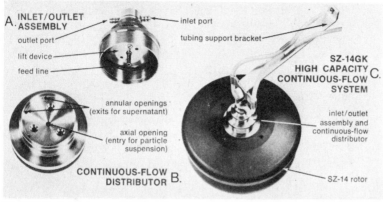

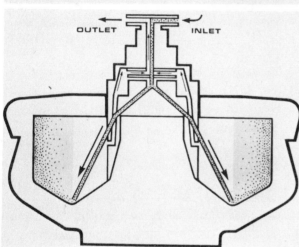

Fig. 6. *Left:*

Inlet/outlet assembly *A* and distributor *B* used for continuous-flow operation of the SZ-14 rotor.

Right: Completely assembled continuous-flow system.

Fig. 7. Diagrammatic representation of continuous-flow operation of the SZ-14 rotor. *Arrows* indicate the direction of flow through the rotor.

fugation may be carried out up to 19,500 rev/min, with flow rates up to 1,400 ml/min; more than 700 ml of sediment can be collected in a single, uninterrupted operation. Rotor speed and flow rate are independently adjusted to provide maximum particle clean-out.

Fig. 7 depicts the movement of liquid into and out of the SZ-14 during continuous flow operation. The suspension is swept through the distributor and septa lines of the core/septa piece to the vertex of the annular V-groove. As it leaves the septa, particles sediment up the outer surface of the V-groove under the combined influences of convection and centrifugal force and are packed against the rotor wall. Simultaneously, the clarified supernatant is displaced centripetally and expelled [15]. It is to be noted that flow through the core and septa is opposite to that which occurs during zonal density gradient centrifugation. The rotor has permitted isolation of a variety of particles from large-volume suspensions including bacteria, blood cells, yeast cells, protozoa, starch grains, and ammonium sulphate precipitates of proteins. By sequential use of the zonal and continuous-flow distributors, Nixon *et al.* [16] have recently employed the SZ-14 to isolate large quantities of calf thymus nuclei in sucrose gradients by continuous-flow-with-banding.

References

1. Fisher, W.D., Cline, G.B. & Anderson, N.G., *Anal. Biochem.* 9 (1964) 477-482.

2. Sheeler, P. & Wells, J.R., *Anal. Biochem.* 32 (1969) 38-47.

3. Vedel, F. & D'Aoust, M.J., *Anal. Biochem.* 35 (1970) 54-59.

4. Anderson, N.G., *Quart. Revs. Biophys.* 1 (1968) 217-263.

5. Anderson, N.G., Price, C.A., Fisher, W.D., Canning, R.E. & Burger, C.L., *Anal. Biochem.* 7 (1964) 1-9.

6. Wells, J.R., Gross, D.M. & Sheeler, P., *Lab. Practice 19* (1970) 497-499.

7. Sheeler, P., Gross, D.M. & Wells, J.R., *Biochim. Biophys. Acta 237* (1971) 28-42.

8. Wells, J.R., Sheeler, P. & Gross, D.M., *Anal. Biochem.* 46 (1972) 7-18.

9. Sheeler, P., *Amer. Lab. 3* (Feb. 1971) 19-25.

10. Wells, J.R. & James, T.W., *Exp. Cell Res.* 75 (1972) 465-474.

11. Wells, J.R., *Exp. Cell Res.*, submitted.

12. Sheeler, P., *(in preparation)*.

13. Carter, C.E., Wells, J.R. & Macinnis, A.J., *Biochim. Biophys. Acta 262* (1972) 135-144.

14. Wilson, M.A. & Cascarano, J., *Biochem. J. 129* (1972) 209-218.

15. Sheeler, P. & Wells, J.R., *Amer. Lab.* (Jan. 1972) 36-40.

16. Nixon, J.C., McCarty, K.S. Jr. & McCarty, K.S. Sr., *Anal. Biochem.* (1973) *(in the press)*.

6 COMPUTER SIMULATION OF RATE-ZONAL CENTRIFUGATION

Jens Steensgaard and Lars Funding
Institute of Medical Biochemistry
University of Aarhus, Denmark

A numerical method for predicting the outcome of rate-zonal centrifugations has been worked out in order to facilitate optimization of experimental conditions in zonal experiments. The principle of this method [1] is as follows. A model zonal rotor is divided into a series of annular segments by imaginary cylindrical planes. The fluxes of sample and gradient materials across each plane during centrifugation are calculated for short periods of time, and the concentrations of materials in each segment are adjusted accordingly after each period. Summations of these changes provide concentration-radius profiles describing the events in the rotor at different timepoints. The present contribution deals mainly with a short description of the numerical method, a series of program tests, and a study of the changes in zone shape produced by centrifugation of the same sample in different sucrose gradients.

Equations describing the pattern of particle movements in a density and viscosity gradient during centrifugation have been used widely to calculate sedimentation coefficients after rate-zonal centrifugation [2, 3]. The opposite use of the same equations, namely to predict the pattern of particle movements in a gradient, has presented some numerical problems, but a numerical solution to these equations can be obtained in two ways. Firstly, programs written for the calculation of sedimentation coefficients can be run with specially constructed data until a desired result has been obtained. Although troublesome, this approach has been found highly valuable in theoretical tests of gradients [4,5]. Secondly, programs performing complete simulations of sedimentation and diffusion processes in sucrose gradients can be written to provide full information on the fate of a zone during gradient centrifugation [1,6].

In addition, the simulation program is valuable because it allows assessment of the extent to which simple flow equations can describe the events in a zonal rotor during centrifugation.

THE MODEL ZONAL ROTOR

A rotor with a cylindrical chamber and without septa has been used as a model rotor, but the geometry of this rotor has been chosen so as to be closely related to that of a genuine B-XIV zonal rotor [7]. Fig. 1 gives a schematic drawing of the model rotor, seemingly a little lower than the B-XIV zonal rotor. This decrease in height has been introduced in order to compensate for the volume of the removed septa. The height of the model rotor, 4.878 cm, has been found by a least-square analysis, and use of this rotor height makes theoretical and practical results directly comparable.

For the purpose of simulation the rotor chamber is divided into a number of annular segments by a number of cylindrical planes. Employing 206 planes the volume of one segment is between 1.5 and 4 ml and depends primarily on its radius. The first and last segments are, however, larger than the others for numerical reasons, because they must act as buffers for incoming and departing materials.

The simplified geometry of the model rotor has the advantage that the radius (r_n) and the volume (v_n) of a segment (n) can be calculated from the following linear expressions:

$$r_n = 2.34 + 0.02n \qquad (1)$$

$$v_n = 4.878\pi (0.0932 + 0.0008n) \qquad (2)$$

and the area (A) of a cylindrical plane from:

$$A_n = 2\pi r_n 4.878 \qquad (3).$$

The flow equations

The flow (J) of a substance across a plane in the model rotor at a given time can be calculated from:

$$J = sC\omega^2 r - D(dC/dr) \qquad (4)$$

and the amount (M) of transported material (sample or gradient molecules) from:

$$M = AJ\Delta t \qquad (5).$$

During gradient centrifugation both sample and gradient molecules will sediment and diffuse through a medium whose density and viscosity are determined by the kind and the concentration of gradient material. Hence, it is necessary to correct equation (4) accordingly by use of the following expressions:

$$s = s_{20.w} \frac{\eta_{20.w}}{\rho_p - \rho_{20.w}} \cdot \frac{\rho_p - \rho_{T.m}}{\eta_{T.m}} \qquad (6)$$

$$D = D_{20.w} \frac{\eta_{20.w}}{293.16} \cdot \frac{T}{\eta_{T.m}} \qquad (7).$$

In the case of sucrose gradients, density and viscosity at given sucrose concentrations and at given termperatures can be calculated by use of Barber's polynomial expressions [8].

Essentially, the equations mentioned above provide all the information

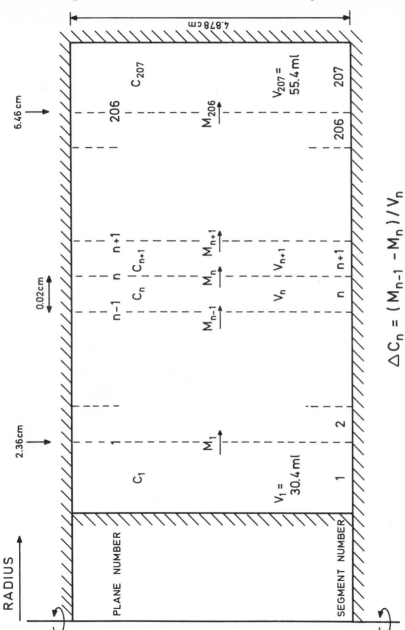

Fig. 1.

$$\Delta C_n = (M_{n-1} - M_n)/V_n$$

Fig. 1. The model zonal rotor. C is the concentration of sample or gradient material, M the amount of transported material, V the volume of a segment.

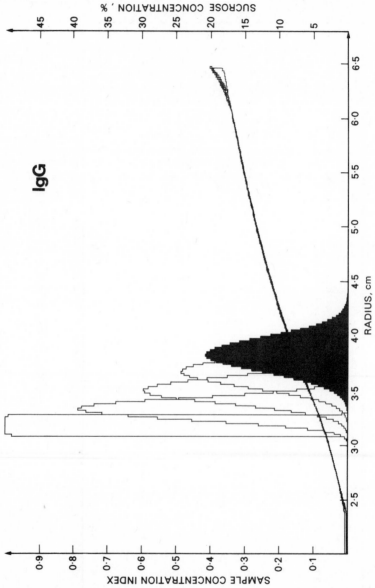

Fig. 2. Computer plot of the result of a simulated run. IgG ($s = 7.0 \times 10^{-13}$ sec, $D = 4.0 \times 10^{-7}$ cm²/sec) was used as sample material, and was centrifuged in an isokinetic gradient. The block zone to *the left* is the initial sample zone, while the dark hatched gaussian zone is the final sample zone. The other gaussian curves show three intermediate zones.

needed to calculate the amount of material which is transported through one of the cylindrical planes in a given short period of time under given experimental conditions, and assuming ideal behaviour of the particles in question.

The simulation program

A rate-zonal centrifugation experiment is conveniently described by the concentration-radius profile at the beginning of the experiment and that at the end of the experiment. For the purpose of simulation this profile may appear as a list of the concentrations of sample and gradient materials, respectively, in each segment. The equations mentioned in the previous paragraph may allow calculation of the amount of material which under given conditions is transported through any of the planes during a short period of time. Then having completed the calculation of transported materials for each plane, the concentrations of sample and gradient materials can be adjusted accordingly, and a new and repeated calculation of transported materials may take place. These iterations may be continued until the desired total time of the centrifugation is reached, providing the final distribution of sample and gradient materials.

A FORTRAN program was written to perform these numerical operations. A comprehensive description of the theory behind the program is given in [1], and the complete program is presented in [6].

Fig. 2 shows as an example the fate of an IgG block zone during centrifugation for 300 min in an isokinetic sucrose gradient.

RESULTS AND DISCUSSION

Fig. 3 gives an example of the type of results which can be obtained with this simulation technique. A 'blood'-sample containing plasma albumin, IgG and IgM has been centrifuged in an isokinetic sucrose gradient at 8° for 300 min. At the beginning of the run the three components of the sample were located at a radius of 3.2 cm corresponding to 100 ml overlay. During centrifugation the three different proteins have migrated with different mobilities, and on completion of the run they are located at the following radius values: albumin 3.6 cm, IgG 3.8 and finally IgM at 4.9 cm. Initially the sample zone was of a rectangular shape, and this shape is changed to the shape of a normal frequency curve. Although purely theoretical, this simulation run demonstrates a well-known practical observation, namely, that macroglobulins may be separated relatively easily from smaller plasma proteins. It has, however, proved difficult to separate albumin from 7 S globulins in serum samples [7,9,10].

Practical *versus* simulated experiment

The extent to which the simulation procedure can predict the outcome of a practical zonal experiment has been checked by performing the same experiment

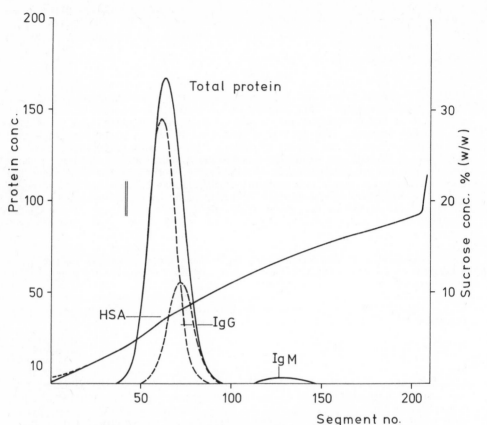

Fig. 3.   Simulated separation of plasma proteins.   The calculations are based on the following hydrodynamic parameters:

human serum albumin, $s = 4.3 \times 10^{-13}$ sec, $D = 5.9 \times 10^{-7}$ cm^2/sec;

IgG,	7.0	"	4.0	"	;
IgM,	19.0	"	2.40	"	;
sucrose,	0.26	"	52.30	"	.

in a centrifuge and on a computer.   A 2 ml sample containing 2 mg/ml serum albumin was centrifuged on a 3 to 20 % (w/w) sucrose gradient for 320 min at 48,000 rev/min at 8°.   The overlay was 100 ml.   The experimental measurements from this run were used to calculate sedimentation coefficients and to plot the results as protein concentration against rotor radius.   In parallel, the same isokinetic sucrose gradient was constructed by another program [5], and

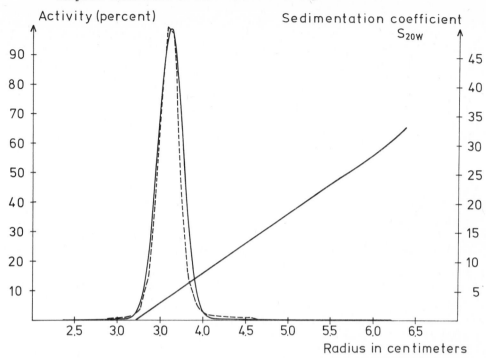

Fig. 4. Simulated versus practical experiment. A 2 ml HSA sample was centrifuged in an isokinetic gradient. In parallel an identical run was simulated
on the computer. The result of the simulated run is shown with the fully
drawn curve. The practical run is shown with the dashed curve. The results
from the two experiments have been converted to the same graph and superimposed.

used as data for the simulation program. The results from the simulation procedure were used as data for the program for calculation of sedimentation
coefficients. The results from the two different experiments were finally
plotted (by hand) in the same graph (Fig. 4).

It appears from Fig. 4 that the agreement between practically obtained
and simulated results is quite good. The ascending parts of the sample zones
are nearly identical while the descending parts of the two zones are separated
with a distance corresponding to about two segments. However, the two curves

are perfectly parallel from the top to the baseline. Hence, it is concluded
that the simulation procedure as tested with a critical sample as a small
soluble protein can be used to predict the outcome of a zonal experiment with
a reasonable accuracy.

The slight zone-broadening effect observed in the simulation procedure
is due mainly to the calculation of sedimentation flux. Previous studies [1]
on the problem have revealed that zone broadening will occur in the simulation
procedure if each segment is not emptied in each iterative cycle. In principle, this effect should be avoidable if either the Δt-value were sufficiently long, or if the distance between two consecutive segments were sufficiently
short to ensure a complete emptying of the segment in question. Unfortunately,
these demands are opposite to the demands for a perfect simulation of diffusion fluxes where the least error occurs when the Δt-value is the shortest
possible. Hence, a compromise is necessary, and the choice of Δt-value and
of a number of segments will depend on the hydrodynamic parameters of the test
particles also.

Three different types of investigations have been carried out to clarify this problem:-
1) Variation of Δt-values with all other conditions kept constant have shown
that there is an upper limit for the size of Δt-values. Runs with serum albumin as test particle showed that Δt could be varied between 1 and 30 sec, and
that the smallest zone widths were found with the highest Δt values [1].
2) Variation of the sedimentation coefficient of test particles with all
other conditions kept constant (including the size of the diffusion coefficient) showed that the zone-broadening effect decreased sharply with increasing sedimentation coefficients [1].
3) Variation of the number of segments. The simulation program, normally
used with 207 segments, was revised to give an option between 43, 207 and
1027 segments in order to study the importance of the number of segments with
respect to the final zone shape. It appeared, however, that a program with
1027 segments could be run only with very short Δt-values (1 to 3 sec). If
the program were used with larger Δt-values it was normally stopped by the
computer because of division by nought arising by an overestimation of gradient movements. Use of this program with Δt-values less than 3 sec proved
highly time-consuming on the computer (a CDC 6400), and use of Δt = 1 sec
used about as much central processor time as the same experiment would require
in a real centrifuge. Hence, a solution employing 1027 segments is regarded
as prohibitive for practical use.

The program version with 43 segments was found to work without problems
in the computer. However, the segments are generally too large especially
with respect to sample volumes, and the final sample zones are broader than
found by use of 207 segments (Fig. 5).

Therefore, it was concluded that use of 207 segments in general is
most suitable.

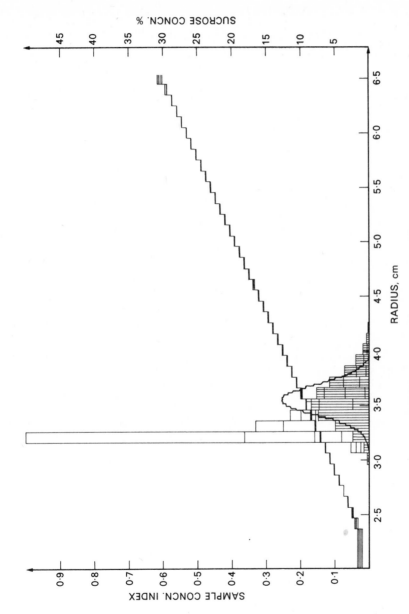

Fig. 5. Comparison of program working with 43 segments and a version of the same program working with 207 segments. The computer-made plot gives the results of a simulated run with the 43-segment program. Inserted by hand drawing are the results of an identical experiment calculated with the 207-segment version.

Zone width and gradient concentration

The zonal simulation program has been written as an aid in optimization work, and with respect to that it has several useful functions. One is that a theoretical separation in a relatively short time and without technical problems can be re-run several times under different conditions until a desired set of conditions is found. Another problem which can be solved with the simulation program is a calculation of the hydrodynamic stability situation throughout a complete experiment. The latter aspect is discussed in the previous paper in this series [1], and reference with respect to these problems should be made to the papers of Meuwissen [11, 12].

Pertinent to the types of problems mentioned first are the following calculations. Two samples (one a soluble protein and another a 70 S ribosome) were centrifuged in different isokinetic gradients. The gradients were constructed so that the sample particles in each case sedimented the same distance. Hence, more concentrated sucrose gradients require a longer run time. The results of these simulated experiments are summarized in Table 1, from which it appears that there is a slight increase in zone width with increasing duration of centrifugation. In generalized form these calculations may suggest that it is advisable to choose the shortest possible run time in planning zonal experiments.

Concluding remarks

In conclusion it may be stated that the numerical method as described above may be used to predict the result of a separation in a zonal centrifuge. However, it should be noted that the method so far represents a first attempt to simulate the sedimentation and diffusion in a zonal rotor, and that future improvements of the method may be expected. Planned work on these problems includes the following aspects: 1) a simplification by introduction of some computer-minded approximative polynomials, 2) an improved method of calculation of sedimentation fluxes, and 3) methods for introduction of means for studying the effect of increasing concentration of particles in the sample zone.

Acknowledgments

This work has been supported in parts by Statens Laegevidenskabelige Forskningsraad and by Fonden til Laegevidenskabens Fremme.

References

1. Steensgaard, J., Funding, L. & Meuwissen, J.A.T.P., *Eur. J. Biochem. 39* (1973) 481-491.

2. Bishop, B.S., *Nat. Cancer Inst. Monogr. 21* (1966) 175-188.

3. Funding, L., in *European Symposium of Zonal Centrifugation* (Chermann, J.C.,

Table 1. Zone width in different isokinetic sucrose gradients

Particle size *Svedbergs*	Sucrose concn. *% w/w*	Run time *min*	Zone width *cm*
4	6.3	300	0.34
4	19.6	600	0.37
4	29.4	1200	0.38
4	34.1	1800	0.39
4	37.1	2400	0.40
4	39.2	3000	0.41
70	14.9	100	0.43
70	36.5	500	0.48
70	39.7	700	0.49

ed.) [*Spectra (2000) 4*], Editions Cité Nouvelle (1973), pp. 45-49.

4. Steensgaard, J. & Hill, R.J., *Anal. Biochem. 34* (1970) 485.

5. Steensgaard, J., *Eur. J. Biochem. 16* (1970) 66.

6. Steensgaard, J., Funding, L. & Meuwissen, J.A.T.P., *this volume*,

7. Anderson, N.G., Waters, D.A., Fisher, W.D., Cline, G.B., Nunley, C.E., Elrod, L.H. & Rankin, Jr. C.T., *Anal. Biochem. 21* (1967) 235.

8. Barber, E.J., *Nat. Cancer Inst. Monogr. 21* (1967) 219.

9. Cooke, K.B. & Apsey, M.E., in *Separations with Zonal Rotors* (Reid, E., ed.), Wolfson Bioanalytical Centre, University of Surrey, Guildford (1971), pp. M-2.1 - M-2.9.

10. Cooke, K.B., at *European Symposium of Zonal Centrifugation* (unpublished contribution; cf. refs. [3] & [12]).

11. Meuwissen, J.A.T.P., in *Methodological Developments in Biochemistry*, Vol. 3 (Reid, E., ed.), Longman, London (1973), pp. 29-44.

12. Meuwissen, J.A.T.P., in *European Symposium of Zonal Centrifugation* (Chermann, J.C., ed.) [*Spectra (2000) 4*], Editions Cité Nouvelle, Paris (1973) pp. 21-31.

7 A FORTRAN PROGRAM FOR SIMULATION OF ZONAL CENTRIFUGATION

Jens Steensgaard, Lars Funding and Jules A.T.P. Meuwissen
Institute of Medical Biochemistry   *Liver Physiopathology Laboratory*
University of Aarhus   *Rega Institute*
Denmark   *Leuven, Belgium*

In order to facilitate the use of the procedure for simulation of rate-zonal centrifugation, the theoretical background of which is described in the preceding article [1] and elsewhere [2], a complete FORTRAN program written for this purpose is now presented, with an example of output.

The FORTRAN version used here is defined by the Control Data FORTRAN extended reference manual [3], but can be accepted directly by most FORTRAN compilers.

The program is initiated by a few functions and subroutines. One of these calculates density of sucrose solutions from sucrose concentration and temperature. The others are introduced to facilitate data input from remote tape terminals (TTY terminals or TTY comparable remote terminals), and these functions may be replaced by other input statements according to the available input possibilities.

Several comments are inserted in the program to illustrate the route of calculations and to separate different parts of the program. Table 1 gives the physical meaning of some important variables and a translation to the physical notation used in refs. [1,2]. Table 2 shows an example of a data file. Following the program, an output example is given. The outputs are designed to yield snapshots of the situation in the rotor at different time-points. An output table therefore shows for each segment the concentration of sample material, the concentration of gradient material, and a massbalance. Moreover, an index of hydrodynamic stability, calculated according to the second stability criterion of Meuwissen [1, 4], is given in the column 'load index'. A zone is regarded as stable if this index is zero or negative.

References

1. Steensgaard, J. & Funding, L., *this volume,* 55-65.

2. Steensgaard, J., Funding, L. & Meuwissen, J.A.T.P., *Eur. J. Biochem.* *39* (1973) 481-491.

3. Control Data: FORTRAN Extended Reference Manual, CDC, Sunneyvale, Calif. 94086, U.S.A. (1970).

4. Meuwissen, J.A.T.P., in *Methodological Developments in Biochemistry, Vol.3* (Reid, E., ed.), Longman, London (1973) 29-44.

Table 1. Meaning of some important symbols. *The Table is continued opposite*

FORTRAN name	Physical notation	Meaning
H	h	Rotor height
VN	v	Volume of segment no. n
RAD	r	Radius of plane n
J1	M'_n	Amount of transported sample material across a plane in t sec

Table 2. Example of data file. *Note that numbers can be given without format, but separated by , or CR.*

600	Experiment number
0,0 0,0.09 .. 2.0,3.7 .. .. 0.0,19.	207 paired values giving the concentration of sample material (mg/ml) and the concentration of sucrose (%, w/w) in each segment at the beginning of an experiment
8	Temperature (degrees C)
10	Acceleration period (min)
300	Duration of centrifugation (min)
10	Deceleration period (min)
48000	Rotor speed (rev/min)
12	Sample volume (ml)
96	Overlay volume (ml)
4.3	$s_{20.w}$ of sample material (Svedberg units)
5.9	$D_{20.w}$ of sample material (Fick units)
1.4	Particle density of sample material (mg/ml)
0.26	$s_{20.w}$ of sucrose (Svedberg units)
52.3	$D_{20.w}$ of sucrose (Fick units)
1.6	Particle density of sucrose (mg/ml)
15	Differential cycle length, Δt (sec)
4	Number of snapshots wanted
*	
*	*Space for insertion of comments on experiment*

Table 1, *continued from opposite*

J2	J''_n	Amount of transported gradient material across a plane in t sec
S1	$s'_{20.w}$	Sedimentation coefficient of sample particles
S2	$s''_{20.w}$	Sedimentation coefficient of gradient particles
C1(N)	C'_n	Concn. of sample material in segment n
C2(N)	C''_n	Concn. of gradient material in segment n
OMEGA	ω	Angular velocity
D1	$D'_{20.w}$	Diffusion coefficient of sample particles
D2	$D''_{2o.w}$	Diffusion coefficient of gradient particles
DC1	dC'	Increase in concn. of sample material across a plane
DC2	dC''	Increase in concn. of gradient material across a plane
T	Δt	Time increment
TEMP	T	Temperature
VISC	$n_{T.m}$	Viscosity
DENS	$\rho_{T.m}$	Density

PROGRAM

```
C     EDITON WITHOUT PLOT. 25/6/1973


      REAL FUNCTION DENFU(SUC)
C     REAL FUNCTION DENFU CALCULATES THE DENSITY OF A SUCROSE SOLUTION
C     AT A GIVEN TEMPERATURE AND OF A GIVEN SUCROSE CONCENTRATION
C     (O/O W/W = CG/G SUCROSE)
      COMMON /DENF/MD2,MD1,MD0
      REAL SUC,MD2,MD1,MD0
      DENFU=(MD2*SUC/100.0 + MD1)*SUC/100.0 + MD0
      END


      INTEGER FUNCTION READCH(DUM)
C     ALGOL-LIKE FUNCTION FOR INPUT OF CHARACTERS FROM DATA TAPE
      COMMON/READCOM/CARD(80),CONV(64),COL,SIGN,CHAR
      INTEGER CARD,CONV,COL,SIGN,CHAR,DUM,T
      COL=COL+1
      IF (COL.LT.81) GOTO 4200
      READ(3,12300) (CARD(T),T=1,80)
      COL=0
      READCH=56B
      GOTO 4210
4200  READCH=CARD(COL)
4210  RETURN
12300 FORMAT(80(R1))
      END                                        [continued
```

[continued

```
      INTEGER FUNCTION READINT(DUM)
C     ALGOL-LIKE FUNCTION FOR INPUT OF INTEGERS FROM DATA TAPE
      COMMON/READCOM/CARD(80),CONV(64),COL,SIGN,CHAR
      INTEGER READCH,CARD,CONV,COL,SIGN,CHAR,DUM,I,VAL
      SIGN=+1
      VAL=0
      DO 4000 I=1,1000
      CHAR=READCH(DUM)+1
      CHAR=CONV(CHAR)
      IF (CHAR.LT.12) GOTO 4010
4000  CONTINUE
4010  IF (CHAR.LE.9) GOTO 4030
      IF (CHAR.EQ.10) GOTO 4020
      SIGN=-1
4020  CHAR=0
4030  DO 4040 I=1,15
      VAL=VAL*10+CHAR
4035  CHAR=READCH(DUM)+1
      CHAR=CONV(CHAR)
      IF(CHAR.EQ.17)GOTO 4035
      IF (CHAR.GT.9) GOTO 4050
4040  CONTINUE
4050  READINT=VAL*SIGN
      RETURN
      END


      REAL FUNCTION READREA(DUM)
C     ALGOL-LIKE FUNCTION FOR INPUT OF REALS FROM DATA TAPE
      COMMON/READCOM/CARD(80),CONV(64),COL,SIGN,CHAR
      INTEGER READCH,READINT,CARD,CONV,COL,SIGN,CHAR,DUM,I
      REAL AP,BP,DIV

      DIV=10.0
      AP=0.0
      BP=FLOAT(READINT(DUM))
      IF (CHAR.NE.12) GOTO 4110
      DO 4100 I=1,15
4105  CHAR=READCH(DUM)+1
      CHAR=CONV(CHAR)
      IF(CHAR.EQ.17)GOTO 4105
      IF (CHAR.GT.9) GOTO 4110
      AP=AP+CHAR/DIV
4100  DIV=DIV*10.0
4110  READREA=BP+SIGN*AP
      RETURN
      END


      SUBROUTINE NEWPAGE
C     SUBROUTINE NEWPAGE PRINTS THE FIRST LINE ON A NEW PAGE
      COMMON/PAGEINF/CDATE,COMTIME,EXPNO,PAGE
      INTEGER EXPNO,PAGE,CDATE,COMTIME
      WRITE(1,12000)EXPNO,PAGE,CDATE,COMTIME
      PAGE=PAGE+1
      RETURN
12000 FORMAT(1H1,//,11X,9HEXP. NO.   ,I3,4H P. ,I2,22X,A10,A10)
      END
```

```
      PROGRAM SIMZON(INPUT,OUTPUT,TAPE1=OUTPUT,TAPE3=INPUT)
C     PROGRAM FOR SIMULATION OF ZONAL CENTRIFUGATION
      INTEGER EXPNO,TEMP,RU,RT,RD,SPEED,SA,OV,T,NUPRINT,N,CYCLES
      INTEGER PRINT,MIN,N1,ENOUGH,B121,B999,COUNT,MINS,N2
      INTEGER TEXT(4),PAGE,I,J,DUM,CONV,CARD,COL,SIGN,CHAR
      INTEGER READCH,READINT,PUT,II,I1
      REAL C1(207),C2(207),DJ1(207),DJ2(207),LOAD(207),MASS(207)
      REAL S1,D1,PDENS1,S2,C2,PDENS2,PI,H,HEPI,INTOM,MAX,TRY,TOTAL1
      REAL TOTAL2,Q1S,Q2S,Q1D,Q2D,F,TABS,K1,DIF1,DIF2,KLOAD
      REAL OMEGA,OMEGA1,OMEGA2,J1,J2,CC1,CC2,RAD,OLDC1,OLDC2,C2G,DENS,L1
      REAL L2,MY,AD,BD,CG,VISC,K,DC1,DC2,OLDJ1,OLDJ2,BAL1,BAL2,VN
      REAL HIGH,RJ,FLYT,X,Y,READREA,MIDRAD,OLDBAL,DENFU,SUC
      LOGICAL FIRSTPL
      COMMON/READCOM/CARD(80),CONV(64),COL,SIGN,CHAR
      COMMON/PAGEINF/CDATE,COMTIME,EXPNO,PAGE
      COMMON/DENF/MD2,MD1,MD0
      REAL MD2,MD1,MD0
      DATA(CONV(I),I=1,64)/14*13,0,12*13,0,1,2,3,4,5,6,7,8,9,10,11,
     *14,5*13,17,16,12,11*13,15,4*13/
C     SUCROSE CONCENTRATIONS IN PER CENT(W/W),SAMPLE IN MG/ML
C     SUCROSE CONC. MUST NOT EXCEED 48 PER CENT

      CDATE=DATE(DUM)
      COMTIME=TIME(DUM)
      COL=80
      PUT=0

C     READ DATA HERE
C     DATA MUST APPEAR AS FOLLOWS:
C     EXPNO(NUMBER OF EXPERIMENTS -INTEGER)
C     SAMPLE AND GRADIENT CONC.(207 PAIRED REALS)
C     TEMP (CENTIGRADES-INTEGER)
C     RUN UP(MINS-INTEGER)
C     RUN TIME(MINS-INTEGER)
C     RUN DOWN(MINS-INTEGER)
C     SPEED(REV/MIN -INTEGER)
C     SAMPLE VOLUME(ML-INTEGER)
C     OVERLAY VOLUME(ML-INTEGER)
C     SEDIMENTATION COEFFICIENTS(SVEDBERGS-REAL)
C     DIFFUSION COEFFICIENTS(FICKS-REALS)
C     PARTICLE DENSITY VALUES(G/ML -REALS)
C     DIFFERENTIAL CYCLE LENGTH(SECONDS-INTEGER)
C     NUMBER OF PRINTOUTS (INTEGERS)

      EXPNO=READINT(DUM)
      DO 90 J=1,207
      C1(J)=READREA(DUM)
   90 C2(J)=READREA(DUM)
      TEMP=READINT(DUM)
      RU=READINT(DUM)
      RT=READINT(DUM)
      RD=READINT(DUM)
      SPEED=READINT(DUM)
      SA=READINT(DUM)
      OV=READINT(DUM)
      S1=READREA(DUM)
      D1=READREA(DUM)
      PDENS1=READREA(DUM)
      S2=READREA(DUM)
      D2=READREA(DUM)
      PDENS2=READREA(DUM)
      T=READINT(DUM)
      NUPRINT=READINT(DUM)
```

```
       FIRSTPL=.TRUE.
       PI=3.1415926535898
       H=4.878
       HEPI=H*PI
       INTOM=((SPEED*PI)**2)*(RT+(RU+RD)/3.0)/15.0
       MC0=(-5.8513271E-6*TEMP + 3.9680504E-5)*TEMP + 1.0003698
       MD1=(1.2392833E-5*TEMP - 1.0578919E-3)*TEMP + 0.38982371
       MD2=(-8.9239737E-6*TEMP + 4.7530081E-4)*TEMP + 0.17097594

C      CALCULATION OF TOTAL AMOUNT OF SAMPLE AND GRADIENT MATERIAL
C      AND IDENTIFICATION OF MAXIMUM SAMPLE CONCENTRATION
       TRY=0.0
       TOTAL1=C1(1)*30.4
       TOTAL2=C2(1)*30.4*DENFU(C2(1))
       DO 100 N=2,206
       TRY=C1(N)
       IF(TRY.GT.MAX)MAX=TRY
       VN=(0.0932+0.0008*N)*HEPI
       TOTAL1=TOTAL1+TRY*VN
100    TOTAL2=TOTAL2+C2(N)*VN*DENFU(C2(N))
       TOTAL1=TOTAL1+C1(207)*55.43
       TOTAL2=TOTAL2+C2(207)*55.43*DENFU(C2(207))

C      OUTPUT TYPE1 HERE. GENERAL DESCRIPTION OF SIMULATED EXPERIMENT.
       WRITE(1,10005)EXPNO,CDATE,COMTIME
       DO 110 J=1,5
110    WRITE (1,10250)
       PAGE=2
       WRITE(1,10000)
       WRITE(1,10010)
       WRITE(1,10020)
       WRITE(1,10030)
       WRITE(1,10040)
       WRITE (1,10250)
       WRITE(1,10050)
       WRITE(1,10060)
5000   COL=80
       CHAR=READCH(DUM)
       CHAR=READCH(DUM)+1
       IF (CONV(CHAR).NE.14) GOTO 5010
       WRITE(1,10070) (CARD(II),II=2,80)
       GOTO 5000
5010   WRITE(1,10080)
       WRITE(1,10090)
       WRITE(1,10100) SA,RU
       WRITE(1,10110) OV,RT
       WRITE(1,10120) SPEED,RD
       WRITE(1,10130) TEMP,T
       WRITE(1,10140) INTOM
       WRITE(1,10150) NUPRINT
       WRITE(1,10160)
       WRITE(1,10170) S1,S2
       WRITE(1,10180) D1,D2
       WRITE(1,10190) PDENS1,PDENS2
       WRITE(1,10200) TOTAL1,TOTAL2
```

```
C      CALCULATION OF CONSTANTS DEPENDING ON THE ACTUAL EXPERIMENT
       COUNT=0
       Q1S=S1*1E-13*1.002/(PDENS1-0.998203)
       Q2S=S2*1E-13*1.002/(PDENS2-0.998203)
       Q1D=D1*1E-7*1.002/293.16
       Q2D=D2*1E-7*1.002/293.16
       F=342.3/18.023
       TABS=273.16+TEMP
       K1=2*PI*H*T
       DIF1=Q1D*TABS*50
       DIF2=Q2D*TABS*50
       KLOAD=-PDENS1*D1/PDENS2/D2
       CYCLES=60*(RU+RT+RD)/T
       PRINT=CYCLES/NUPRINT

150    DO 1000 N1=1,3
C      THE OUTER LOOP TREATS RUNUP- RUN- AND RUNDOWN PERIODS SEPARATELY
       MIN=RU
       OMEGA=((2*PI*SPEED/60.0)**2)/3.0
       IF(N1.EQ.2)MIN=RT
       IF(N1.EQ.3)MIN=RD
       IF(N1.EQ.2)OMEGA=3*OMEGA
       ENOUGH=60*MIN/T

200    DO 900 N2=1,ENOUGH
C      THIS LOOP CALCULATES DJ-VALUES,NEW CONCENTRATIONS
C      AND SNAPSHOT OUTPUTS
       OMEGA1=Q1S*OMEGA
       OMEGA2=Q2S*OMEGA
       J1=0
       J2=0
       COUNT=COUNT+1
       CC1=C1(1)
       CC2=C2(1)

       DO 300 N=1,206
C      CALCULATION OF NET FLOW ACROSS ANY PLANE(N)
       RAD=2.34+N*0.02
       OLDC1=CC1
       OLDC2=CC2
       CC1=C1(N+1)
       CC2=C2(N+1)
       C2G=(CC2+OLDC2)/2
       DENS=(MD2*C2G/100+MD1)*C2G/100+MD0
       L1=OMEGA1*OLDC1*RAD*(PDENS1-DENS)
       L2=OMEGA2*OLDC2*RAD*(PDENS2-DENS)
       MY=C2G/(C2G+(100-C2G)*F)
       AD=((((((4.5921911E9*MY-1.1028981E9)*MY+1.0323349E8)*MY
      *    -4.6927102E6)*MY+1.0504137E5)*MY-1.1435741E3)*MY+9.4112153)*MY
      *    -1.5018327
       BD=((((((-5.4970416E11*MY+1.3532907E11)*MY-1.2985834E10)*MY
      *    +6.0654775E8)*MY-1.4184371E7)*MY+1.6911611E5)*MY+1.6077073E3)
      *    *MY+2.1169907E2
       CG=146.06635-25.251728*SQRT(1+(MY/0.070674842)**2)
       VISC=10.0**(AD+BD/(TEMP+CG))
       K=K1*RAD/VISC
       DC1=CC1-OLDC1
       DC2=CC2-OLDC2
       OLDJ1=J1
       OLDJ2=J2
       J1=(L1-DC1*DIF1)*K
       J2=(L2-DC2*DIF2)*K
```

```
C       CALCULATICN OF NET INCREMENT OF SAMPLE AND GRADIENT SUBSTANCE IN
C       EACH SEGMENT ABOVE A PLANE AND CALCULATION OF MEUWISSENS
C       STABILITY CRITERION.
        DJ1(N)=OLDJ1-J1
        DJ2(N)=OLCJ2-J2
        IF (DC1.EQ.0) GO TO 290
        IF(DC2.EQ.0) GOTO 295
        LOAD(N)=KLOAD-DC1/(10.0*DC2*DENFU(C2G))
        GO TO 300
290     LOAD(N)=0
        GOTO 300
295     LCAD(N)=9999
300     CONTINUE

        DJ1(207)=J1
        DJ2(207)=J2
        LOAD(207)=0

C       ADJUSTMENT OF CONCENTRATIONS IN EACH SEGMENT
        C1(1)=C1(1)+DJ1(1)/30.4
        C2(1)=C2(1)+DJ2(1)/(30.4*DENFU(C2(1)))
        BAL1=C1(1)*30.4/TOTAL1
        MASS(1)=BAL1
        BAL2=C2(1)*30.4*DENFU(C2(1))
        DO 500 N=2,206
        VN=(.0932+.0008*N)*HEPI
        C1(N)=C1(N)+DJ1(N)/VN
        C2(N)=C2(N)+DJ2(N)/(VN*DENFU(C2(N)))
        OLDBAL=BAL1
        BAL1=BAL1+C1(N)*VN/TOTAL1
        IF ((BAL1.GE.0.5).AND.(OLDBAL.LT.0.5)) MIDRAD=2.34+(N-1)*.02
       *+(((.5-OLDBAL)/(BAL1-OLDBAL))**2)*.02
        BAL2=BAL2+C2(N)*VN*DENFU(C2(N))
500     MASS(N)=BAL1
        C1(207)=C1(207)+DJ1(207)/55.43
        C2(207)=C2(207)+CJ2(207)/(55.43*DENFU(C2(207)))
        MASS(207)=BAL1+C1(207)*55.43/TOTAL1
        BAL2=(BAL2+C2(207)*55.43*DENFU(C2(207)))/TOTAL2

C       OUTPUT OF TABLES WITH SNAPSHOTS
        B121=MOD((COUNT-1+PRINT),PRINT)
        IF (B121.EQ.0) GO TO 357
        IF (COUNT.EQ.CYCLES) GO TO 357
        GO TO 900
357     MINS=COUNT*T/60
C       OUTPUT OF TYPE2 HERE.
        PUT=PUT+1
        CALL NEWPAGE
        WRITE(1,10215)
        WRITE(1,10210)MINS,PUT
        WRITE(1,10250)
        WRITE(1,10220)
        WRITE(1,10230)
        WRITE(1,10215)
        DO 5020 J=1,207
        IF(MOD(J,52).NE.0)GOTO 5020
        CALL NEWPAGE
        DO 5019 I1=1,3
5019    WRITE(1,10250)
5020    WRITE(1,10240)J,C1(J),C2(J),LOAD(J),MASS(J)
        WRITE(1,10215)
        WRITE(1,10260)BAL2,COUNT
        WRITE (1,10265) MIDRAD
        WRITE(1,10215)
```

```
 900  CONTINUE

1000  CONTINUE

10000 FORMAT(1X,37X,46HSIMULATION OF ZONAL CENTRIFUGATION EXPERIMENTS)
10005 FORMAT(1H1,//,1X,21X,9HEXP. NO. ,I4,2X,4HP. 1,37X,A10,A10)
10010 FCRMAT(1X,37X,46(1H=))
10020 FORMAT(/,1X,35X,49H3Y J.STEENSGAARD, L.FUNDING AND J.A.T.P.MEUWISS
     *EN)
10030 FORMAT(1X,43X,33HINSTITUTE OF MEDICAL BIOCHEMISTRY)
10040 FORMAT(1X,45X,29HUNIVERSITY OF AARHUS, DENMARK)
10050 FORMAT(1X,37X,46HPROGRAM IN CDC FORTRAN EX. VERSION OF 12/02/73)
10060 FORMAT(/,1X,52X,16HB-XIV LIKE ROTOR,///)
10070 FORMAT(1X,21X,79R1)
10080 FORMAT(///,1X,44X,26HCENTRIFUGATION PARAMETERS!)
10090 FORMAT(1X,44X,26(1H-))
10100 FCRMAT(1X,21X,13HSAMPLE VOLUME,12X  ,I5,3H ML,6X,19HACCELERATION P
     *ERIOD,7X,I4,8H MINUTES)
10110 FORMAT(1X,21X,14HOVERLAY VOLUME,11X  ,I5,3H ML,6X,15HDURATION OF R
     *UN,9X,I6,8H MINUTES)
10120 FORMAT(1X,21X,11HROTOR SPEED,12X  ,I7,4H RPM,5X,19HDECELERATION PE
     *RIOD,7X,I4,8H MINUTES)
10130 FORMAT(1X,21X,11HTEMPERATURE,14X  ,I5,7H CEGR C,2X,17HDIFFERENTIAL
     *CYCLE,8X,I5,8H SECONDS)
10140 FORMAT(/,1X,31X,27HINTEGRATED FORCE-TIME FIELD,3X,G10.4,13H RAD.**
     *2*SEC.)
10150 FCRMAT(/,1X,21X,19HNUMBER OF PRINTOUTS,6X,I3)
10160 FORMAT(///,1X,49X,21HOLECULAR PARAMETERS!,/,1X,49X,21(1H-),/,1X,5
     *7X,6HSAMPLE,15X, 8HGRADIENT)
10170 FORMAT(1X,21X,26HSEDIMENTATION COEFFICIENTS, 5X,F7.2,10H SVEDBERGS
     *,5X,F7.2,10H SVEDBERGS)
10180 FORMAT(1X,21X,22HDIFFUSION COEFFICIENTS,9X,F7.2,6H FICKS,9X,F7.2,6
     *H FICKS)
10190 FORMAT(1X,21X,18HPARTICLE DENSITIES,13X,F7.3,5H G/ML,10X,F7.3,5H G
     */ML)
10200 FORMAT(1X,21X,18HAMOUNT OF MATERIAL,13X,F7.3,3H MG,12X,-2PF7.3,2H
     *G)
10210 FORMAT(1X,10X,12HRUN DURATION, 3X,I5,6H MINS.,16X,11HOUTPUT NO  ,
     *I3)
10220 FCRMAT(1X,10X, 7HSEGMENT, 6X,6HSAMPLE, 6X, 8HGRADIENT, 6X, 4HLOAD
     *,6X,11HMASSBALANCE)
10230 FORMAT(1X,10X, 7H NO  , 5X, 8HD/D*10E3,5X,8HW/W D/D , 6X,5HINDEX
     *,4X,11H   INDEX  )
10215 FORMAT(1X,10X,60(1H-))
10240 FORMAT(1X,10X,2X,I3,2X,3X,3PF9.4,6X,2X,0PF4.1,2X,5X,G10.4,
     *4X,F6.4,4X,F6.2)
10250 FORMAT(/)
10260 FORMAT(1X,10X,23HMASSBALANCE OF GRADIENT, 3X,F6.4,   6X,12HNO OF C
     *YCLES,3X,0PI5)
10265 FORMAT (1X,10X,23HRADIUS AT MASS MIDPOINT,3X,F6.4
     *,3H CM)


      END
```

SIMULATION OF ZONAL CENTRIFUGATION EXPERIMENTS
===

BY J.STEENSGAARD, L.FUNDING AND J.A.T.P.MEUWISSEN
INSTITUTE OF MEDICAL BIOCHEMISTRY
UNIVERSITY OF AARHUS, DENMARK

PROGRAM IN CDC FORTRAN EX. VERSION OF 12/02/73

B-XIV LIKE ROTOR

ALBUMIN-LIKE SAMPLE ON LINEAR GRADIENT.

CENTRIFUGATION PARAMETERS:

SAMPLE VOLUME	12 ML	ACCELERATION PERIOD	10 MINUTES
OVERLAY VOLUME	96 ML	DURATION OF RUN	300 MINUTES
ROTOR SPEED	48000 RPM	DECELERATION PERIOD	10 MINUTES
TEMPERATURE	8 DEGR C	DIFFERENTIALCYCLE	15 SECONDS

INTEGRATED FORCE-TIME FIELD   .4649E+12 RAD.**2*SEC.

NUMBER OF PRINTOUTS   4

MOLECULAR PARAMETERS:

	SAMPLE	GRADIENT
SEDIMENTATION COEFFICIENTS	4.30 SVEDBERGS	.26 SVEDBERGS
DIFFUSION COEFFICIENTS	5.90 FICKS	52.30 FICKS
PARTICLE DENSITIES	1.400 G/ML	1.600 G/ML
AMOUNT OF MATERIAL	23.539 MG	75.491 G

RUN DURATION   320 MINS.   OUTPUT NO   5

SEGMENT NO	SAMPLE 0/0*10F3	GRADIENT W/W 0/0	LOAD INDEX	MASSBALANCE INDEX
1	.0000	.5	-.9871E-01	.0000
2	.0000	.6	-.9871E-01	.0000
3	.0000	.6	-.9871E-01	.0000
4	.0000	.7	-.9871E-01	.0000
5	.0000	.7	-.9871E-01	.0000
6	.0000	.8	-.9871E-01	.0000
7	.0000	.8	-.9871E-01	.0000
8	.0000	.9	-.9871E-01	.0000
9	.0000	1.0	-.9871E-01	.0000
10	.0000	1.0	-.9871E-01	.0000
11	.0000	1.1	-.9871E-01	.0000
12	.0000	1.2	-.9871E-01	.0000
13	.0000	1.2	-.9871E-01	.0000
14	.0000	1.3	-.9871E-01	.0000
15	.0000	1.4	-.9871E-01	.0000
16	.0000	1.5	-.9871E-01	.0000
17	.0000	1.5	-.9871E-01	.0000
18	.0000	1.6	-.9871E-01	.0000
19	.0000	1.7	-.9871E-01	.0000
20	.0001	1.8	-.9871E-01	.0000
21	.0001	1.9	-.9871E-01	.0000
22	.0002	1.9	-.9871E-01	.0000
23	.0004	2.0	-.9871E-01	.0000
24	.0009	2.1	-.9871E-01	.0000
25	.0017	2.2	-.9871E-01	.0000
26	.0032	2.3	-.9871E-01	.0000
27	.0061	2.4	-.9872E-01	.0000
28	.0112	2.5	-.9872E-01	.0000
29	.0204	2.5	-.9873E-01	.0000
30	.0364	2.6	-.9874E-01	.0000
31	.0640	2.7	-.9876E-01	.0000
32	.1107	2.8	-.9880E-01	.0000
33	.1883	2.9	-.9885E-01	.0000
34	.3148	3.0	-.9894E-01	.0001
35	.5176	3.1	-.9906E-01	.0001
36	.8368	3.2	-.9926E-01	.0002
37	1.3300	3.2	-.9954E-01	.0003
38	2.0784	3.3	-.9994E-01	.0004
39	3.1929	3.4	-.1005	.0007
40	4.8219	3.5	-.1013	.0011
41	7.1582	3.6	-.1023	.0017
42	10.4456	3.7	-.1037	.0025
43	14.9828	3.8	-.1054	.0038
44	21.1239	3.9	-.1076	.0056
45	29.2731	4.0	-.1103	.0080
46	39.8725	4.0	-.1135	.0114
47	53.3804	4.1	-.1171	.0159
48	70.2416	4.2	-.1212	.0220
49	90.8472	4.3	-.1255	.0298
50	115.4877	4.4	-.1300	.0398
51	144.3016	4.5	-.1345	.0524

52	177.2243	4.6	-.1386	.0679
53	213.9439	4.7	-.1420	.0868
54	253.8594	4.8	-.1445	.1094
55	296.1180	4.9	-.1457	.1358
56	339.5279	5.0	-.1454	.1663
57	382.6973	5.0	-.1434	.2009
58	424.0522	5.1	-.1396	.2394
59	461.9372	5.2	-.1341	.2817
60	494.7241	5.3	-.1269	.3272
61	520.9279	5.4	-.1185	.3753
62	539.3189	5.5	-.1091	.4254
63	549.0184	5.6	-.9925E-01	.4768
64	549.5691	5.7	-.8940E-01	.5284
65	540.9706	5.8	-.8006E-01	.5796
66	523.6770	5.9	-.7167E-01	.6294
67	498.5574	6.0	-.6459E-01	.6770
68	466.8236	6.0	-.5909E-01	.7219
69	429.9336	6.1	-.5530E-01	.7634
70	389.4812	6.2	-.5326E-01	.8012
71	347.0839	6.3	-.5286E-01	.8351
72	304.2791	6.4	-.5393E-01	.8650
73	262.4384	6.5	-.5622E-01	.8909
74	222.7041	6.6	-.5945E-01	.9130
75	185.9532	6.7	-.6330E-01	.9315
76	152.7858	6.8	-.6751E-01	.9469
77	123.5367	6.9	-.7182E-01	.9593
78	98.3045	7.0	-.7602E-01	.9693
79	76.9921	7.0	-.7994E-01	.9771
80	59.3534	7.1	-.8349E-01	.9832
81	45.0406	7.2	-.8661E-01	.9878
82	33.6475	7.3	-.8926E-01	.9913
83	24.7472	7.4	-.9147E-01	.9939
84	17.9207	7.5	-.9326E-01	.9957
85	12.7783	7.6	-.9468E-01	.9971
86	8.9726	7.7	-.9578E-01	.9980
87	6.2047	7.8	-.9662E-01	.9987
88	4.2258	7.9	-.9724E-01	.9991
89	2.8348	8.0	-.9770E-01	.9994
90	1.8733	8.1	-.9802E-01	.9996
91	1.2195	8.1	-.9825E-01	.9998
92	.7821	8.2	-.9841E-01	.9999
93	.4943	8.3	-.9851E-01	.9999
94	.3078	8.4	-.9858E-01	.9999
95	.1889	8.5	-.9863E-01	1.0000
96	.1142	8.6	-.9866E-01	1.0000
97	.0681	8.7	-.9868E-01	1.0000
98	.0400	8.8	-.9869E-01	1.0000
99	.0232	8.9	-.9870E-01	1.0000
100	.0132	9.0	-.9870E-01	1.0000
101	.0074	9.1	-.9871E-01	1.0000
102	.0041	9.1	-.9871E-01	1.0000
103	.0023	9.2	-.9871E-01	1.0000

104	.0012	9.3	-.9871E-01	1.0000
105	.0006	9.4	-.9871E-01	1.0000
106	.0003	9.5	-.9871E-01	1.0000
107	.0002	9.6	-.9871E-01	1.0000
108	.0001	9.7	-.9871E-01	1.0000
109	.0000	9.8	-.9871E-01	1.0000
110	.0000	9.9	-.9871E-01	1.0000
111	.0000	10.0	-.9871E-01	1.0000
112	.0000	10.1	-.9871E-01	1.0000
113	.0000	10.2	-.9871E-01	1.0000
114	.0000	10.3	-.9871E-01	1.0000
115	.0000	10.3	-.9871E-01	1.0000
116	.0000	10.4	-.9871E-01	1.0000
117	.0000	10.5	-.9871E-01	1.0000
118	.0000	10.6	-.9871E-01	1.0000
119	.0000	10.7	-.9871E-01	1.0000
120	.0000	10.8	-.9871E-01	1.0000
121	.0000	10.9	-.9871E-01	1.0000
122	.0000	11.0	-.9871E-01	1.0000
123	.0000	11.1	-.9871E-01	1.0000
124	.0000	11.2	-.9871E-01	1.0000
125	.0000	11.3	-.9871E-01	1.0000
126	.0000	11.4	-.9871E-01	1.0000
127	.0000	11.4	-.9871E-01	1.0000
128	.0000	11.5	-.9871E-01	1.0000
129	.0000	11.6	-.9871E-01	1.0000
130	.0000	11.7	-.9871E-01	1.0000
131	.0000	11.8	-.9871E-01	1.0000
132	.0000	11.9	-.9871E-01	1.0000
133	.0000	12.0	-.9871E-01	1.0000
134	.0000	12.1	-.9871E-01	1.0000
135	.0000	12.2	-.9871E-01	1.0000
136	.0000	12.3	-.9871E-01	1.0000
137	.0000	12.4	-.9871E-01	1.0000
138	.0000	12.5	-.9871E-01	1.0000
139	.0000	12.6	-.9871E-01	1.0000
140	.0000	12.6	-.9871E-01	1.0000
141	.0000	12.7	-.9871E-01	1.0000
142	.0000	12.8	-.9871E-01	1.0000
143	.0000	12.9	-.9871E-01	1.0000
144	.0000	13.0	-.9871E-01	1.0000
145	.0000	13.1	-.9871E-01	1.0000
146	.0000	13.2	-.9871E-01	1.0000
147	.0000	13.3	-.9871E-01	1.0000
148	.0000	13.4	-.9871E-01	1.0000
149	.0000	13.5	-.9871E-01	1.0000
150	.0000	13.6	-.9871E-01	1.0000
151	.0000	13.7	-.9871E-01	1.0000
152	.0000	13.8	-.9871E-01	1.0000
153	.0000	13.8	-.9871E-01	1.0000
154	.0000	13.9	-.9871E-01	1.0000
155	.0000	14.0	-.9871E-01	1.0000

EXP. NO. 600 P. 21   27/06/73 17.14.36.

156	.0000	14.1	-.9871E-01	1.0000
157	.0000	14.2	-.9871E-01	1.0000
158	.0000	14.3	-.9871E-01	1.0000
159	.0000	14.4	-.9871E-01	1.0000
160	.0000	14.5	-.9871E-01	1.0000
161	.0000	14.6	-.9871E-01	1.0000
162	.0000	14.7	-.9871E-01	1.0000
163	.0000	14.8	-.9871E-01	1.0000
164	.0000	14.9	-.9871E-01	1.0000
165	.0000	15.0	-.9871E-01	1.0000
166	.0000	15.1	-.9871E-01	1.0000
167	.0000	15.1	-.9871E-01	1.0000
168	.0000	15.2	-.9871E-01	1.0000
169	.0000	15.3	-.9871E-01	1.0000
170	.0000	15.4	-.9871E-01	1.0000
171	.0000	15.5	-.9871E-01	1.0000
172	.0000	15.6	-.9871E-01	1.0000
173	.0000	15.7	-.9871E-01	1.0000
174	.0000	15.8	-.9871E-01	1.0000
175	.0000	15.9	-.9871E-01	1.0000
176	.0000	16.0	-.9871E-01	1.0000
177	.0000	16.1	-.9871E-01	1.0000
178	.0000	16.2	-.9871E-01	1.0000
179	.0000	16.3	-.9871E-01	1.0000
180	.0000	16.4	-.9871E-01	1.0000
181	.0000	16.5	-.9871E-01	1.0000
182	.0000	16.5	-.9871E-01	1.0000
183	.0000	16.6	-.9871E-01	1.0000
184	.0000	16.7	-.9871E-01	1.0000
185	.0000	16.8	-.9871E-01	1.0000
186	.0000	16.9	-.9871E-01	1.0000
187	.0000	17.0	-.9871E-01	1.0000
188	.0000	17.1	-.9871E-01	1.0000
189	.0000	17.2	-.9871E-01	1.0000
190	.0000	17.3	-.9871E-01	1.0000
191	.0000	17.4	-.9871E-01	1.0000
192	.0000	17.6	-.9871E-01	1.0000
193	.0000	17.7	-.9871E-01	1.0000
194	.0000	17.8	-.9871E-01	1.0000
195	.0000	17.9	-.9871E-01	1.0000
196	.0000	18.0	-.9871E-01	1.0000
197	.0000	18.1	-.9871E-01	1.0000
198	.0000	18.3	-.9871E-01	1.0000
199	.0000	18.4	-.9871E-01	1.0000
200	.0000	18.5	-.9871E-01	1.0000
201	.0000	18.7	-.9871E-01	1.0000
202	.0000	18.8	-.9871E-01	1.0000
203	.0000	19.0	-.9871E-01	1.0000
204	.0000	19.1	-.9871E-01	1.0000
205	.0000	19.3	-.9871E-01	1.0000
206	.0000	19.5	-.9871E-01	1.0000
207	.0000	19.6	0.	1.0000

MASSBALANCE OF GRADIENT   1.0001   NO OF CYCLES   1280
RADIUS AT MASS MIDPOINT   3.6040 CM

----

8   EFFECT OF CENTRIFUGATION ON SUBCELLULAR STRUCTURES

Robert Wattiaux
Laboratoire de Chimie Physiologique
Facultés Universitaires Notre-Dame de la Paix
Namur, Belgium

One cannot centrifuge subcellular particles in a density gradient with impunity at any speed in any rotor: subcellular structures markedly deteriorate when the hydrostatic pressure they are exposed to in the centrifuge tube becomes too high. This is shown for rat-liver mitochondria by the distributions of reference enzymes after isopycnic centrifugation in a density gradient and by electron-microscopic examination of the fractions. One factor which appears to influence the behaviour of mitochondria during high-speed centrifugation in a density gradient is the temperature at which the run is performed.

Preparative centrifuges able to develop g values of the order of hundreds of thousands with high performance rotors are now available. It might seem interesting to make use of such centrifuges and rotors to separate cellular organelles. However, recent works indicate that certain subcellular structures suffer seriously when the hydrostatic pressure they are exposed to in the centrifuge tube becomes too high. We illustrate this damaging effect of centrifugation by showing some results of experiments performed in our laboratory, mainly on rat-liver mitochondria, by S. Wattiaux-De Coninck, M.F. Ronveaux-Dupal, M. Collot and F. Dubois.

In the first part we will show that these organelles may undergo marked deterioration during isopycnic centrifugation, and that the main causal agent of the phenomenon is the hydrostatic pressure. Then we will consider some factors which should accordingly be taken into account in centrifugation experiments.

RESULTS AND DISCUSSION

Fig. 1 shows the distribution of several mitochondrial enzymes after isopycnic centrifugation of a rat-liver mitochondrial fraction in a sucrose gradient. These enzymes have been selected for their different submitochondrial localization. In this experiment, centrifugation has been performed at 39,000 rev/min with the Spinco rotor SW 65. All the enzymes exhibit a similar distribution. They are mostly recovered in a relatively narrow region centred around a density of about 1.19 g/ml. The distribution pattern is similar to that described for cytochrome oxidase by Beaufay *et al.* [1] under comparable experimental conditions. In the following experiment (Fig. 2) the centri-

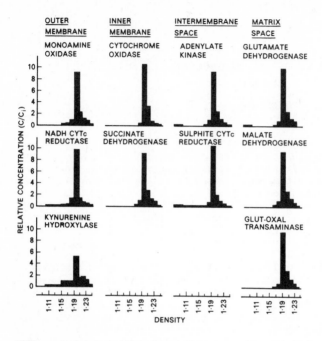

Fig. 1. Distribution of mitochondrial enzymes after isopycnic centrifugation of a rat-liver mitochondrial fraction in a sucrose gradient. SW 65 Spinco rotor; speed of centrifugation: 39,000 rev/min; time integral of the square angular velocity; 144 rad^2/n-sec; limits of the gradient: 1.09 to 1.26 g/ml. Mitochondrial fraction: M + L according to de Duve *et al.* [2]. The granules were layered at the top of the gradient. *Ordinate:* relative concentration, i.e. the ratio of the observed activity to that which would have been found if the enzyme had been homogeneously distributed throughout the gradient.

MAO, monoamine oxidase; cyt.oxid., cytochrome oxidase; AK, adenylate kinase; GDH, glutamate dehydrogenase; SD, succinate dehydrogenase; sulph. red., sulphite cytochrome *c* reductase; MD, malate dehydrogenase; kyn. hydrox., kynurenine hydroxylase; GOT, glutamate oxalo-acetate transaminase.

fugation was performed at 65,000 rev/min with the same rotor, the time integral of the square angular velocity being identical. Obviously the results are quite different. The distribution curves of mitochondrial enzymes after centrifugation at 65,000 rev/min differ in two respects from those observed after centrifugation at 39,000 rev/min. First, for a given enzyme, the distribution pattern is considerably more heterogeneous. Secondly, different mitochondrial enzymes do not exhibit the same general distribution pattern.

The difference is particularly impressive when a comparison is made between the distribution of the membrane enzymes and those of the matrix enzymes. The membrane enzymes show an almost trimodal distribution: a small peak of activity is observed where the bulk of the mitochondria are recovered after centrifugation at 39,000 rev/min, i.e. in a zone of mean density of about 1.19 g/ml; a second peak is found in a zone of higher density (≈1.22 g/ml), and a third main peak in a zone of lower density (1.16 - 1.17 g/ml)

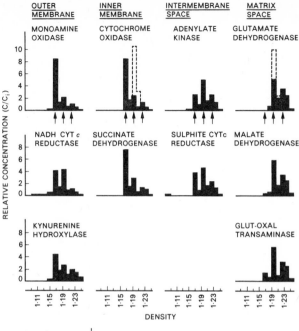

Fig. 2. Distribution of mitochondrial enzymes after isopycnic centrifugation of a mitochondrial fraction in a sucrose gradient. Conditions similar to those of the experiment of Fig. 1 (with same abbreviations) except that the speed of centrifugation is 65,000 rev/min.

To facilitate comparison the distributions of cytochrome oxidase and glutamate dehyrogenase illustrated in Fig. 1 are indicated by dotted lines on the graph.

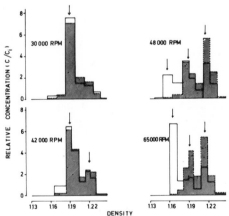

Fig. 3. Distribution of cytochrome oxidase (open blocks) and glutamate dehydrogenase (hatched blocks) after isopycnic centrifugation of a mitochondrial fraction in a sucrose gradient: influence of the speed of centrifugation. (From [3].)

The matrix enzymes are mainly recovered in two regions: in the zone of mean density 1.19 g/ml and in that of mean density 1.22 g/ml. No activity, or only a little, is recovered in the region of density 1.16-1.17 g/ml. The enzymes of the intermembrane space are characterized by an intermediate distribution.

The distribution changes observed when various centrifugation speeds are used are illustrated in Fig. 3. We represent only the distribution of a typical mitochondrial membrane enzyme, cytochrome oxidase, and of a matrix enzyme, glutamate dehydrogenase. The time integral of the square angular velocity is, of course,

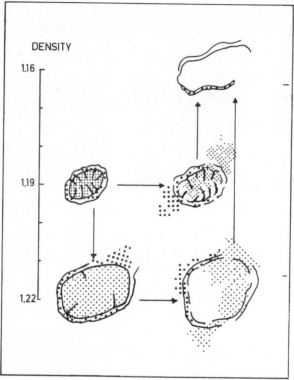

DENSITY

1.16

1.19

1.22

Fig. 4. Interpretation of the mitochondrial enzyme distributions observed after isopycnic centrifugation of a mitochondrial fraction in a sucrose gradient at increasing centrifugation speeds.

See text for explanation.

the same in each experiment. At 30,000 rev/min the enzymes exhibit a homogeneous distribution: no difference in distribution is observed between the membrane enzyme and the matrix enzyme. At 42,000 rev/min the enzymes are found predominantly in a zone between densities 1.18-1.20 g/ml; but a peak of activity is also seen in a fraction with a mean density of about 1.22 g/ml. When the centrifugation speed reaches 48,000 rev/min a dissociation is apparent between the distribution curve of cytochrome oxidase and that of glutamate dehyrogenase. The distribution of cytochrome oxidase becomes trimodal: a peak of activity is observed at a mean density of 1.16-1.17 g/ml, a second at 1.19 g/ml and a third at 1.22 g/ml. Glutamate dehydrogenase exhibits a bimodal distribution. No activity is found in the zone of density 1.16 g/ml but two peaks are distinctly visible at 1.19 and 1.22 g/ml. At 65,000 rev/min cytochrome oxidase is chiefly recovered in the fraction of mean density 1.16 g/ml; the distribution of glutamate dehydrogenase is comparable to the former distribution.

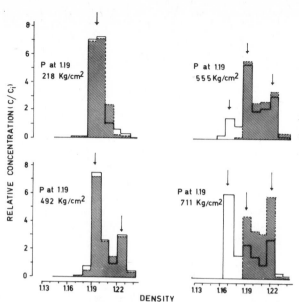

Fig. 5. Distribution of cytochrome oxidase (open blocks) and glutamate dehydrogenase (hatched blocks) after isopycnic centrifugation of a rat-liver mitochondrial fraction in a sucrose gradient: influence of hydrostatic pressure. (From Wattiaux *et al.* [3].)

The more plausible explanation of these results is represented schematically in Fig. 4. At a relatively low centrifugation speed, the well preserved mitochondria equilibrate in a relatively narrow region situated around a density of 1.19 g/ml. The mitochondrial enzymes are recovered together in that region. At a higher centrifugation speed, certain mitochondria become denser, and are found in a zone of density 1.22 g/ml; this could be explained if, following a deterioration of their inner membrane, mitochondria become more permeable to sucrose. If so, it is to be expected that the granules swell with consequent disruption of their outer membranes. At a still higher centrifugation speed, a true disruption of the two membranes occurs, with a release of the matrix content; then the mitochondrial ghosts migrate to their equilibrium position, at about 1.16 - 1.17 g/ml.

Morphological evidence agrees well with the hypothesis deduced from enzyme distribution curves [3]. The main components of fraction of mean density 1.19 g/ml are mitochondria in a condensed form with an intact double membrane. These structures stand out against a dark and relatively homogeneous background, probably released matrix proteins which have been denatured by the fixation procedure and which consequently are unable to pass through the millipore filter.

The fraction of density 1.22 g/ml consists chiefly of swollen mitochondria with a disrupted outer membrane; the dark background is also observed near the surface of the pellicle, in contact with the millipore filter. Mitochondrial ghosts are the major components of the fraction of mean density 1.16 g/ml. No dark background is seen as in the two other fractions.

The hydrostatic pressure to which the granules are subjected in the gradient seems to be the main cause of these phenomena. This is illustrated in Fig.5. We show the distributions of cytochrome oxidase and glutamate dehydrogenase, after isopycnic centrifugation in a sucrose gradient, at the same speed in the same rotor but under different hydrostatic pressures. The pressure given is that exerted on the zone of density 1.19 g/ml around which mitochondria normally equilibrate. The influence of hydrostatic pressure on enzyme distributions is evident. The first important change observed consists in an increase in the activity recovered beyond density 1.20 g/ml, the enzyme distributions becoming bimodal. At a higher pressure, the distribution of cytochrome oxidase becomes trimodal with an additional peak appearing between densities 1.16 and 1.17 g/ml. The bimodality of the glutamate dehydrogenase distribution pattern becomes more marked, but no activity is measurable in the 1.16-1.17 g/ml region.

The experiments we have described have been performed in sucrose gradients. During such centrifugation, the granules are exposed to hypertonic conditions. Other experiments, not described here, were performed in iso-osmotic medium by making use of a glycogen gradient with 0.25 M aqueous sucrose as solvent according to Beaufay *et al.* [1]. They show that rat liver mitochondria deteriorate in such conditions when the hydrostatic pressure during centrifugation becomes too high. Pelleting mitochondria in iso-osmotic sucrose at high speed leads to deterioration of the outer and the inner membrane of the organelles [4]. Rat-liver peroxisomes and lysosomes may also be damaged by centrifugation [5].

In agreement with our results, Bronfman and Beaufay [6] have shown that rat liver mitochondria, lysosomes and peroxisomes markedly deteriorate when they are subjected to a high hydrostatic pressure generated by a hydraulic press.

Evidently, then, hydrostatic pressure appears to be an important factor that must be taken into account in centrifugation experiments.

During centrifugation, a zone located at a distance x from the axis of rotation is subjected to an hydrostatic pressure P given by the following relationship:

$$P = \omega^2 \rho \left(\frac{x^2 - x_0^2}{2} \right)$$

when the density of the liquid is constant and:

$$P = \omega^2 a \left(\frac{x^3 - x_0^3}{3} \right) + (\rho_0 - ax_0)\left(\frac{x^2 - x_0^2}{2} \right)$$

when the density linearly varies as a function of the radial distance.

In these relationships, ω is the angular velocity, ρ the density of the liquid, x the radial distance at the selected zone from the axis of rotation, x_0 the radial distance at the meniscus from the axis of rotation, a the slope of the gradient, and ρ_0 the density at x_0 (lower density limit).

The second relationship particularly needs consideration in isopycnic centrifugation experiments. For a given rotor, the maximum hydrostatic pressure to which a particle will be subjected depends on the centrifugation speed and on the position of the isopycnic zone of the particle in the gradient.

Fig. 6 shows, for example, how strikingly the hydrostatic pressure increases as a function of the distance from the meniscus at various speeds of rotation in the Spinco Rotor SW 50. For the calculations, we have supposed a sucrose gradient extending from 1.09 to 1.26 g/ml. Obviously the position of the isopycnic zone of a particle in a gradient depends on the limits of the gradient. Therefore, it is to be expected, that the damaging effect of the centrifugation will decrease at a given speed, if the limits of the gradients are selected so that the isopycnic zone is located not too far from the meniscus.

This is illustrated by the results of the following experiment. We have established the distribution of cytochrome oxidase and malate dehydrogenase after isopycnic centrifugation of a mitochondrial fraction in five sucrose gradients, with use of the SW 50 Spinco rotor. The gradients had the same slope but the limits were chosen so that the zone of density 1.19 g/ml was more and more shifted toward the meniscus. As shown by Fig. 7, the distributions differ strikingly. They show that mitochondria deteriorate markedly when the gradient extends from 1.07 to 1.24 g/ml but are probably well preserved when the gradient extends from 1.16 to 1.33 g/ml.

The effect of the hydrostatic pressure has also to be taken into account when one makes use of different rotors. Marked differences exist amongst certain rotors running at a speed generating the same average centrifugal field. For example the same g_{av} generates a markedly higher hydrostatic pressure in a rotor like the Spinco SW 40 than in the SW 65. This influence of the rotor is illustrated in Fig. 8. Striking differences are observed in the distribution of mitochondrial enzymes according to whether the centrifugation has been performed in the SW 65 rotor or in the SW 40 rotor, the g_{av}

being identical.   The centrifugation in the SW 40 rotor has caused marked
deterioration of mitochondria; these seem well preserved after centrifugation
in the SW 65 rotor.   Attention is here drawn to the quality, in this respect,
of the zonal rotor with a small annular cell designed by Beaufay [7].   At
identical *g* values, the hydrostatic pressure is considerably lower than in
the swinging-bucket rotor.

Centrifugations in the experiments we have described were performed
near 0°.   Subsequently, we observed that the behaviour of mitochondria could
be greatly changed if the temperature of centrifugation were increased.   In
these experiments, we have to distinguish between two activities for sulphite
cytochrome *c* reductase (an intermembrane-space enzyme [8]) and for malate dehydrogenase (a matrix enzyme):   the *free activity*, i.e. the activity measured
without addition of the detergent Triton X-100, and the *total activity* determined in the presence of Triton.   When mitochondrial membranes are intact,
the free activity is low, because the enzyme has no access to external substrate; it increases when the mitochondrial membranes are altered.   There are
no differences in enzyme distributions according as the centrifugation is
performed at 0° or at 15° when the centrifugation speed is 39,000 rev/min.
The results are quite different when the run is done at 65,000 rev/min.   At
0°, one observes heterogeneous distributions like those we have previously
described.   The free activity of malate dehydrogenase and sulphite cytochrome
c reductase is high.   When the temperature of centrifugation is increased,
enzyme distributions progressively return to normal, that is to a pattern
like that seen after centrifugation at lower speed.   First the proportion of
membrane enzymes recovered in the zone of density 1.16-1.17 g/ml decreases;
this means that fewer and fewer mitochondria have their membranes disrupted.
Then, the enzymatic activity present in the zone of density 1.22-1.23 g/ml
decreases whilst there is an increase in the activity recovered in the zone
of density 1.19-1.20 g/ml.   At the same time the free activities of sulphite
cytochrome *c* reductase and of malate dehydrogenase decrease.   Finally with
centrifugation at 15°; the distributions of the four mitochondrial enzymes
seem normal, the free activities of sulphite cytochrome *c* reductase and of
malate dehydrogenase are comparable with those measured after centrifugation
at lower speed [Fig. 9].   Morphological observations are in good agreement
with the biochemical results [9].   The components of the mitochondrial fraction centrifuged in a sucrose gradient at 0° are mainly swollen mitochondria
and mitochondrial ghosts.   On the contrary, after centrifugation at 15°, the
preparation consists chiefly of mitochondria in a condensed form with the
double membrane intact.   The effect of temperature is interesting from two
points of view; first it shows that it is possible by adjusting the temperature to prevent the damaging effects of centrifugation on mitochondria, and
secondly, it suggests that the pressure acts either on the physical state of
the mitochondrial membrane lipids or by disrupting hydrophobic bonds essential for the mitochondrial membrane stability [9].

The above results came from experiments on rat-liver mitochondria.
Subsequent studies have indicated that the effect of hydrostatic pressure

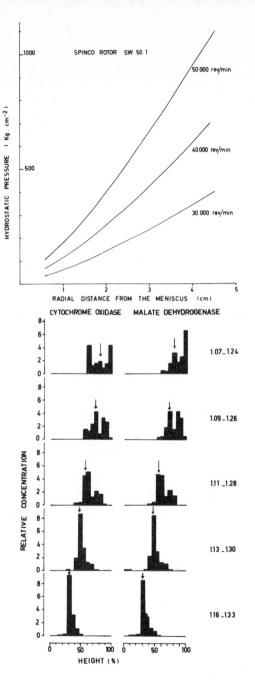

Fig. 6. Hydrostatic pressure during centrifugation as a function of the radial distance in the Spinco rotor SW 50. There is a notional sucrose gradient of 5 ml extending from 1.09 to 1.26 g/ml, over which is a layer of 0.25 M sucrose, 0.65 cm in height.

Fig. 7 *(below, left)*. Distribution of cytochrome oxidase and malate dehydrogenase after isopycnic centrifugation of a mitochondrial fraction in sucrose gradients with different density limits. SW 50.1 Spinco rotor; centrifugation at 50,000 rev/min; time integral of the square angular velocity 144 rad^2/n-sec. The density limits of the gradient in g/cm^3 are given at the right of the Figure. *In such experiments, the hydrostatic pressure exerted on the zone of density 1.19 g/ml varies from about 850 to 250 kg/cm^2.*

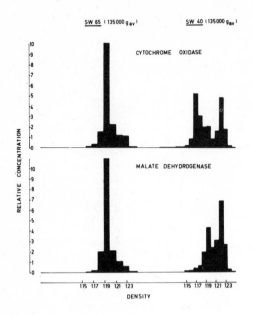

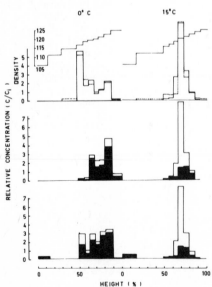

Fig. 8. Distribution of cytochrome oxidase and malate dehydrogenase after isopycnic centrifugation of a mitochondrial fraction in a sucrose gradient, at the same g_{av} (135,000) in the SW 65 and the SW 40 Spinco rotors. The gradient extended from 1.09 to 1.26 g/ml density. The time of centrifugation was 118 min.

In such experiments, the hydro-static pressure exerted on the zone of density 1.19 g/ml was about 450 (SW 65) or 780 (SW 40) kg/cm².

Fig. 9 *(below, left)*. Distribution of mitochondrial enzymes after iso-pycnic centrifugation of a mitochon-drial fraction in a sucrose gradient at 0° and 15°. SW 65 Spinco rotor; speed of centrifugation 65,000 rev/min; time integral of the square angular velocity: 144 rad²/n-sec.

Top: Cytochrome oxidase (——)
   and monoamine oxidase (••••••).
Middle: Malate dehydrogenase, free
   activity *(solid blocks)* and total
   activity *(open blocks)*.
Bottom: Sulphite-cytochrome c reductase, free and total *(as above)*.

can vary according to the origin of the particles. In a study of sub-cellular structures of Morris hepa-tomas, we have found that with one of these tumours, hepatoma 16, the mitochondria exhibit a considerable resistance to hydrostatic pressure during centrifugation [10]. Striking differences are also seen when com-parison is made between mitochon-dria isolated from foetal liver just before birth and mitochondria from the liver of an animal one or two days after birth. Foetal mito-chondria seem more resistant to centrifugation than mitochondria of newborn animals (Mertens-Strijthagen & De Schrijver, unpublished results). We cannot yet explain such differences.

Such results indicate that perhaps one should sometimes be more cautious in interpreting observations made on membrane systems isolated by centrifugation. It must be emphasized that the deterioration of the subcellular structures that we have observed is of marked extent. It is possible that more subtle damage could occur under apparently safe centrifugation conditions.

References

1. Beaufay, H., Jacques, P., Baudhuin, P., Sellinger, O.Z., Berthet, J. & de Duve, C., *Biochem. J.* 92 (1964) 184-205.

2. de Duve, C., Pressman, B.C., Gianetto, R., Wattiaux, R. & Appelmans, F., *Biochem. J., 60* (1955) 604-612.

3. Wattiaux, R., Wattiaux-De Coninck, S. & Ronveaux-Dupal, M.F., *Eur. J. Biochem.* 22 (1971) 31-39.

4. Ronveaux-Dupal, M.F., Collot, M., Wattiaux-De Coninck, S. & Wattiaux, R., *Arch. Intern. Physiol. Biochim.* 80 (1972) 406-407.

5. Wattiaux, R., Wattiaux-De Coninck, S. & Collot, M., *Arch. Intern. Physiol. Biochim.* 79 (1971) 1050-1051.

6. Bronfman, M. & Beaufay, H., *Abst. 9th Int. Congr. Biochem. (Stockholm)* (1973) 21.

7. Beaufay, H., *Thesis, University of Louvain,* Centerick, Louvain (1966).

8. Wattiaux-De Coninck, S. & Wattiaux, R., *Eur. J. Biochem.* 39 (1973) 93-99.

9. Wattiaux-De Coninck, S., Ronveaux-Dupal, M.F., Dubois, F. & Wattiaux, R., *Eur. J. Biochem.* 39 (1973) 93-99.

10. Wattiaux-De Coninck, S., Collot, M., Wattiaux, R. & Morris, H.P., *Eur. J. Cancer* 8 (1972) 415-420.

9   GRADIENT MATERIAL INTERFERENCE IN THE ASSAY OF FRACTIONS

G.C. Hartman, N. Black, R. Sinclair and *R.H. Hinton
*Biochemistry Department and *Wolfson Bioanalytical Centre
University of Surrey, Guildford GU2 5XH, U.K.*

The various ways in which density gradient materials can affect the properties of biological particles[†] are discussed. Especial attention is paid to the interference by density-gradient solutes of low molecular weight in the assay of enzymes and of protein, and in the measurement of ^{3}H-radioactivity by scintillation counting. These effects are proportional to the concentration of the gradient solute and are reversed on dilution. It is suggested that this interference is caused by the binding of low molecular weight solutes within the hydration spheres of macromolecules. Density-gradient materials of high molecular weight, such as Ficoll, were found not to interfere in the assay of a model enzyme.

Although density-gradient centrifugation has been widely used in biochemical research, relatively little attention has been paid to the possible interactions of the density-gradient solute with the particles which are being separated. Indeed, there has been little systematic study of density-gradient solutes in any respect. An ideal density-gradient solute should meet the following criteria:- it should be an uncharged species of high molecular weight; having the possibility of forming aqueous solutions covering a wide density range with low viscosity and stable over the physiological pH range. Such a solute could be used to generate diverse iso-osomotic density gradients, and Meuwissen [1] has pointed out that high molecular weight solutes should permit greater loading of sample into the starting zone. This ideal solute should be readily available and of reproducible quality, innocuous in the sense of being inert towards the tissue under investigation, be readily removable from the separated particles, and should not interfere with subsequent analyses. The development of high-capacity rotors has led to the need for large amounts of the density-gradient material; thus the solute should either be relatively inexpensive or be recoverable without undue manipulation. The presence of a physical property which facilitates monitoring of the concentration is desirable. From these criteria it may well be that a solute which approaches ideality in one experimental situation may prove unsatisfactory with, say particles from a different tissue as is the case with homogenization media [2].

[†] *The term 'particles' is used in a general sense, and includes macromolecules.*

Deviations from ideality include the following :-
(1) Interaction with the material under investigation;
(2) reaction with reagents (including substrate) used in the estimation;
(3) interference with the performance of equipment used in the assay;
(4) interference with the sensitivity of the assay without noticeable reaction with the reagents.

Interference under any of these headings may be due either to the density gradient materials themselves or to contaminants therein.

The commonly used density gradient materials listed in Table 1 may be divided into three main classes :-
(i)   Salts of the alkali metals;
(ii) small hydrophilic organic molecules;
(iii) hydrophilic macromolecules.
A number of other compounds have been used for special applications which fall into none of these classes, notably the iodinated compounds Urografin (Schering A/G, Berlin) and Metrizamide (Nyegaard and Co. A/S, Oslo), and colloidal silica (Ludox; Du Pont, Wilmington, Del.). As mentioned earlier, gradient solutes of high molecular weight have a number of theoretical advantages over other materials [1]; but otherwise the choice between density gradient materials is largely determined by price and by the necessity to minimize damage to the material which is being separated.

Solutions with the widest density range and the lowest viscosities are obtained with solutes in the first class (salts of alkali metals). However, the high ionic strength of solutions of these compounds is lethal to living cells and will severely damage subcellular structures. Thus even the RNA and protein of ribosomes are dissociated by concentrations of CsCl lower than those required for the isopycnic banding of such particles [3]. The ionic density-gradient solutes are restricted to the banding of macromolecules and of the few multimolecular complexes such as serum lipoproteins that are stable in solutions of high ionic strength, and to the separation of nucleoprotein structures that have been stabilized by 'fixation' with formaldehyde [3].

Sucrose is an example of a compound in the second class and is by far the most widely used of the non-ionic density-gradient solutes of low molecular weight. Certain other low molecular weight sugars are soluble enough and cheap enough to warrant consideration if it could be shown that they have a significant advantage over sucrose; but, as is mentioned below, no such advantage seems to exist in respect of inhibitory effects. Glycerol has been used for some specialised applications [8, 9].

All the high molecular weight solutes, with the exception of Ludox, are relatively expensive and tend to form very viscous solutions at high concentrations. Of those listed in Table 1, Ficoll is by far the most widely used. This polymer was expressly designed for density-gradient centrifugation.

Table 1. Properties of the more commonly used density gradient materials.

Material	Mol. wt.	Max. density of aqueous solution	Ionic strength of solutions	Viscosity of 20% w/v solution[1]	UV absorbance	Approx. price[2] £/100g	Ref. for further information on properties
Caesium chloride	168.4	1.918	high	+	low	11.15[3,4]	4
Caesium sulphate	361.9	>1.8	high	+	low	8.98[3,4]	6
Sodium bromide	102.9	>1.5	high	+	low	0.15[5]	4
Sodium iodide	149.9	1.9	high	+	high	0.55[5]	7
Potassium tartrate	235.3	1.485	high	+	low	0.37[5]	5
Sucrose	342.3	1.3	low	2.954 (5°)	low[6]	0.03[5], 0.54[7]	4, 5
Glycerol	92.09	1.26	low	++	low	0.13[3]	4
Ficoll[8]	400,000	1.23	low	43.19 (4°)	low	6.05[3]	5
Dextran	(72,000)[9]	1.05	low	+++	low	6.62	4
Bovine serum albumin	69,000	1.12	low	+++	high	15.90[10]	-
Ludox[11]	-[12]	1.219	low	+		negligible[11]	-13
Urografin[14]	809[14]	1.6	?[14]	++	high	3.56	-13, 15
Metrizamide[16]	789	1.46	low	1.7 (20°)	high	≯30	-13
MGU[17]	599	1.614	?	?++	high	n.a.[17]	16[15]
Chloral hydrate	165.4	1.91	low	++	quite low	0.24	11

[1] It is difficult to locate exact figures for the viscosities of many density-gradient solutes; + indicates a solution almost as mobile as water, ++ a solution with viscosity similar to that of a sucrose solution of the same concentration, +++ a solution similar in viscosity to Ficoll at same concn.

[2] The approx. prices as given merely for comparison represent those listed by major manufacturers in Jan. 1974; no effort has been made to locate 'best buys'.

[3] Analytical grade reagent

[4] as a 60% w/w aqueous solution

[5] General-purpose reagent

[8] Manufacturer: Pharmacia & Co.

[10] Cohn Fraction V.

[11] Manufacturer: Du Pont Inc.; if some needed, *inquire.*

[13] Data available from the manufacturer.

[6] Traces of UV-absorbing material may be removed by activated charcoal treatment [19].

[7] Especially purified RNase-free grade.

[9] A relatively low mol. wt. preparation; dextrans of mean mol. wts. between ∿50,000 and 300,000 are available.

[12] 'Ludox' is an especially finely divided form of colloidal silica; hence 'mol. wt.' meaningless.   *[CONTINUED overleaf*

High-molecular weight dextrans, which ante-dated Ficoll, tend to form solutions with significantly higher viscosity. Bovine serum albumin is extremely expensive and forms very viscous solutions, but is probably the least damaging of all density gradient solutes for use in the separation of living cells. Glycogen sediments significantly even at low centrifugal force; indeed this polysaccharide has been superseded by Ficoll.

Of the iodinated aromatic compounds listed, Urografin and MGU have been available much longer and thus much more widely used. However, Metrizamide is, in the authors' experience [10], easier to use. The properties of Metrizamide are discussed in detail later in this book (Articles 10 & 12). Chloral hydrate has been used for the isopycnic banding of chromatin [11].

In spite of the introduction of density-gradient materials such as Metrizamide and Ludox (colloidal silica) [12], sucrose is by far the most commonly used density-gradient solute, followed, at a considerable distance, by Ficoll. Therefore, in considering the deleterious effects of density-gradient materials on subcellular particles we will largely be concerned with these two compounds. We will now consider the problems under the four headings listed earlier.

1. Direct interference of the gradient material or contaminants with the separating particles

Both sucrose and Ficoll are fairly innocuous towards subcellular particles. The high osmotic strength of the sucrose solutions used to form density gradients is, however, lethal to living cells [13] and is damaging to respiratory control in mitochondria [14]. In view of the high molecular weight of Ficoll, the osmolarity of even very concentrated solutions is low, and, indeed, when Ficoll gradients are employed it is normal to include a low molecular weight solute to maintain isotonicity with the separating particles. Nevertheless Ficoll has been reported to cause aggregation of nucleated erythrocytes [15]. Moreover, damage to spermatozoa has been reported [16]; but other workers in this field have used Ficoll successfully in separating viable spermatozoa, and the problems may well be in the materials which are added to the gradient to maintain isotonic conditions (R.A.P. Harrison, personal communication). It is, in fact, likely that the quality of commercially available Ficoll has improved with the passage of time. Damage both by sucrose and Ficoll seems to involve only the most organized systems, with

---

FOOTNOTES to Table 1, continued from previous page

[14] A mixture of sodium and methylglucamine salts of 3,5-diacetylamino-2,4,6-triiodobenzoic acid (the mol. wt. given refers to the latter salt); manufactured by Schering AG, Berlin.   [16] See Article 10.

[15] These complex salts may form ionized solutions, but whether they would then produce effects similar to those produced by alkali metal ions is not clear.

[17] 3,5-Di iodo-4-pyridone-N-acetic acid Me-glucamine salt; the diethanolamine salt is commercially available as 'Diodrast' or 'Diodone' (May & Baker; Pharmacia).

sucrose having considerably more effect than Ficoll.

It is difficult to generalize about contaminants in density-gradient materials. We do, however, emphasize the contamination of most ordinary grades of sucrose with traces of ribonuclease [17, 18] and the fact that this ribonuclease can be removed merely by treatment with activated charcoal [19]. The danger of heavy metal contaminants has been pointed out by Gardy and Cash [20].

2. Reaction of the gradient material with the reagents used in the assay of separated constituents

All carbohydrate gradient materials will tend to interfere with the analysis of sugars, notably the determination of glycogen, and the estimation of RNA ribose by the orcinol procedure [21]. Solutes such as NaI, Metrizamide and Urografin which absorb strongly in the ultraviolet will interfere when this parameter is measured for example in the estimation of nucleic acids by procedures based on precipitation and extraction followed by reading the extinction at 260 nm. Even though the nucleic acids can usually be separated from the bulk of the density-gradient material by precipitation with perchloric acid (an exception is Urografin which is precipitated by acid), the pellet must still be washed several times to remove all traces of density-gradient solute. Ficoll severely interferes with the estimation of protein by the Lowry procedure [22]. Sucrose does not itself react with the Lowry reagents, but does affect the sensitivity of the assay *(see below)*. Sucrose does, however, interfere in a micro-biuret procedure [23], reacting with the copper reagent to form a complex with a very high ultraviolet absorbance [18].

3. Interference with the performance of analytical equipment

We have shown that sucrose affects the performance of AutoAnalyzers quite markedly [24]. The cross-contamination between samples is markedly increased due to a broadening of the trailing edge of each peak. The leading edge is actually sharpened so that the effect is not due to simple mixing. We have suggested that the cause is, in fact, density inversion in the sample probe, but have not examined other density-gradient solutes to test this hypothesis.

4. Interference of the gradient material in the sensitivity of assays

Several authors have shown that the sensitivity of the Lowry procedure for the assay of protein is reduced in the presence of sucrose [18, 25, 26]. We have also found that there is considerable inhibition of all the liver enzymes which we have tested when they are assayed in the presence of high concentrations of sucrose [18]. The percentage inhibition is typically linear with sucrose concentration, so that if the inhibition at a standard sucrose concentration is known, it is simple to correct results. The inhibition is fully reversed upon dilution [18]. The percentage inhibitions as given in Table 2 for a wide range of liver enzymes show no particular pattern. The three

Table 2. Inhibition of rat-liver enzymes by sucrose. The figures given in this Table are the percentage of the activity in very dilute sucrose (<0.02 M) that is lost when the assay is carried out in the presence of 1 M sucrose. The figures are derived from a least squares fit to an inhibition curve similar to that shown in Fig. 1, a linear relationship being assumed between enzyme activity and sucrose concentration.

Enzyme activity	Inhibition, %
Plasma membrane enzymes	
5'-Nucleotidase	65[1,2], 60[3]
ADPase	30
Alkaline phosphatase	59[4]
Alkaline phosphodiesterase	56[5,6]
L-leucyl-β-naphthylamidase	51[7]
ATPase	41
UTPase	19
Alkaline ribonuclease	20[1]
Mitochondrial enzymes	
Succinate dehydrogenase	62[1]
Monoamine oxidase	51[7]
Peroxisomal enzyme	
Urate oxidase	58[6]
Lysosomal enzymes	
Acid β-glycerophosphatase	16[1]
Acid ribonuclease	23[1]
Acid phosphodiesterase	28[5,6]
Endoplasmic reticulum enzymes	
Glucose-6-phosphatase	50[1]
UDPase	38

[1] See ref. [18].
[2] Substrate 5'-AMP.
[3] Substrate 5'-UMP.
[4] Substrate *p*-nitrophenylphosphate.
[5] Substrate *bis-p*-nitrophenylphosphate.
[6] K.A. Norris, *unpublished observations*.
[7] See ref.[27].

lysosomal enzymes examined do, however, seem to be among the least affected. Metrizamide was found to cause an inhibition similar to that caused by sucrose (*see* Article 10); but with the three enzymes which we tested there was no simple relationship between the inhibition with sucrose and that with Metrizamide. Townsend and Lata [28] found that uricase was inhibited to about the same extent by glycerol as by sucrose.

In an attempt to obtain a mechanism of interference we investigated purified alcohol dehydrogenase, on the grounds that the use of a purified soluble enzyme would avoid complications due to variations in osmotic pressure on membrane-bounded structures. The alcohol dehydrogenase was similar in inhibition pattern to the liver enzymes mentioned earlier [18]. Examination of the effects of glycerol and of low molecular weight carbohydrates on alcohol dehydrogenase showed that all were inhibitory, and that the percentage inhibition was proportional to the concentration of the material in g/ml (Fig. 1). Urea showed a biphasic effect, at low concentrations behaving like the low molecular weight carbohydrates, but becoming much more inhibitory at concentrations of greater than

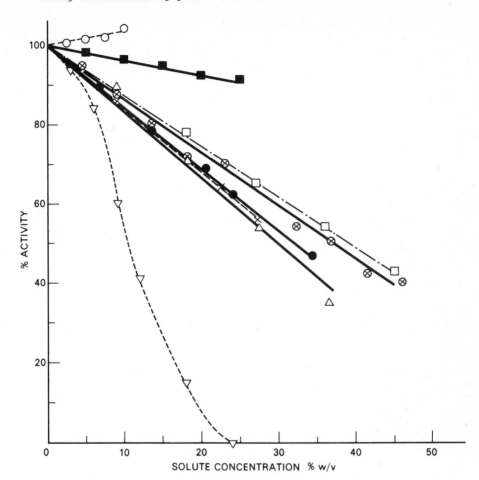

Fig. 1. Alcohol dehydrogenase activity as a function of solute concentration. --o--, Dextran (mean mol. wt. 115,000); ——■——, Ficoll (mean mol. wt. 400,000); ——△——, sorbitol; ——⊗——, glycerol; –·–□–·–, fructose; ——●——, sucrose; --×--, glucose; ···▽··, urea.

0.5 g/ml. This transition point probably marks the beginning of the unfolding of the enzyme protein. In contrast, the high molecular weight solutes Ficoll (average mol. wt. 400,000) and dextran (average mol. wt. 115,000) did not inhibit alcohol dehydrogenase at all.

Apart from its inhibitory effects on enzymes, we found that sucrose (and caesium chloride) will quench the counting of ^{3}H-labelled RNA and ribosomes in an aqueous scintillant based on Triton X-100, while having little effect on the counting of ^{14}C-labelled RNA or ribosomes and none on the counting of small molecules labelled with either ^{3}H or ^{14}C [29]. All the other low molecular weight solutes that we examined caused some quenching of the counting of ^{3}H-labelled RNA, but it was not possible to determine the effect of Ficoll or dextran as, at quite low concentrations, these caused the counting mixture to separate into two phases. There was considerable variation in the percentage quenching caused by the various carbohydrates, unlike the similarity in the effect on enzyme activity. However, the physical structure of aqueous scintillation mixtures using Triton X-100 as a solubilizer is very complex [30, 31] and different low molecular weight compounds are known to have widely varying effects on the phase diagram [30]. Quenching varies from one scintillation mixture to another [32], but it seems inescapable that density-gradient solutes of low molecular weight will quench significantly at high concentrations.

It is not yet clear why low molecular weight hydrophilic solutes should inhibit enzyme activities but not high molecular weight ones (tested on one enzyme). The explanation cannot be simply withdrawal of water, as compounds such as Ficoll are heavily hydrated and are likely to immobilize similar amounts of water to the low molecular weight gradient materials. Nor can the effect be due to the viscosity of the solutions, as the viscosities of glucose and fructose solutions are markedly different, while their inhibitory effects are very similar. One possible explanation is that small hydrophilic molecules may become bound within the hydration spheres of the enzyme molecules. The molecular weights of the density-gradient solutes are sufficiently greater than that of water for their inertia to present a significant obstacle to the approaching substrate molecules or to departing products. In addition, the presence of hydrophilic carbohydrates in the hydration sphere is likely to cause increased binding of water, thus increasing the width of this sphere. This would explain the trapping of weak electrons arising from the decay of ^{3}H, while the more energetic electrons from ^{14}C, which can penetrate well beyond the limits of the hydration sphere, escape. This hypothesis is consistent with other work on the physical state of macromolecules in a scintillation counting mixture containing Triton X-100 [31].

We conclude that both the inhibition of enzyme activities by density-gradient solutes of low molecular weight and the quenching of the counting of ^{3}H-labelled macromolecules can be explained by the solute interacting with the hydration sphere of the macromolecules. The lack of effect of high molecular weight solutes can be explained by assuming that they gather their own hydration sphere and move as independent particles. We have, however, no direct measurements in support of our explanation, and further experiments will be needed to establish its veracity.

Acknowledgements

The authors thank Mrs. H. Spencer for technical assistance in the experiments on the quenching of RNA radioactivity. This portion of the work was carried out as part of a project supported by the Cancer Research Campaign.

References

1. Meuwissen, J.A.T.P., in *Separations with Zonal Rotors* (Reid, E., ed.), Wolfson Bioanalytical Centre, University of Surrey, Guildford (1971) pp. B-2.1—B-2.8.

2. Bowers,W.E., Finkenstaedt, J.T. & de Duve, C., *J. Cell Biol. 32* (1967) 325-337.

3. Spirin, A.S., Belitsina, N.V. & Lerman, M.I., *J. Mol. Biol. 14* (1965) 611-615.

4. Wolf, A.V. & Brown, M.G., in *Handbook of Chemistry and Physics* (Weast, W.C., ed.), Chemical Rubber Co., Cleveland (1964) pp. D-127—D-166.

5. Dobrota, M., *as for ref. 1,* pp. Z-6.1—Z-6.2.

6. Szybalski, W., *Methods in Enzymology, 12B* (1968) 330-360.

7. Anet, R. & Strayer, D.R., *Biochem. Biophys. Res. Commun.* 37 (1969) 52-58.

8. Wallach, D.F.H., in *The Specificity of Cell Surfaces* (Davis, B.D. & Warren, L., eds.), Prentice Hall, Englewood Cliffs, N.J. (1967) pp. 129-163.

9. Schreier, M.H. & Staehelin, T., *Nature New Biol. 242* (1973) 35-38.

10. Mullock, B.M. & Hinton, R.H., *Trans. Biochem. Soc. 1* (1973) 577-581.

11. Hossainy, E., Zweidler, A. & Bloch, D.P., *J. Mol. Biol. 74* (1973) 283-289.

12. Pertoft, H. & Laurent, T.C., in *Modern Separation Methods of Macromolecules and Particles* (Gerritsen, T., ed.), Wiley, New York (1969) pp. 71-90.

13. Mateyko, G.M. & Kopac, M.J., *Ann. N.Y. Acad. Sci. 105* (1963) 183-286.

14. Zimmer, G., Keith, A.D. & Packer, L., *Arch. Biochem. Biophys. 152* (1972) 105-113.

15. Mathias, A.P., Ridge, D. & Trezona,N. St. G., *Biochem. J. 111* (1969) 583-591.

16. Benedict, R.C.,Schumaker, V.N. & Davies, R.E., *J. Reprod. Fertility 13* (1967) 237-249.

17. Barlow, J.J., Mathias, A.P., Williamson, R. & Gammack, D.B., *Biochem. Biophys. Res. Commun. 13* (1963) 61-66.

18. Hinton, R.H., Burge, M.L.E. & Hartman, G.C., *Anal. Biochem. 29* (1969) 248-256.

19. Steele, W.J. & Busch, H., in *Methods in Cancer Research,* Vol. 3 (Busch, H., ed.), Academic Press, New York (1967) pp. 61-152 (*vide* p. 104).

20. Gardy, M. & Cash, W.D., *Fed. Proc. 25* (1966) 737 *(Abst. 3087).*

21. Slater, T.F., *Biochim. Biophys. Acta 27* (1958) 201-202.

22. Lo, C.-H. & Stelson, H., *Anal. Biochem. 45* (1972) 331-336.

23. Itzhaki, R.F. & Gill, D.M., *Anal. Biochem. 9* (1964) 401-410.

24. Hinton, R.H. & Norris, K.A., *Anal. Biochem. 48* (1972) 247-258.

25. Schuel, H. & Schuel, R., *Anal. Biochem. 20* (1967) 86-93.

26. Gerhardt, B. & Beevers, H., *Anal. Biochem. 24* (1968) 337-339.

27. Burge, M.L.E., Ph.D. Thesis, University of Surrey, Guildford (1973).

28. Townsend, D. & Lata, G.F., *Arch. Biochem. Biophys. 135* (1969) 166-172.

29. Dobrota, M., & Hinton, R.H., *Anal. Biochem. 56* (1973) 270-274.

30. Fox, B.W., in *Liquid Scintillation Counting,* Vol. 2 (Crook, M.A., Johnson, P. & Scales, B., eds.), Heyden & Son, London (1972) pp. 184-204.

31. Paus, P.M. *as for ref. 30,* pp. 205-212.

32. McDowell, R.E. & Copeland, J.C., *Anal. Biochem. 41* (1971) 338-343.

33. Neal, W.K., Hoffman, H.P., Avers, C.J. & Price, C.A., *Biochem. Biophys. Res. Commun. 38* (1970) 414-422.

ADDENDUM

Sorbitol has been used as a gradient material in rate-sedimentation studies on yeast mitochondria with a zonal rotor [33]. MGU (methylglucamine salt of umbradilic acid) was effective in buoyant-density studies on spermatozoa [16].

10 THE USE OF METRIZAMIDE FOR THE FRACTIONATION OF RIBONUCLEOPROTEIN PARTICLES

Richard H. Hinton, Barbara M. Mullock, and Carl-Christian Gilhuus-Moe
Wolfson Bioanalytical Centre   Nyegaard & Co. A/S
University of Surrey   PO Box 4220
Guildford GU2 5XH, U.K.   Oslo 4, Norway

Metrizamide [(2-(3-acetamido-5-N-methylacetamido-2.4.6-triiodobenzamido)-2-deoxy-D-glucose] has been investigated as a medium for the isopycnic banding of unfixed ribonucleoprotein particles. Metrizamide forms stable solutions with densities up to 1.45 g/ml and has apparently no more damaging effect on biological material than sucrose. Comparison of the banding of fixed and unfixed ribonucleoprotein particles on Metrizamide gradients shows that the compound does not dissociate RNA and protein. Isolated liver polysomes and native and derived large and small subunits all band at a density of 1.26 g/ml. Subsidiary bands were sometimes found at higher densities. The ribonucleoprotein particles are clearly separated from the proteins, glycoproteins and ferritin which contaminate the preparations of native ribosome subunits, but the various classes of ribonucleoprotein particle are not separated from each other. The reasons for the apparent uniformity of banding density of ribonucleoprotein particles in Metrizamide gradients, in contrast with their heterogeneity in CsCl gradients, are discussed.*

As discussed in the article by G.C. Hartman and co-authors[†], an ideal density gradient medium should (a) form solutions covering the density range needed for the particular application; (b) form solutions of low viscosity; (c) possess some property, such as refractive index, by which its concentration may be measured simply; (d) not damage the material which is being separated; (e) be readily removable after the separation and (f) not interfere with the analysis of the separated fractions. The density gradient solute conventionally used for the isopycnic banding of ribonucleoprotein particles, caesium chloride, satisfies all the above criteria except for (a). The RNA and proteins of ribonucleoprotein particles are dissociated by the high ionic strength of the caesium chloride solutions required to band the particle [1]. It is therefore necessary to fix ribonucleoprotein particles with formaldehyde prior to isopycnic banding on caesium chloride gradients, and although

---

* *manufactured by Nyegaard and Co. A/S, Oslo 4.*
† *p. 93 (i.e. preceding Article)*

it is possible to recover the RNA of such fixed particles by pronase digestion [2], it is absolutely impossible to recover the proteins in their original state. As cytoplasmic particles carrying messenger RNA are distinguished from ribosome subunits by their lower density [3], it is obviously desirable to find a density gradient medium on which to band unfixed ribonucleoprotein particles. As Metrizamide appeared, at first examination, to satisfy many of the above criteria, we have tested its suitability for the fractionation of cytoplasmic ribonucleoprotein particles. Some of the results to be presented have already been published [4, 5].

MATERIALS AND METHODS

Cytoplasmic ribonucleoprotein particles were separated from the livers of rats which had been injected 24 h before death with [^{3}H-]orotic acid and 45 min before death with [^{14}C-]orotic acid. Native ribosome subunits were prepared by zonal centrifugation as described earlier [6, 7]. Polysomes were prepared by the magnesium precipitation method of Leytin and Lerman [8]. Derived ribosome subunits were prepared by treating polysomes with 2.5 μmoles EDTA/mg polysomes and separating the subunits by centrifugation on a titanium B-XIV zonal rotor (M.S.E. Ltd., Crawley, Sussex) in the same way as the native subunits. Fractions separated from the zonal rotor were concentrated by filtration using Diaflo PM 30 membranes (Amicon Ltd., High Wycombe, Bucks). The purity of the preparations was checked by analysis of the RNA on SDS-containing sucrose gradients [9]. The native 60 S particles were heavily contaminated by particles containing 18 S RNA, probably deriving from dimers of the small ribosome subunit (B.M. Mullock and R.H. Hinton, *unpublished experiments*), but the native 40 S subunits and the derived 40 S and 60 S subunits were essentially pure.

Metrizamide gradients extending from 20% w/v to 80% w/v were prepared in the tubes of a 3 x 6.5 ml titanium swing-out rotor (M.S.E.) using an M.S.E. small-tube gradient maker. All gradient solutions contained 5 mM MgCl$_2$ and either 5 mM Tris HCl or 5 mM triethanolamine pH 7.8. Aliquots of the ribonucleoprotein particles were layered over the gradients and centrifuged for 22 h at 54,000 rev/min (240,000 × g) in an M.S.E. SuperSpeed 65 Ultracentrifuge. The temperature of the rotor was controlled at 9°. After centrifugation, the gradient was fractionated using an M.S.E. gradient displacement device. Metrizamide absorbs strongly in the ultraviolet, so that it was not possible to monitor the positions of the ribonucleoprotein bands directly; but a general idea of the distribution of material through the gradient could be obtained by measuring the extinction at 400 nm. Fractions of about 0.2 ml were collected automatically, and the refractive index of a small aliquot of every fifth fraction was measured on an Abbé refractometer (Bellingham and Stanley Ltd., London); finally aliquots were counted, after dilution with 1.5 ml of water, in a Packard scintillation counter using a butyl-PBD-based scintillant [10].

RESULTS AND DISCUSSION

In preliminary experiments we examined how well Metrizamide met the criteria
for density gradient materials laid down at the beginning of this article.
(a) Metrizamide readily forms solutions with density up to 1.45.
(b) Metrizamide solutions are less viscous than sucrose solutions of the
same density (at ρ = 1.30 the viscosity of Metrizamide solutions is one-tenth
of that of sucrose)*. Nevertheless, the viscosity of Metrizamide solutions
rises markedly at densities greater than 1.3 g/ml, so that particles more
dense than this will band more slowly.
(c) The concentration, and thus the density, of Metrizamide solutions can
readily be estimated from their refractive index, using the following formula:
$$\rho_{20°} = \frac{3.350}{3.462} \eta_{20°} .$$
(d) Two types of damage to particles were conceivable. Firstly, Metrizamide
could dissociate the RNA and the protein. This was tested by banding ribonucleoprotein particles, either untreated or fixed with formaldehyde, on
Metrizamide gradients. Experiments which have been presented elsewhere [4]
showed there to be no difference in the banding pattern, which indicates that
there is no dissociation of the RNA and protein of the unfixed particles.

The second type of damage that we envisaged was denaturation of the
RNA or protein of the particles. This was not tested directly. We did find
that a number of enzymes were inhibited in the presence of Metrizamide [5].
This inhibition was, however, comparable to the inhibition found in the presence of other gradient materials (Table 1; see also the article by G.C.
Hartman and co-authors), and was readily reversed on dilution (Table 2). Hence
there is no evidence that Metrizamide is a protein denaturing agent.
(e) Metrizamide is not precipitated by 5% perchloric acid or 75% ethanol.
Hence ribonucleoprotein particles can be readily precipitated from solution.
Alternatively, Metrizamide can be removed by dialysis in Visking tubing or
by diafiltration using a PM 10 Diaflo membrane.
(f) Metrizamide causes some quenching of counting of [^{14}C-] and [^{3}H-]
labelled ribonucleoprotein particles even when using a quench-resistant scintillant such as butyl-PBD. The quenching appears to be a combination of the
chemical quench found generally with halogenated organic molecules and the
specific quenching of [^{3}H-]labelled macromolecules and small particles, which
is probably due to the binding of hydrophilic gradient solutes to the hydration sphere [see the article by G.C. Hartman *et al.*]. Metrizamide also reacts
with the reagents used in the estimation of protein by the Lowry method [12],
a 36% solution of Metrizamide giving the same colour as a 1 mg/ml solution
of bovine serum albumin. However, this does not prohibit the use of this
method in the estimation of protein in Metrizamide solutions. Metrizamide,
as would be expected from its chemical structure, reacts strongly with the
phenol-sulphuric acid reagent used in the determination of glycogen [13].

---

* *See the article by A. Hell, D. Rickwood & G.D. Birnie - Art. 12 (p. 117)*

Table 1. Inhibition of enzymic activity: comparison of sucrose and Metrizamide. Values represent % inhibition with 34% (w/v) in the assay system.

	Sucrose (1M) *	Metriz-amide
5'-Nucleotidase	65	38
Glucose-6-phosphatase	50	41
Succinate dehyd-rogenase (INT)	62	77

* from ref. [10]

Table 2. Reversibility of the Metriz-amide inhibition, studied with a 20% (w/v) rat-liver post-nuclear fraction.
(a) To 1 ml, add 9 ml of 83% (w/v) Metri-zamide, and leave 4 h at 0°. *(This con-centration in an assay system would have inhibited 5'-nucleotidase by 85% and suc-cinate dehydrogenase 100%.)* Then add 9 ml water, and assay (+ 2.5 ml of assay medium).
(b) To 1 ml, add 9 ml of water, and leave 4 h at 0°. Then add 9 ml 83% Metrizamide and assay as for (a). *Partial inhibition (= Table 1 % value) is expected; but if* (a) = (b) *below, then no irreversible effect of standing in very high Metrizamide concn.*

	Reading: pre-exposure to-	
	(a) Metrizamide	(b) Water
5'-Nucleotidase	0.098	0.110
Succinate dehyd-rogenase	0.385	0.39

Thus Metrizamide would ap-pear to satisfy all the prelimi-nary criteria for a density gra-dient material for the isopycnic separation of ribonucleoprotein particles. Polysomes were found to band in Metrizamide gradients at a density of 1.26 g/ml, with an additional band sometimes appearing at a density of 1.36 g/ml (Fig. 1). Very similar results were found with derived ribosome subunits. Again the major band was at a density of 1.26 with subsidiary bands sometimes appearing at higher densities (Fig. 2). These subsidiary bands were not reproducible and appear to be due to an interaction of the particles and the gradient material during centrifugation rather than to changes in the particles themselves. One possibility was that the subsidiary bands were due to aggregation, some water being removed from the hydration spheres. However, native ribosome subunits gave very similar patterns to the derived subunits (Fig. 3) except that an extra band was found at a density of 1.14 g/ml coincident with the largest peak of light-scattering. This peak is almost certainly due to aggregation of some ribonucleoprotein particles with other material, probably lipoprotein. Otherwise the ribonucleoprotein partic-les were well separated from the peaks of light-scattering material, which can tentatively be ascribed to protein (ρ = 1.24), to glycoproteins or glyco-gen (ρ = 1.4) and, in the case of the 60 S subunits, to ferritin (ρ = 1.42).

The results presented in the previous paragraph show that Metrizamide is an efficient material for banding rat-liver ribonucleoprotein particles, but not for fractionating them. After fixation in formaldehyde, the 40 S and 60 S subunit preparations used in the experiment of Fig. 2 banded at densities of 1.55 and 1.59 g/ml respectively in caesium chloride gradients, so that, *a priori,* one would have expected a difference in the banding densities on

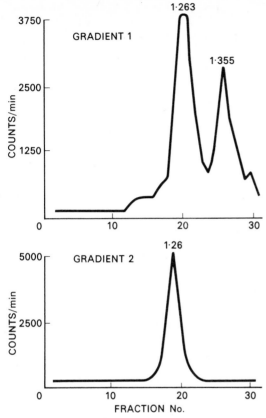

Fig. 1.   Separation of polysomes on 20-80% (w/v) Metrizamide gradients containing 5 mM
triethanolamine pH 7.8 and 5 mM
$MgCl_2$, after centrifugation for
22 h at 54,000 rev/min (240,000
× g) at 9° in an M.S.E. 3 ×
6.5 ml swing-out rotor.
*The results come from two sepa-
rate experiments and illustrate
the variability in the amount
of material recovered at densities greater than 1.3 g/ml.*

Metrizamide gradients. As native
ribosome subunits band at the
same density as derived ribosome
subunits (which are essentially
free from any non-ribosomal proteins), the explanation cannot be
aggregation of RNA and protein.
The much lower banding densities
in Metrizamide gradients must,
therefore, be due to variations
in the amount of water associated
with the particles. Assuming that
the density of anhydrous RNA is
1.895 and that of anhydrous protein is 1.245 g/ml, and that polysomes contain 53% RNA and 47% protein [14], then 1.27 mg of water
must be bound per mg of polysomes
if the hydrated particles are to
band at a density of 1.26 g/ml.
This is not unexpected, for early experiments on the viscosity of ribosome
preparations indicated that in a dilute salt solution between 2.7 and 3.7 g
of water were bound per g of ribosomes [15]. Assuming that the small and
large ribosome subunits contain respectively 55.3% and 50.5% RNA and 44.7%
and 49.5% protein [14] and that the proportion of bound water is the same as
in the polysomes, then one would expect banding densities of 1.252 and 1.266
g/ml respectively for the large and small subunits in Metrizamide gradients.
These would not have been distinguished under the present conditions but
might have been resolved if a shallower gradient had been employed. However,
Metrizamide gradients are self-forming [see the article by A. Hell, D. Rickwood & G.D. Birnie], so that a less steep gradient would necessitate slower
and longer centrifugation, which would increase the risk of enzymic break-

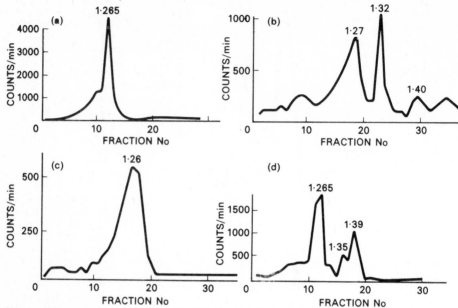

Fig. 2.   Separation of (a) and (b) derived 40 S subunits, and (c) and (d) derived 60 S subunits, on 20 - 80% Metrizamide gradients. The conditions for centrifugation are given in the legend to Fig. 1.

down of the RNA. One possible solution would be to use angle-head rotors, in which the equilibrium gradients are more shallow than in swing-out rotors [16].

We thus conclude that the lack of resolution of ribonucleoprotein particles on Metrizamide gradients is due to the large amounts of water associated with the particles. It is also reasonable to suggest that the appearance of subsidiary bands at higher density is due to a loosening of the structure, possibly due to ribonuclease action [2] which allows Metrizamide to penetrate into the particle. Metrizamide, in itself, would thus appear to have no ill effects on ribonucleoprotein particles and may still have a role in separating ribonucleoprotein particles which differ considerably in density, such as RNA virus and host ribosomes. One would expect a similar lack of resolution in other non-ionic density gradient materials. From this point of view, it would be interesting to re-examine the reported banding of light ribonucleoprotein particles on sucrose-D_2O gradients [17].

Acknowledgements

We thank Dr. E. Reid for his support in this work and Mr. M. Dobrota for his

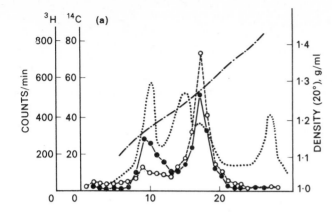

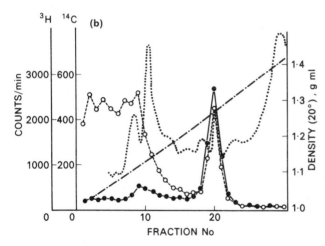

Fig. 3. Separation of (a) native 40 S subunits, and (b) native 60 S subunits, on 20-80% Metrizamide gradients. Conditions for centrifugation are given in the legend to Fig. 1. *Part (a) of this figure has already been published in ref. [4].*
——— RNA [^{3}H]cts/min; ------- rapidly labelled RNA [^{14}C]cts/min;.......... A_{400}; – ·– ·–· – ·–· – Density (20°), g/ml.

help with zonal centrifugation.   Financial support was given by the Cancer Research Campaign and by Nyegaard A/S.

References

1.   Spirin, A.S., *Eur. J. Biochem. 10* (1969) 20-35.

2.   Dessev, G.N. and Grancharov, K., *Eur. J. Biochem. 31* (1972) 112-118.

3.   Henshaw, E.C. and Loebenstein, J., *Biochim. Biophys. Acta 190* (1970) 405-420.

4.   Mullock, B.M. and Hinton, R.H., *Trans. Biochem. Soc. 1* (1973) 518-521.

5.   Hinton, R.H., Mullock, B.M., Reid, E. and Gilhuus-Moe, C.C., *Abstracts 9th Internat. Congr. Biochem.,* Stockholm (1973) p.31.

6.   Mullock, B.M., Hinton, R.H., Dobrota, M., Froomberg, D. and Reid, E., *Eur. J. Biochem. 18* (1971) 485-495.

7.   Hinton, R.H. and Mullock, B.M., in *Separations with Zonal Rotors* (Reid, E., ed.) Wolfson Bioanalytical Centre, University of Surrey, Guildford (1971) pp. P-7.1 - P-7.6.

8.   Leitin, V.L. and Lerman, M.J., *Biokhimiya 34* (1969) 839-848.

9.   McConkey, E.H., in *Methods in Enzymology* (Moldave, K. and Grossman, L. eds.) Vol. 12A, Academic Press, New York, pp. 620-634.

10. Dobrota, M. and Hinton, R.H., *Anal. Biochem. 56* (1973) 270-274.

11. Hinton, R.H., Burge, M.L.E. and Hartman, G.C., *Anal. Biochem. 29* (1969) 248-256.

12. Lowry, O.H., Rosebrough, N.J., Farr, A.L. and Randall, R.J., *J. Biol. Chem. 193* (1951) 265-275.

13. Montgomery, R., *Arch. Biochem. Biophys. 67* (1957) 378-386.

14. Hamilton, M.G. and Ruth, M.E., *Biochemistry 8* (1969) 851-856.

15. Petermann, M.L., *The Physical and Chemical Properties of Ribosomes,* Elsevier, Amsterdam (1964).

16. Flamm, W.G., Birnstiel, M.L. and Walker, P.M.B. in *Subcellular Components, Preparation and Fractionation,* 1st edn. (Birnie, G.D. and Fox, S., eds.) Butterworths, London, pp. 125-156.

17. Kempf, J.F., Egly, J.M., Stricker, C., Schmidt, M. and Mandel, P., *FEBS Lett. 26* (1972) 130-134.

11 BUOYANT DENSITY SEPARATION OF NUCLEIC ACIDS IN SODIUM IODIDE

Anna Hell, Elizabeth MacPhail and G.D. Birnie
The Beatson Institute for Cancer Research
132 Hill Street, Glasgow G3 6UD, Scotland

Solutions of caesium salts are the most frequently used media for buoyant density analyses of nucleic acids. However, the separation of native and denatured DNA in either CsCl or Cs_2SO_4 is not entirely satisfactory even when advantage is taken of the increased resolution obtainable with fixed-angle rotors [1]. Recently it has been shown that shallow density gradients are generated when a solution of NaI is centrifuged in a fixed-angle rotor, and that DNA, DNA-RNA hybrid and RNA all band isopycnically in these gradients [2-4]. If care is taken to remove heavy metal ion contaminants, for many purposes NaI is an acceptable, and much cheaper, substitute for CsCl and Cs_2SO_4, while for some purposes it is superior to both these salts. The major drawback to the use of NaI is the very high absorbance of its solutions at 260 nm.

Solutions of caesium salts are suitable for separating nucleic acid species which differ quite markedly in buoyant density [5] but are much less satisfactory for resolving mixtures of native and denatured DNA. Anet and Strayer [2] introduced the use of NaI as a gradient solute for the preparative-scale fractionation of DNAs, and showed that DNAs of closely related base composition were better resolved in NaI than in CsCl, mainly because of the shallower gradients generated in solutions of NaI. DeKloet and Andrean [3] also investigated the use of NaI and KI for the separation of native and denatured DNA and RNA; they concluded that, although resolution was poorer in KI than in NaI, KI was more versatile because RNA precipitated in NaI.

In our experiments we have found that NaI is suitable for the isopycnic banding of native and denatured DNA, DNA-RNA hybrids and RNA [4]. The precipitation of RNA in NaI appears to be due to contamination of the salt with heavy metal ions. In aqueous 5 M NaI about 75% of RNA precipitates in 24 h at 20°; addition of 10 mM EDTA reduces this to 40%, while if the NaI solution is first treated with Chelex-100 (BioRad Laboratories) no significant proportion of the RNA is precipitated [4].

BANDING OF DNA

A comparison of the gradients formed in NaI, CsCl and Cs_2SO_4 (Table 1) shows that NaI solutions centrifuged under the same conditions form much shallower gradients than either of the other two salts. Fig. 1 illustrates the excellent

Table 1.   Comparison of gradients formed in sodium iodide,caesium chloride and caesium sulphate solutions. Each solution (4.6 ml), overlaid with liquid paraffin in a 10 ml polyallomer tube, was centrifuged at 45,000 rev/min for 66 h at 20° in the 10 × 10 ml titanium fixed-angle rotor of an MSE Superspeed 65 Mark II ultracentrifuge. The gradients were unloaded by upwards displacement [10] and 0.2 ml fractions were collected; densities were calculated from refractive index measurements.

	Initial density g/ml	Gradient range g/ml	Δ g/ml
NaI	1.545	1.514-1.584	0.070
CsCl	1.750	1.702-1.820	0.118
Cs_2SO_4	1.540	1.430-1.670	0.240

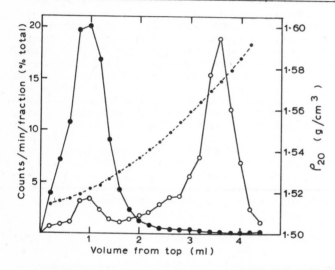

Fig. 1.   Isopycnic banding of native ($\bullet$——$\bullet$) and denatured ($\circ$——$\circ$) LS-cell [³H]DNA in NaI gradients (initial density 1.546 g/ml) centrifuged for 66 h as described for Table 1; $\bullet$---$\bullet$ shows the density gradient. (From ref. [4].)

separation of native and denatured DNA in such an NaI gradient. Not only is the separation enhanced by the shallower gradient but also the difference in buoyant densities in NaI is greater than in CsCl (0.052 g/ml in NaI compared to 0.015 in CsCl). This increase may be due to an ionic interaction of the bases in single-stranded DNA with the anion, making it more dense in the presence of iodide than chloride ions. However, the increased density of denatured DNA in NaI is not due to iodination of bases, as was shown by recovering the DNA from an NaI gradient, removing free iodide by passage through

Sephadex G-25, and re-banding the DNA in CsCl; the density of this DNA was identical to that of DNA which had not previously been exposed to NaI.

NaI appears to fractionate DNA mainly on the basis of secondary structure, with a smaller effect of base composition, which makes NaI an excellent tool for investigating the re-annealing of DNA. Two methods have been used to determine the degree of double-strandedness of re-annealed DNA and they give quite different results. First, hydroxyapatite has a greater affinity for double-stranded DNA than for denatured DNA [6]. However, concentrations of phosphate buffer which elute denatured DNA from hydroxyapatite allow retention of material which is up to 80% single-stranded, thus giving a high estimate of the degree of re-annealing. The second method measures the proportion of re-annealed DNA which is resistant to digestion with a single-strand specific nuclease [7,8]. These enzymes digest all single-stranded DNA tails as well as loops and regions of poor base complementarity, and so give the minimum estimate of re-annealing. For example, a sample of LS-cell DNA which was sheared to an average molecular weight of 5×10^4, then denatured and re-annealed to a C_0t * of 445 was found by hydroxyapatite analysis to contain 42% of material with some degree of double-strandedness while a nuclease assay indicated that only 16% was in well-matched duplexes without tails of single-stranded material. However, both the quality and the quantity of the re-annealed duplexes were readily assessed by isopycnic banding of the mixture in NaI. In Fig. 2, region A consists of well-matched duplexes, region B of re-annealed material with increasing proportions of single-stranded stretches, and region C of unannealed DNA.

BANDING OF RNA

Both total nuclear and cytoplasmic RNAs band in NaI gradients, though much more heterogeneously than DNA (Fig. 3). In contrast, globin messenger RNA (9 S RNA isolated from mouse reticulocytes) bands as a relatively sharp and dense peak (Fig. 4). Since RNAs are known to have considerable secondary structure it seems that, as in the case of DNA, the buoyant density of RNA is more dependent on secondary structure than on base composition. Since the usual methods of fractionating RNA on sucrose gradients or polyacrylamide gels depend mainly on the size and shape of the RNA molecules, fractionation by isopycnic sedimentation in NaI, which depends on a different parameter, could prove to be a valuable extra dimension for the fractionation of RNA.

BANDING OF DNA-RNA HYBRIDS, AND GENERAL COMMENTS

DNA-RNA hybrids [4] also band isopycnically in NaI gradients. The banding densities of DNA, RNA and hybrids in NaI and CsCl are summarized in Table 2. One particularly interesting point illustrated by Table 2 is the difference in the order in which these nucleic acid species band in CsCl and NaI. In

*C_0t = *initial concentration of DNA (moles of nucleotide/litre)* × *time (sec).*

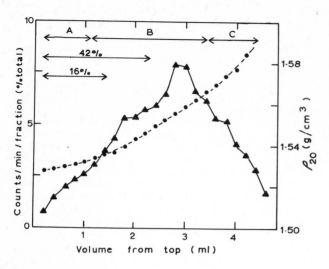

Fig. 2. Isopycnic banding of partly re-annealed low molecular-weight LS-cell [^{14}C]DNA in NaI. Centrifugation was in NaI solution (initial density 1.544 g/ml) as described for Table 1.

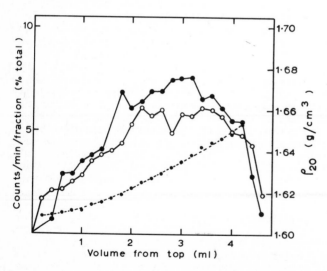

Fig. 3. Isopycnic banding of [^{3}H]RNA from nuclei (●——●) and cytoplasm (○——○) of LS-cells in NaI fradients (initial density 1.630 g/ml) centrifuged for 65 h as described for Table 1; ●---● shows the density gradient. (From ref. [4].)

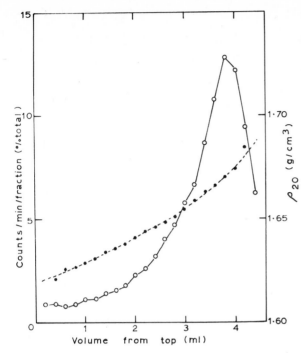

Fig. 4. Isopycnic banding of [^{32}P]-labelled globin messenger RNA (9S RNA from mouse reticulocytes) in NaI. The solution (initial density 1.646 g/ml) was centrifuged as described for Table 1; ●---● shows the density gradient.

CsCl, the buoyant density of a DNA-RNA hybrid is intermediate between those of denatured DNA and RNA while in NaI it is intermediate between those of native and denatured DNA. This again illustrates the dependence of buoyant density in NaI on secondary rather than primary structure. It also indicates that NaI is superior to CsCl for the buoyant density-gradient analysis of DNA-RNA hybrids. In CsCl, a hybrid molecule with unpaired regions of RNA and DNA may band close to the position of perfect hybrid since RNA is heavier and DNA lighter than the hybrid molecule. In NaI, on the other hand, any unpaired stretches of RNA and/or DNA increase the buoyant density of a hybrid molecule, so that imperfect hybrids are displaced from the isopycnic banding position of a perfect hybrid.

These data indicate that NaI has some considerable advantages over CsCl and Cs$_2$SO$_4$ as a solute for buoyant density-gradient analyses. Unfortunately, full characterisation of NaI density gradients is hampered by the strong UV-absorption of NaI solutions, which precludes its use in the analytical ultracentrifuge. It is necessary to use either labelled nucleic acids, or to have sufficient nucleic acid in the gradient to allow detection either by the fluorescence of their complexes with ethidium bromide [3] or by microchemical analyses [9].

Table 2. Buoyant densities of LS-cell nucleic acids in sodium iodide and caesium chloride. Labelled DNA, RNA and hybrid were prepared [4] and solutions of them were mixed with a concentrated NaI solution which had been treated with Chelex-100 [4]. The densities of the mixtures were adjusted as appropriate and 4.6 ml portions were centrifuged as described for Table 1.

	Buoyant density g/ml	
	in NaI	in CsCl
Native DNA	1.522	1.700
Denatured DNA	1.574	1.715
Nuclear RNA	1.63	>1.9
DNA-RNA Hybrid	1.540	1.775

Acknowledgements

This work has been supported by grants to the Beatson Institute from the Medical Research Council and the Cancer Research Campaign. We are grateful to Dr. G. Threlfall for preparing DNA-RNA hybrids and to Mr. G. Lanyon for providing the reticulocyte 9 S RNA. We acknowledge with thanks permission from the Editors of *FEBS Letters* to republish Figs. 1 and 3.

References

1. Flamm, W.G., Birnstiel, M.L. & Walker, P.M.B., in *Subcellular Components, Preparation and Fractionation,* 2nd edn. (Birnie, G.D., ed.), Butterworth, London (1972) pp. 279-310.

2. Anet, R. & Strayer, D.R., *Biochem. Biophys. Res. Commun. 37* (1969) 52-58.

3. DeKloet, S.W. & Andrean, B.A.G., *Biochim. Biophys. Acta 247* (1971) 519-527.

4. Birnie, G.D., *FEBS Lett. 27* (1972) 19-22.

5. Szybalski, W., in *Methods in Enzymology, Vol. 12B* (Colowick, S.P. & Kaplan, N.O., eds.), Academic Press, New York (1968) pp. 330-360.

6. Bernardi, G., *Biochim. Biophys. Acta 174* (1969) 423-448.

7. Linn, S. & Lehman, I.R., *J. Biol. Chem. 240* (1965) 1287-1293.

8. Sutton, W.D., *Biochim. Biophys. Acta 240* (1971) 522-531.

9. Klevecz, R.R. & Kapp, L.N., *J. Cell Biol. 58* (1973) 564-573.

10. Hell, A., *MSE Application Information Sheet A6/6/72* (1972).

12 BUOYANT DENSITY-GRADIENT CENTRIFUGATION IN SOLUTIONS OF METRIZAMIDE

Anna Hell, D. Rickwood and G.D. Birnie

The Beatson Institute for Cancer Research
132 Hill Street, Glasgow G3 6UD, Scotland

Metrizamide is very soluble in water and dilute buffer solutions, and forms dense solutions of relatively low viscosity. It is chemically inert and non-ionic, and so is potentially useful for isopycnic banding of biological macro-molecules. Both ribonucleoprotein particles [1] and chromatin [2] can be banded in Metrizamide in their native state without prior fixation. Metriza-mide readily forms gradients in a high centrifugal field. The shape of a gradient, as well as the rate at which it is formed, depends on a number of parameters including the rotor used, the speed, time and temperature of cen-trifugation, and the initial concentration of Metrizamide. This gives Metri-zamide considerable versatility as a buoyant density-gradient solute, since it enables the shape of the gradient to be manipulated to a much greater ex-tent than is possible with, for example, CsCl, so facilitating optimum reso-lution of particles.

Much effort has been expended in searches for gradient solutes which would be suitable for the isopycnic banding of nucleoprotein complexes in their native states. Several gradient systems have been described, including chloral hydrate [3], Urographin [1,4] and sucrose-glucose mixtures [5], but, for a variety of reasons, none have been very successful. A new compound, Metriza-mide [2-(3-acetamido-5-N-methylacetamido-2,4,6-tri-iodobenzamido)-2-deoxy-D-glucose], recently introduced by Nyegaard & Co. A/S, Oslo, Norway, has pro-perties which suggested that it might have considerable potential as a gra-dient solute for the isopycnic sedimentation of native nucleoproteins. Indeed, it has now been shown that ribonucleoprotein particles [1], chromatin [2] and nuclei [6] will band isopycnically in gradients of Metrizamide, as also will RNA, DNA and proteins [7].

FORMATION OF METRIZAMIDE GRADIENTS

Metrizamide is non-ionic and chemically inert. It is very soluble in water and dilute buffer solutions in which it forms dense solutions. The viscosity of these solutions is greater than that of CsCl solutions, but much less than that of dense sucrose solutions [2]. The refractive index of solutions of Metrizamide bears a linear relationship to both density and concentration, so that it is easy to obtain accurate estimates of density from refractive index

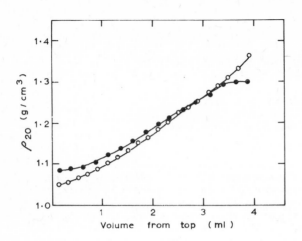

Fig. 1. The gradient
formed by centrifuging
6 ml of 29% (w/v) Metrizamide solution in the MSE
× 6.5 ml swing-out rotor;
------, initial density,
o———o, density after cen-
trifugation at 40,000 rev/
min for 44 h at 4°. Gra-
dients were unloaded by
upwards displacement [8]
with Fluorochemical FC 43
(3M Co. Ltd.).

readings:

$$\rho_{20°} = 3.350\ n_{20°} - 3.462$$

Density gradients readily form in solutions of Metrizamide which are
centrifuged at high speed. The rate of formation of a gradient, and the shape
of the gradient obtained, are both very dependent on a number of parameters
including the type of rotor used, the initial concentration of Metrizamide,
and the speed, temperature and time of centrifugation. However, swing-out
rotors are not suitable for forming gradients in Metrizamide solutions. For
example, when a uniform solution of Metrizamide of initial density 1.145 g/ml
was centrifuged at 40,000 rev/min for 45 h in the MSE 3 × 6.5 swing-out rotor,
steep gradients were found only at the top and bottom of the tube, while the
remainder remained unchanged (Fig. 1). However, the slow rate at which gradients are formed in swing-out rotors means that they can be used for centrifuging pre-formed gradients of Metrizamide. Even after centrifugation at
34,000 rev/min for 34.5 h there is little change in the shape of a preformed
14.5% to 58% (w/v) gradient (Fig. 2). In contrast, gradients are formed much
more rapidly in solutions of Metrizamide centrifuged in fixed-angle rotors
(Fig. 3).

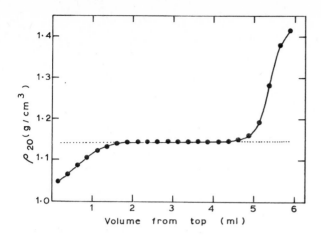

Fig. 2. Change in a pre-formed gradient caused by centrifugation in a swing-out rotor. Gradients were pre-formed by equilibration at 4° for 24 h of 1.0 ml layers of 14%, 29%, 43% and 58% (w/v) Metrizamide solutions; density gradient shown before (●———●) and after (○———○) centrifugation at 34,000 rev/min for 34.5 h at 2° in the MSE 3 × 5 swing-out rotor.

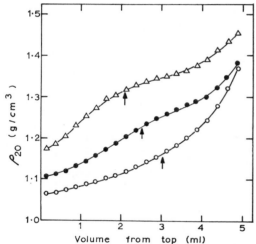

Fig. 3. The effect of initial density on the formation of a Metrizamide gradient. Tubes containing 5 ml of Metrizamide solution of initial density 1.158 g/ml (○———○), 1.224 g/ml (●———●) and 1.312 g/ml (△———△) and overlaid with liquid paraffin were centrifuged in the MSE 10 × 10 Al fixed-angle rotor at 35,000 rev/min for 40 h at 5°. The *arrows* indicate the initial densities.

The effects of initial density and of speed and time of centrifugation on gradient formation are summarized in Figs. 3-7. In Fig. 3, three 5 ml gradients are shown; all were centrifuged under the same conditions, the only variation being in the initial concentration of the solution of Metrizamide. The marked differences in the *shapes* of the gradients are due to the differences in the viscosities of the solutions, upon which depend the relative rates of diffusion and sedimentation of Metrizamide. For the same reason the speed of centrifugation has a marked effect on the shape of the gradient formed (Fig. 4). Also for the same reason, similar differences will be seen when a solution of Metrizamide is centrifuged at different temperatures.

These observations indicate that equilibrium gradients in Metrizamide are slow to form as compared with gradients in less viscous solutions such as those of CsCl. This conclusion is confirmed by Fig. 5, which shows the gradual changes in the shape of the gradient formed in a 29% (w/v) solution of Metrizamide by centrifugation at 35,000 rev/min for between 14.5 h and 68 h. Expressing this data in a different form (Fig. 6) clearly shows that the gradient is approaching equilibrium by 40 h, though there still may be slight changes even beyond 68 h.

The slow formation of equilibrium gradients in Metrizamide solutions can be exploited to vary the shape of the gradient obtained, as shown by Fig.7. In this way, gradients with steep and shallow regions can readily be obtained. In contrast, the rapid attainment of equilibrium with CsCl gradients precludes such adjustments as this to the shape of the gradient.

The dependence of the shape of a Metrizamide gradient on such a variety of parameters means that considerable care has to be taken in choosing conditions of centrifugation. However, it also means that Metrizamide is a versatile buoyant density-gradient material since the shape of the gradient can frequently be adjusted to obtain maximum resolution simply by careful manipulation of the conditions under which the Metrizamide solution is centrifuged.

ISOPYCNIC BANDING OF NUCLEIC ACIDS, PROTEINS AND NUCLEOPROTEINS IN METRIZAMIDE

DNA, RNA and proteins band isopycnically in Metrizamide gradients [7]. Proteins band at the same density (approx. 1.28 g/ml) in Metrizamide as in CsCl. However, in sharp contrast, both DNA and RNA band at much lower densities in Metrizamide (1.118 g/ml and 1.168 g/ml, respectively) than in CsCl. Calculations reveal that the nucleic acids and proteins are fully hydrated in Metrizamide whereas only protein is fully hydrated in CsCl; both DNA and RNA are grossly dehydrated in the latter. Ribonucleoprotein particles [1] and deoxyribonucleoproteins [2] band in Metrizamide at densities intermediate between those of the nucleic acid and protein. In the case of deoxyribonucleoproteins it has been shown that the buoyant density is dependent on the ratio of protein to DNA [7], and preliminary experiments indicate that the same rule holds for ribonucleoproteins. In addition, there is good evidence that Metrizamide neither dissociates nucleic acid-protein complexes [1, 7], nor binds irrever-

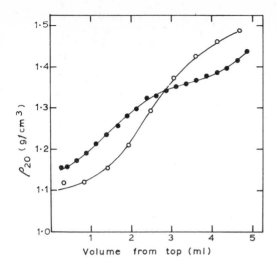

Figure 4. The effect of speed of rotation on the formation of a Metrizamide gradient. Tubes containing 5 ml of 58% (w/v) Metrizamide solution (density 1.312 g/ml) and overlaid with liquid paraffin were centrifuged in the MSE 10 × 10 Al fixed-angle rotor for 68 h at 35,000 rev/min (●————●) or 45,000 rev/min (○————○) at 5°. [From ref. 2].

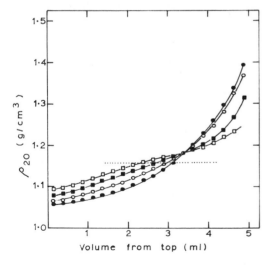

Fig. 5. The effect of time of centrifugation on the formation of a Metrizamide gradient. Tubes containing 5 ml of 29% (w/v) Metrizamide solution (density 1.158 g/ml) and overlaid with liquid paraffin were centrifuged in the MSE 10 × 10 Al fixed-angle rotor at 35,000 rev/min at 5° for 14.5 h (□————□), 22,5 h (■————■), 40 h (○————○) and 68 h (●————●). [From ref. 2.]

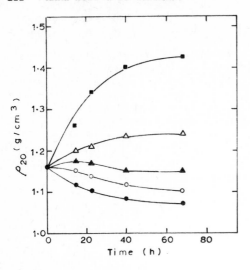

Fig. 6. The rate of attainment of equilibration of a Metrizamide gradient. Centrifugation data as in Fig. 5. Density at 5 ml from top (■———■), 4 ml (△———△), 3 ml (▲———▲), 2 ml (○———○), and 1 ml (●———●).

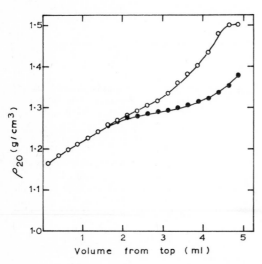

Fig. 7. Variation in gradient shape caused by inclusion of a dense cushion of Metrizamide. Gradients were centrifuged in the MSE 10 × 10 Al fixed-angle rotor at 30,000 rev/min for 42.5 h at 2°; ●———●, 5 ml of 50% (w/v) Metrizamide; ○———○, 4 ml of 50% Metrizamide + 1 ml cushion of 90% Metrizamide.

sibly to nucleoproteins [7].

These observations indicate that Metrizamide has considerable potential for the fractionation of nucleoproteins such as chromatin in their native state, and also in studies of the interaction of specific proteins with nucleic acids.

Acknowledgements

This work was supported by grants to the Beatson Institute from the Medical Research Council and Cancer Research Campaign and by a generous gift from Nyegaard and Co. A/S, Oslo, Norway. We are grateful to the Editors of FEBS Letters for permission to reproduce Fig. 4 and 5, and we are pleased to acknowledge the valued technical assistance of Mr. R. McFarlane.

References

1. Mullock, B.M. & Hinton, R.H., *Trans. Biochem. Soc. 1* (1973) 579-581.

2. Rickwood, D., Hell, A. & Birnie, G.D., *FEBS Lett. 33* (1973) 221-224.

3. Hossainy, E., Zweidler, A. & Bloch, D.P., *J. Mol. Biol. 74* (1973) 283-289.

4. Benedict, R.C., Schumaker, V.N. & Davies, R.E., *J. Reprod. Fert.13* (1967) 237-249.

5. Raynaud, A & Ohlenbusch, H.H., *J. Mol. Biol. 63* (1972) 523-537.

6. Mathias, A.P. & Wynter, C.V.A., *FEBS Lett. 33* (1973) 18-22.

7. Birnie, G.D., Rickwood, D. & Hell, A., *Biochim. Biophys. Acta 331* (1973) 283-294.

8. Hell, A., *MSE Application Information Sheet A6/6/72* (1972).

13 THE USE OF ZONAL ROTORS IN SEPARATING FRAGMENTS OF RIBONUCLEIC ACIDS

R.A. Cox, W. Hirst, Elizabeth Godwin and Piroska Huvos
National Institute for Medical Research
London NW7 1AA, U.K.

The need for methods of fractionating macromolecules of rather similar sizes is widespread. Polyacrylamide gel electrophoresis has been successfully used on the analytical scale to separate RNA fragments. We have needed larger amounts of material than are given by this technique, and have used zonal centrifugation to effect the separation of products of partial digestion of rabbit reticulocyte ribosomal RNA, as illustrated by the properties of 4 sets of pooled fractions. These had molecular weight ranges lying between 2×10^4 and 4.4×10^5, each fraction containing several bands of similar R_f values. The largest fraction had a nucleotide composition of approx. 80% G + C (compared with 67% G + C for the undigested RNA) and a relatively stable partly double helical secondary structure, judged by T_m. The properties of the fragments indicate a very uneven distribution of nucleotides throughout the RNA moiety of the larger subribosomal particle of rabbit reticulocytes.

There is a need for fractionating RNA fragments within the range 100,000 - 500,000 daltons mass, in amounts sufficient for further physicochemical and chemical studies. For microanalytical work, gel electrophoresis has been successfully used for studies of the primary sequence of rRNA [1]. The separation is achieved by using very low loads of very highly radioactive rRNA. The demand for additional techniques is illustrated by a problem concerning the structure of the rRNA of the larger subribosomal particle of mammalian ribosomes. In earlier spectrophotometric studies of the secondary structure of this rRNA species it was shown that the molecule forms double-helical secondary structure comprising short hairpin loops as might be expected for single-stranded RNA [2]. However it was found that the nucleotide composition of the double-helical regions was different from the overall nucleotide composition. The rRNA species was found by chemical analysis to comprise approx. 67% G + C, whereas approx. 40% of double-helical structure was found to be stabilized by approx. 55% G + C base-pairs and a further approx. 40% to be stabilized by approx. 80% G + C base-pairs. This observation suggests a very uneven distribution of nucleotides along the polynucleotide chain.

Further information about the structure of 28 S rRNA of rabbit reticulocytes was sought by means of partial nuclease digests. The previous results did not distinguish between the possibility that the uneven distribution of nucleotides might arise from regions that are as long as or little longer

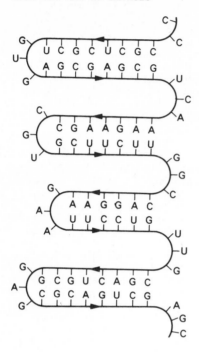

Fig. 1. Hypothetical structure of ribosomal RNA [2].

The ribosomal RNA is believed to consist of short hairpin loops. A hypothetical sequence is illustrated here. Although overall 65% of the nucleotides in the hypothetical sequence are G + C, some loops contain 50% G + C base-pairs while others contain 75% G + C base-pairs. This represents the possibility that these GC-rich regions are sufficiently long to form hairpin loops and that many such sequences are widely distributed throughout the polynucleotide chain. Another contrasting possibility is that there are a few long sequences containing the majority of the GC-rich hairpin loops. In the first case, fragments of 100,000 daltons might be expected to have much the same nucleotide composition as the intact molecule. In the second case, there is a much greater likelihood of isolating fragments of >100,000 daltons that differ appreciably in nucleotide composition from the intact molecule.

than individual double-helical regions (e.g. Fig. 1) and the possibility that there may be a few long tracts rich in G + C. There were indications that there might be at least one region of high G + C that was approx. 600 nucleotides long [3-5]. We have shown that fragments of $\sim$ 0.5 x 10^6 daltons mass ($\sim$ 1500 nucleotides) or 30% of the rRNA of 80% G + C can be isolated from rabbit 28 S RNA after partial digestion with RNase T_1.

PARTIAL DIGESTION

RNA of the reticulocyte larger subparticle (245 E_{260} units) in 0.01 M potassium phosphate buffer pH 7 (4ml) was digested with RNase T_1 (73.8 units) for 1 h at 20°. The solution was made up to 5% sucrose, layered onto a Sephadex G-200 column (50 × 1.2 cm diameter) and eluted with 0.01 M potassium acetate buffer pH 7 at a flow rate of 15 ml/h. Fractions of 1 ml were collected and the E_{260} measured against the appropriate blank (see below, Fig. 5). The void volume (7 ml) containing about half the initial load of RNA was collected and loaded onto a 15-40% sucrose gradient in a B-XIV rotor. Centrifugation was for 5 h at 43,000 rev/min in anM.S.E. Superspeed 65 centrifuge at thermostat setting 7. Four 50ml and 96 5ml fractions were collected and the E_{260} recorded. Fractions 24, 30, 36 and 40 were dialyzed against 0.01 M potassium phosphate buffer pH 7 at 4°, and the melting profiles (Pye Unicam SP 700) and $S_{20,w}$ (Beckman Model E analytical ultracentrifuge fitted with ultraviolet optics) of the sucrose-free samples were measured (Fig. 2). As shown in the Figure, other fractions were pooled, and the RNA precipitated by making up to 2% potassium acetate and adding two volumes of ethanol. The precipitate was dissolved in 0.5 ml of potassium acetate buffer pH 7, and the mobility on polyacrylamide gel electrophoresis (Fig. 3), sedimentation coefficient, melting profiles (e.g. Fig. 8, below) and nucleotide composition were assayed (Table 1 - *see p. 132*).

CONDITIONS OF ZONAL CENTRIFUGATION

The zonal separation experiment described above showed that there were particles in the void volume from the G-200 column that varied in size from 3 S to 12 S. Our usual sucrose gradient conditions allowed full separation of 18 S and 28 S particles, but smaller molecules stayed near the top of the gradient.

Further improvement in the separation was sought by lengthening the distance travelled by the RNA fragments along the gradient. A satisfactory separation of the model compounds tRNA and 16 S RNA of *E. coli* was achieved by centrifuging for 16 h at 33.5 k-rev/min so that the larger RNA species had traversed 90% of the gradient while the tRNA had only penetrated about one-third of this distance into the gradient.

The conditions used in later experiments were a B-XIV rotor filled with a 15-40% sucrose gradient centrifuged for 17 h at 33.5 k-rev/min. The sample was loaded at 4 ml/min using a LKB peristaltic pump; neither sample nor overlay contained sucrose. These loading conditions were chosen as they were found to give the sharpest initial zone (Fig. 4).

In the second experiment, the void volume from the G-200 column (Fig.5) was applied to a sucrose gradient as described above. After 17 h, the rotor was unloaded and 4 x 50 ml and 96 x 5 ml fractions were collected. The E_{260} of each fraction was measured against the appropriate buffer. The resulting profile was much sharper than that previously obtained (Fig. 2), with evidence for at least three distinct but overlapping peaks (Fig. 6).

The fractions shown were combined and precipitated with potassium acetate (final concentration 2%) and 2 volumes of ethanol. The precipitates were dissolved in about 0.5 ml buffer and analyzed by polyacrylamide gel electrophoresis. The gels (Fig. 7) show evidence for a clear separation of RNA fragments of different sizes on the sucrose gradient. The melting profiles of combined fractions III & IV are compared with that of the intact molecule in Fig. 8. Nucleotide compositions, including those for fragments, are in Table 1.

DISCUSSION

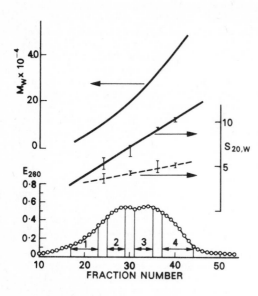

Fig. 2. Zonal separation of high molecular weight fragments from the G-200 column. The volume marked in Fig. 5 was applied directly to a 15-40% sucrose gradient in a B-XIV rotor, overlaid with 0.01 M phosphate buffer pH 7 and spun at 43,000 rev/min for 5 h. The fractions marked 1-4 were combined and the RNA precipitated for gel analysis *(Fig. 3)*. Other single fractions were dialyzed overnight against buffer and the sedimentation constant found using a Spinco Model E analytical ultracentrifuge equipped with UV optics. The same fractions were 'melted' and the sedimentation constants again found after cooling. The drop in S-value (----) indicates about three hidden breaks per RNA fragment.
The melting curves gave information concerning the double helix content of the fragments. Chemical analysis was also performed (Table 1).

We have isolated from partial RNase T_1 digests of rabbit 28 S RNA, a fraction of 0.36-0.44 x 10^6 daltons mass that has a nucleotide composition of approx. 78% G + C. The significance of the presence of such a long stretch of G + C rich sequences within this rRNA species is discussed elsewhere [10]. The separation that was achieved by zonal centrifugation was acceptable for the high molecular weight fragments, even though resolution was sacrificed in our procedure for the analysis of the material from the zonal rotor. As indicated in Fig. 6, several fractions were combined before the RNA was precipitated. Since each fraction contained up to 0.3 mg of RNA, it should be possible to improve the resolution by using single (5 ml) fractions. The separation by zonal centrifuging of a mixture of lower molecular weight species ranging from 300,000 to 100,000 appears difficult. We conclude that zonal centrifugation is a satisfactory means of isolating fragments of 200,000 daltons mass or greater.

ACKNOWLEDGEMENT

We thank the Medical Research Council for the award of a Scholarship for Training in Research Methods to Elizabeth Godwin and the Wellcome Trust for the award of a Fellowship to Piroska Huvos.

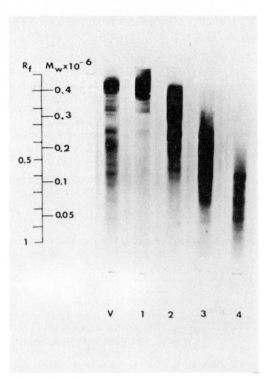

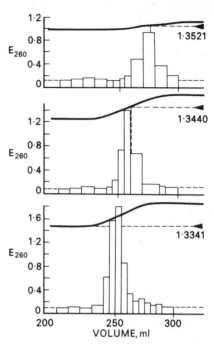

Above, left
Fig. 3. Fractions from the zonal separation shown in Fig. 2 were combined and the RNA precipitated by adding potassium acetate (final concentration 2%) and two volumes ethanol. The precipitate was dissolved in ∿ 0.5 ml buffer and the RNA separated by electrophoresis on 5% polyacrylamide gels.

Above, right
Fig. 4. Change in sample width due to loading conditions. The sample (6 ml) (*E. coli* 30 S, 0.2 mg/ml m M_4) was layered over a 15% sucrose solution in a B-XIV rotor spinning at 2000 rev/min using an LKB peristaltic pump to minimize mixing. After adding 250 ml overlay, the centrifuging was continued for a further 10 min at the same speed before unloading and finding the spread in the absorption at 260 nm. The sucrose content of both the sample and the overlay was varied to see which combination led to the narrowest initial zone:-

Expt.	Sucrose concentration (%)	
	sample	*overlay*
1	12.5	10
2	6	0
3	0	0

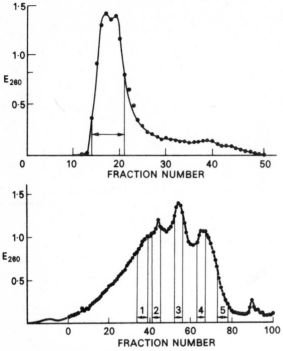

Fig. 5. Separation of RNase-digested reticulocyte 28 S RNA on a G-200 column. RNA of the reticulocyte large particle (245 E_{260} units in 4 ml) was digested with RNase T_1 for 1 h at 20°. The solution was made up to 5% sucrose, layered on a G-200 column and eluted with 0.01 M phosphate buffer pH 7.

Fig. 6. Zonal separation of fragments of hydrolyzed reticulocyte RNA. After digestion of reticulocyte 28 S RNA with RNase T_1, small molecular weight components were removed by gel filtration on Sephadex G-200. The void volume was applied directly to a 15-40% sucrose gradient, overlaid with buffer and spun for 17 h at 33 k-rev/min- 4 × 50 and 96 × 5 ml fractions were collected and the extinction at 260 nm of each fraction was found relative to buffer.

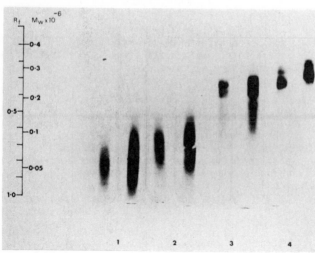

Fig. 7. The fractions marked in Fig. 6 were combined, precipitated with potassium acetate and ethanol and applied to a 5% polyacrylamide gel (cf. Fig. 3). Duplicate gels were run for each sample, using two different volumes of RNA solution so as to allow both a good separation of the major species and a visualization of minor bands.

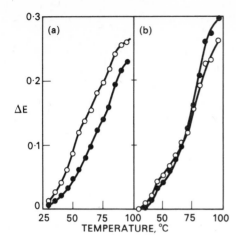

ΔE

TEMPERATURE, °C

Fig. 8. Melting profiles of (a) RNA of the larger subribosomal particles, & (b) of rabbit reticulocyte RNA fragments rich in G & C. ——●——, Change in absorbance at 280 nm, ——o——, *ditto* at 260 nm. (a) *It appears that the population of double helical segments melting over the range 25-60° are stabilized by approximately 55% G-C base-pairs overall, whereas double-helical secondary structure melting over the range 65-95% is stabilized by approx. 80% base-pairs overall.* (b) *The fragments (fractions 3 & 4) have a double-helical secondary structure stabilized mainly by G-C base-pairs. The more labile double-helical regions 'melting' over the range 65-95° comprise approx. 81% G-C base-pairs. The yield of fractions 3 & 4 was approx. 33% (wt. fraction), this to be compared with an estimate from the melting profiles that approx. 40% of the mass of the intact 30 S-rRNA comprises regions of approx. 80% G & C. Thus the fragments isolated account for approx. 85% of the regions rich in G & C whose presence was inferred from the melting profiles.*

References

1. Fellner, P., Ehresmann, C., Stiegler, P. & Ebel, J.P., *Nature New Biol. 239* (1972) 1-5.

2. Cox, R.A., *Biochem. J. 98* (1966) 841-857.

3. Delihas, N., *Biochemistry 6* (1967) 3356-3362.

4. Hadjiolov, A.A., Venkov, P.V., Delapchiev, L.B. & Genchev, D.D., *Biochim. Biophys. Acta 142* (1967) 111-127.

5. Gould, H.J., *J. Mol. Biol. 29* (1966) 307-313.

6. Hall, B.D. & Doty, P., in *Microsomal Particles and Protein Synthesis* (Roberts, R.B., ed.), Pergamon, New York (1958) p. 27

7. McPhie, P., Hounsell, J. & Gratzer, W.B., *Biochemistry 5* (1966) 988-993.

8. Boedtker, H., *J. Mol. Biol. 2* (1960) 171-188.

9. Loening, U.E., *J. Mol. Biol. 38* (1968) 355-365.

10. Cox, R.A., Huvos, P. & Godwin, E.A., *Isr. J. Chem. 11* (1973) 407-422.

Table 1.   Estimated molecular weight ranges of pooled fractions of RNA fragments separated in a zonal rotor.

Values of $S_{20,w}$ were interpolated from Fig. 2 for the appropriate combinations of fractions, and the relation

$$S_{25,w} = 0.021\ M_w^{0.49} \qquad (1)$$

was used to calculate M . Equation (1) was obtained by Hall & Doty [6] who related $S_{20,w}$ (solvent: 0.01 M phosphate buffer, pH 7) to M_w (obtained by light scattering) of degraded calf-liver rRNA. Values for R_f in 5% polyacrylamide gels were related directly to $S_{20,w}$ (solvent: 0.2 M NaCl) of fragments of yeast rRNA by McPhie *et al.* [7] who reported the equation:

$$S_{20,w}^{\circ} = 13.75 - 11.25\ R_f \qquad (2).$$

These authors used Eq. (3) given by Boedtker [8]:

$$S_{20,w}^{\circ} = 0.020\ M_w^{0.5} \qquad (3)$$

to relate R_f and M_w by the equation

$$M_w = (6.88 - 5.63\ R_f)^2 \times 10^4 \qquad (4).$$

Moles of fragment per mole of intact RNA were calculated by dividing the observed weight fraction by the ratio mean M_w of fragment/M_w of intact RNA. The M_w of intact RNA was taken as 1.72×10^6 [9].

The further expression is used: % G-C base-pairs = $\dfrac{1}{1 + df_{AU}/df_{GC}}$

where df_{AU} and df_{GC} respectively are the changes in the mole fraction of A-U and G-C base-pairs on heating over the temperature range 25°-95° (solvent: 0.01 M potassium phosphate buffer pH 7.0 *(see Fig. 7)*. The ratio df_{AU}/df_{GC} was calculated from the appropriate value of E_{260}/E_{280} by means of the equation [2]:

$$df_{AU}/df_{GC} = 0.75\ (E_{260}/E_{280}) - 0.36$$

	$S_{20,w}$	$M_w \times 10^{-5}$	R_f	$M_w \times 10^{-5}$	Yield $\overline{M_w \times 10^{-5}}$ (mean)	Wt. fraction	Moles of fragment /mole intact RNA	% G+C *(chem. analysis)*	%GC of double-helical regions
	(Fig. 2)	*(Eq. 1)*	*(Fig. 3)*	*(Eq. 4)*					
Intact RNA	26.0	17.6	-	-	-	-	-	67.5	67.6
Pool 1	3.0-4.8	0.2-0.8	0.70-0.95	0.2-0.7	0.45	0.10	∿ 2.0	60.8	-
Pool 2	5.4-6.8	0.1-1.6	0.55-0.74	0.75-1.5	1.13	0.13	∿ 2.0	64	71
Pool 3	7.4-8.7	2.0-2.8	0.13-0.41	2.1-3.8	2.95	0.14	∿ 0.8	77.8	76
Pool 4	9.3-11.6	3.6-4.4	0.06-0.14	3.7-4.2	3.95	0.19	∿ 0.8	76.8	77

14 SPECIFIC PROTEINS ASSOCIATED WITH ANIMAL MESSENGER RNA

Robert Williamson
The Beatson Institute for Cancer Research
Hill Street
Glasgow, G3 6UD, U.K.

Specific proteins are found in association with messenger RNA in the nucleus, in free cytoplasmic particles, in polysomes and bound to the poly-A sequence at the 3'-terminus. In each case the proteins appear to be different. There is considerable controversy as to whether the proteins might be non-specifically bound contaminants, and their homogeneity is also in doubt. This arises in part from the multiplicity of tissues and species used in these experiments, and in part from the non-standard technology employed in protein analysis.

PROTEINS ASSOCIATED WITH NUCLEAR MESSENGER RNA

A rapidly labelled ribonucleoprotein sedimenting as a broad band in the polysome region can be isolated from pulse-labelled nuclei from liver and cultured cells [1, 2]. Georgiev and his colleagues have designated the protein component of the complex 'informofer' protein [3]. Messenger RNP was extracted with a low ionic strength wash from rat liver nuclei which had been purified by sedimentation through sucrose [4]. RNase inhibitor could be used to obtain high molecular weight material. After either digestion with low levels of ribonuclease or sonication, ribonucleoprotein particles sedimenting at approximately 30 S are obtained, containing 20 - 25% RNA [5]. Sonication releases more 'mRNP' than washing [5]. Gel electrophoresis has been performed at pH 4.5 in urea [4], at pH 6.8 in urea [6], and at pH 7.0 in sodium dodecyl sulphate (SDS) [5]. Free-flow electrophoresis has also been used [7]. Each of these procedures gives different results. Rat liver nuclei have been generally used as the starting material, but similar informofer protein is obtained from rat liver, rabbit liver and ascites cells.

In the absence of mercaptoethanol a more complex electrophoretic pattern is obtained. In SDS gels a very complex pattern of protein bands is seen, but the dominant components are of molecular weight 40,000 [5], very similar to that found for informofer protein isolated and analyzed by the more usual techniques of Georgiev and his colleagues [1]. Although much of the analysis of the messenger RNP was performed with formaldehyde-fixed material banded on a caesium chloride gradient, this technique can cause non-specific binding of proteins [8].

Both nuclease and poly-A synthetase activities have been shown to occur in the monomeric 30 S nuclear mRNP particles [9, 10]. They have not yet been shown to be related to similar nuclear activities postulated *in vivo,* which may process heterogeneous nuclear RNA to messenger RNA.

PROTEINS ASSOCIATED WITH POLYSOMAL MESSENGER RNA

Attempts have also been made to follow nuclear mRNA-associated proteins into cytoplasmic mRNP. In rat liver there is a protein similar to informofer protein which occurs not in the polysomes but in the endoplasmic reticulum [7,11]. Since liver is an extremely heterogeneous tissue, with regard both to cell types present and to messenger RNAs synthesized, it is difficult to evaluate the significance of a single protein band separated only on one-dimensional gel electrophoresis. The proteins associated with polysomal mRNA which can be dissociated from the ribosomal subunits by EDTA do not appear to be identical to informofer protein in either duck or mouse reticulocytes [12, 13].

Both puromycin and chelating agents have been used to isolate messenger RNA from polysomes [12, 14, 15], particularly from reticulocytes. Puromycin treatment of rabbit reticulocyte polysomes in conditions of high salt dissociates a particle containing two proteins of molecular weights 78,000 and 52,000 [15]. Duck reticulocyte polysomes release a particle after EDTA treatment containing two proteins of molecular weights 73,000 and 49,000 [12], but a similar mRNP from rabbit reticulocytes contains proteins of molecular weights 130,000 and 68,000 [14]. These proteins appear to be quite different from those associated with mRNA in the puromycin-derived particle; in this case the isolations and electrophoresis were performed under similar conditions. There have been many other descriptions of proteins associated with cytoplasmic mRNP from various tissues and cells which synthesize many messenger RNAs for many proteins, and therefore the heterogeneity which is found is not surprising.

Although it is known that proteins are associated with the 'free' mRNA found, for instance, in the post-ribosomal supernatant of reticulocytes [16], no reports have been published of their characterization. Proteins are also found firmly bound to the poly-A region of messenger RNAs. This has been demonstrated for messenger RNAs from mouse sarcoma and L cell polysomes [17,18]; in the latter case a protein of molecular weight 78,000, thought to be identical with that found in puromycin-dissociated mRNP from rabbit reticulocytes, was characterized.

Even from this very brief summary it is apparent that non-standard methodology, particularly with regard to protein analysis, has made comparison of data gathered by different groups difficult.

CONCLUSIONS

In view of the remarkable advances in technology for analyzing messenger RNA, it is surprising that so little accurate information exists concerning proteins associated with mRNA in the cell. Density gradient materials are now available [19] which will band ribonucleoprotein, but do not require prior fixation with formaldehyde. Although preliminary experiments with Urographin, angiocardin and Metrizamide have not yet succeeded in separating mRNP from ribosomal RNP, separations of a similar kind have been achieved for deoxyribonucleoproteins (cf. Articles 10-12).

In many differentiated cells a spectrum of messenger RNAs exist for related proteins. Examples are mRNAs for actin and myosin in muscle cells, for α and β globins in reticulocytes, and for the various viral proteins in virus-infected cells. It is not known whether the proteins associated with these mRNAs are identical, similar or unrelated. The only role postulated to date for polysomal mRNA-associated proteins, the mediation of binding to deoxycholate-treated ribosomes, could be reinvestigated with profit in systems utilizing purified ribosomal subunits.

Experimental procedures now exist which can radioactively label mRNA to several million counts/min, in particular chemical labelling with iodine [20]. This should permit a study of the protein components associated with added heterologous mRNA in cell-free systems. For example, it is not known whether added rabbit mRNA binds to messenger-specific duck proteins before interacting with ribosomes in the duck reticulocyte lysate system. Similar experiments could be performed using heterogeneous nuclear RNA, which also directs protein synthesis in heterologous systems [21]. The oocyte system would be a particularly interesting one to analyze in this regard; it is very active, the mRNA is stable and the distance between the two species involved in the heterologous cross is great.

A great deal of careful experimentation has been performed using HeLa cells, liver or other cell types which synthesize large numbers of different proteins (and hence a wide spectrum of messenger RNAs). Until recently it was very difficult to obtain and assay specific mRNAs, but with the development of techniques utilizing the poly-A region of messenger RNA to permit isolation [22], most experiments could now be performed with specific messenger RNAs. This should permit a direct study of the proportion of the RNA in nuclear and non-polysome-bound ribonucleoprotein which is truly mRNA. It is known that the great majority of the rapidly labelled nuclear RNA is not processed to the polysomes, but no analysis has been carried out to determine whether it is protein-bound and, if so, whether the proteins are of the informomer type.

One hopes that potentially applicable technical advances will soon clarify these and other central aspects of mRNA processing and function.

Acknowledgement

The Beatson Institute for Cancer Research is supported by grants from the Medical Research Council and the Cancer Research Campaign.

References

1. Lukanidin, E.M., Zalmanzon, E.S., Komaromi, L., Samarina, O.P. & Georgiev, G.P., *Nature New Biol. 238* (1972) 193-197.

2. Henshaw, E.C., *J. Mol. Biol. 36* (1968) 401-411.

3. Samarina, O.P., Asrijan, I.S. & Georgiev, G.P., *Dokl. Akad.Nauk, U.S.S.R. 163* (1965) 1510.

4. Samarina, O.P., Lukanidin, E.M., Molnar, J. & Georgiev, G.P., *J.Mol. Biol. 33* (1968) 251-263.

5. Albrecht, C. & Van Zyl, I.M., *Exp. Cell Res. 76* (1973) 8-14.

6. Tomcsanyi, T. and Tigyi, A., *Acta Biochim. et Biophys. Acad. Sci.Hung.6* (1971) 149-151.

7. Schweiger, A. & Hannig, K., *Biochim. Biophys. Acta 204* (1970) 317-324.

8. Woodcock, D.M. & Mansbridge, J.N., *Biochim. Biophys. Acta 240* (1971) 218-232.

9. Niessing, J. & Sekeris, C.E., *Nature New Biol. 243* (1973) 9-12.

10. Niessing, J. & Sekeris, C.E., *Biochim. Biophys. Acta 209* (1970) 484-492.

11. Olsnes, S., *Eur. J. Biochem. 15* (1970) 464-471.

12. Morel, C., Kayibanda, B. & Scherrer, K., *FEBS Lett. 18* (1971) 84-88.

13. Lukanidin, E.M., Georgiev, G.P. & Williamson, R., *FEBS Lett. 19* (1971) 152-156.

14. Lebleu, B., Marbaix, G., Huez, G., Temmerman, J., Burny, A. & Chantrenne, H., *Eur. J. Biochem 19* (1971) 264-269.

15. Blobel, G., *Biochem. Biophys. Res. Comm. 47* (1972) 88-95.

16. Spohr, G., Kayibanda, B. & Scherrer, K., *Eur. J. Biochem. 31* (1972) 194-208.

17. Kwan, S.W. & Brawerman, G., *Proc. Natl. Acad. Sci. U.S. 69* (1972) 3247-3250.

18. Blobel, G., *Proc. Natl. Acad. Sci. U.S. 70* (1973) 924-928.

19. Rickwood, D., Hell, A. & Birnie, G.D., *FEBS Lett. 33* (1973) 221-224.

20. Getz, M.J., Altenburg, L.C. & Saunders, G.F., *Biochim. Biophys. Acta 287* (1972) 485-494.

21. Williamson, R., Drewienkiewicz, C.E. & Paul, J., *Nature New Biol. 241* (1973) 66-68.

22. Aviv, H. & Leder, P., *Proc. Natl. Acad.Sci.(Wash.) 69* (1972) 1408-1412.

15 STUDY ON BOVINE THYROID NUCLEI ISOLATED WITH AN A-XII ZONAL ROTOR

H.J. Hilderson
Laboratory of Human Biochemistry
RUCA - University of Antwerp
Belgium

Nuclei of adult bovine thyroid have been isolated in moderate yield and purity (15% based on DNA recovery). The nuclei are mostly spherical and have an average diameter of 8.4μm with normal and of 9.0μm with hypertrophic glands. Centrifugation of these nuclei in a zonal A-XII rotor gave a single peak of density varying with the composition of the medium, probably due to their permeability to sucrose solutions. By extrapolation of the $\rho_p - \rho_m$ versus ρ_m curves, values of 3.86% and 3.88% for the sucrose-impermeable space and isopycnic densities of 1.41 and 1.43 are obtained respectively for normal and hypertrophic thyroid nuclei. These similar results differ from values found for adult bovine liver nuclei, viz. 7.76% and 1.30 and agree better with those for rat liver nuclei (10.2% and 1.35). Bovine liver nuclei do not resolve into separate bands as do young rat liver nuclei.

Respirometric analysis of crude nuclear fractions (centrifugation in the zonal rotor omitted) and of purified nuclear fractions was performed manometrically and polarographically. In the former fractions low activities (less than 1% in comparison to a mitochondrial fraction subjected to similar experiments) are found. In the latter fraction no respiratory capacity can be detected. It thus seems likely that a further purification occurs during the centrifugation in the zonal rotor. The lipid composition of the purified nuclear fractions reveals that there are no other contaminating membranous fractions.

ISOLATION OF THE NUCLEI

Bovine nuclei (both liver and thyroid) can be prepared by a modification of the method of Widnell and Tata [1]. Thyroid glands are collected immediately after slaughtering, stored at 0° during transportation to the laboratory and treated as soon as possible. Subsequent manipulations are performed at 4°. Connective and fat tissue are removed first. The thyroids are subsequently cut into small cubes ($\sim$ 2.5 mm side) and washed repeatedly with cold isotonic sucrose (0.25M) to remove contaminating blood. Portions of 20 g are made up to 80 ml with 0.32M sucrose (3mM with respect to $MgCl_2$, adjusted to pH 7.4 with $NaHCO_3$) and homogenized in a Ten Broeck hand homogenizer. Per 25 ml filtered homogenate (filtration through two layers of cheese cloth) 15 ml of chilled 0.32M sucrose solution ($3mM$ $MgCl_2$, pH 7.4) and 12 ml of chilled glass-distilled water are added. Portions (25 ml) of dilute homogenate, now 0.25 M with respect to sucrose, are layered on top of 25 ml 0.32M sucrose solution and spun at 700 g for 10 min. The resulting pellet consists mainly of red blood cells and nuclei with some mitochondrial and lysosomal contamination. Most of the post-nuclear fraction and other cellular debris are held up in the top layer. Finally the nuclei are purified by centrifugation through 2.2 M sucrose (1 mM $MgCl_2$, pH 7.4). Isolation of bovine liver nuclei proceeds in a similar way.

Three kinds of nuclear suspensions are prepared. *Nuclear fraction A* is the pellet obtained after centrifugation through 2.2 M sucrose. To avoid damage during re-suspension this pellet is left standing overnight at 4° in 0.32 M sucrose ($3mM$ $MgCl_2$, pH 7.4). Next morning the nuclei are re-suspended. After centrifugation (1,000 g, 20 min) the nuclear pellet *(nuclear fraction B)* is re-suspended in a solution appropriate for zonal centrifugation studies (Table 1, Sample zone). *Nuclear fraction C* is the nuclei recovered after centrifugation in the zonal rotor.

Table 1: Sedimentation behaviour of bovine thyroid and liver nuclei in sucrose solutions: experimental conditions. *Each gradient was used several times with different centrifugation times.*

Sucrose gradient	Sample zone	Overlay	Underlay	Running speed	Pumping speed	Centrifugation time
No.	% w/w sucrose 1 mM $MgCl_2$, pH 7.4		% w/w sucrose	rev/min	ml/min	
1	5-17	3	H_2O	55	600	40	15-45 min
2	20-30	15	10	55	1200	40	15-45 min
3	20-35	15	10	55	1200	40	15-45 min
4	20-50	15	10	55	1200	40	15-45 min
5	35-50	15	10	55	1200	40	15-45 min
6	40-50	15	10	55	2000	20	2-3 h
7	50-60	40	30	65	3000	20	2-3 h

CENTRIFUGATION OF THE NUCLEI IN THE ZONAL ROTOR

Nuclear fraction *B* is subjected to a series of centrifugations in an A-XII zonal rotor through different gradients according to Johnston and co-workers [2]. The experimental data are summarized in Table 1. The runs are carried out at 4°. All gradients are linear with respect to volume and 1mM with respect to $MgCl_2$ (pH 7.4). They are introduced (pumping speed 40 ml/min for sucrose solutions up to 40%) into the MSE A-XII rotor while running at 1200 rev/min (600 rev/min for the 5-17% gradient). For higher sucrose concentrations the pumping speed is reduced to 20 ml/min. The sample of nuclei examined in the zonal rotor is equivalent to $\sim$ 20 g tissue. The sample (12 ml; for sucrose concentration see Table 1) is introduced through the core line followed by an overlay (± 100 ml) until the leading edge of the sample is at least 6 cm away from the axis of rotation. A stroboscopic lamp is used to observe the positions of the peak against a scale marked on the underside of the rotor. This also permits an exact adjustment of the rotor speed. After centrifugation the contents of the rotor are displaced by an appropriate underlay (Table 1). The effluent is passed through a flow cell of 2.5 cm pathlength, monitored at 660 nm and collected manually into 50 ml centrifuge tubes cooled in ice-water.

During the centrifugations the nuclei form one single peak. This holds for all kinds of gradients used and for both liver (adult bovine) and thyroid (normal and hypertrophic). In the typical run shown for thyroid nuclei (Fig. 1), the first peak at the left marks the position of the interface between the sample zone and the top of the gradient. The second peak is the nuclear peak, from which 88.5 ±2.3% of nuclear fraction B can be recovered. This profile coincides with the curves

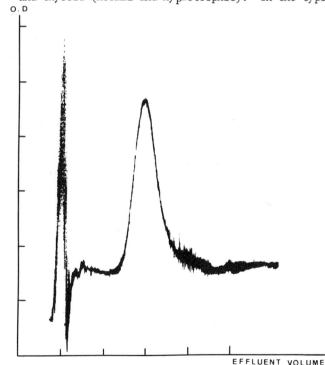

O.D

EFFLUENT VOLUME

Fig. 1. Light-scattering profile obtained on emptying the rotor. The direction of sedimentation is from left to right.

Ordinate: Absorbance at 660 nm.

Abscissa: Effluent vol. Both coordinates are arbitrary units.

for nuclear counts (Celloscope 401) and for protein content. There was no difference in the histograms (nuclear sizes) between the leading and the tailing edge of the peak. The protein content per nucleus was also identical in all parts of the peak. It therefore seems likely that these nuclei behave as a single population. Of the gradients used the most convenient for purification purposes are 20-30% and 20-50% sucrose. On the assumption that the nuclear radii are unaffected by sucrose concentration (this is the case for isotonic and hypertonic sucrose solutions; however they swell in hypotonic solutions), it is possible to compute from the law of Stokes the density of the nuclei at any given point in the gradient from the rate of sedimentation at that point.*

In Fig. 2, $(\rho_p - \rho_m)$ values are plotted versus ρ_m values. The curve is linear only for isotonic and hypertonic media. The intersection of the abscissa represents the isopycnic density ($\rho_p - \rho_m = 0$, hence $\rho_p = \rho_m = \rho_e$). For normal thyroid nuclei $\rho_e = 1.41$. Hypertrophic thyroid nuclei have a similar value of 1.43 whereas for liver nuclei an isopycnic density of 1.30 is found. There is a different slope for thyroid and liver nuclei. In both isotonic and hypertonic media it is reasonable (although a crude approximation) to treat the nuclei as comprising two compartments, one sucrose-permeable and the other sucrose-impermeable. Suppose that x is the nuclear proportion impermeable to sucrose (ρ_e) and $(1 - x)$ is the nuclear proportion permeable to sucrose (ρ_m), then the total density is $\rho_p = x \cdot \rho_e + (1 - x)\rho_m$. The value of x is obtained by writing two simultaneous equations containing the data derived from Fig. 2 and solving for x. Hence x is 3.86% for normal thyroid nuclei, 3.88% for hypertrophic thyroid nuclei and 7.76% for liver nuclei. These values are lower than for, e.g., rat liver nuclei (10.2%) and mouse liver nuclei (23.7%) [2].

PURITY OF NUCLEAR FRACTION *c*

There was no detectable loss of RNA nor DNA on centrifuging nuclear fraction *B* in the zonal rotor. However in fraction *c* the enzymatic activities of phenylphosphatase (Pase) and of monoamine oxidase (MAO) are reduced when compared to nuclear fraction *A* (ratio *c/A*: 1/1.6). Moreover it is appropriate to stress that less than 1% of the original homogenate-Pase and MAO gets into nuclear fraction *A*. No or almost no cytochrome oxidase could be detected in nuclear fraction *c*. It therefore seems likely that centrifugation in a zonal

---

* For a spherical particle, $\rho_p = \rho_m + \dfrac{dX/dt \; 9\eta}{2r^2\omega^2 X}$ where ρ_p = density of the particle, ρ_m = density of the medium, X = radial distance from axis of rotation, dX/dt = rate of sedimentation, η = viscosity, r = nuclear radius, ω = angular velocity. The sedimentation rate can be detected by following the peak through the transparent rotor. Corresponding values for η, X, ρ_m can be found for the different positions in the gradient. For determination of r, microscopic measurements are used.

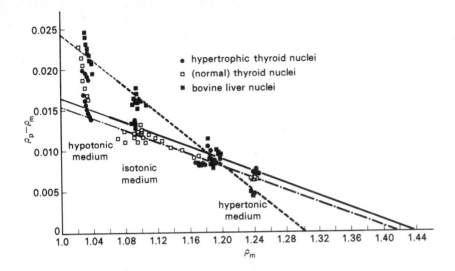

Fig. 2. Variation of nuclear density with density of the medium.
ρ_p = density of nuclei; ρ_m = density of the sucrose solution.

rotor gives rise to a further purification. However some contamination, undetectable by light microscopy, may still be present in nuclear fraction *c*.
Finally no oxygen uptake could be detected by either manometric or polaro-
graphic methods [3].

NUCLEAR SHAPE, SIZE AND CONTENT

The nuclei are well shaped (mostly spherical), undamaged and have RNA-poly-
merase activity. As can be seen from Fig. 3, the average nuclear diameter is
8.4 μm for normal thyroid nuclei, 9.0 μm for hypertrophic thyroid nuclei (shift-
ing nuclear volume from 310 to 381 μm^3) and 8.4 μm for bovine liver nuclei. Normal and hypertrophic thyroid nuclei have different protein contents with a
ratio of 1/1·7. There is no clear explanation for this high protein content
in hypertrophic thyroid nuclei. The values found for RNA and DNA content are
similar in both kinds of thyroid nuclei. Table 2 shows the percentage
composition of hypertrophic thyroid nuclei: it is similar that of rat liver
nuclei. However, it must be stressed that the high %-value found for protein
in hypertrophic thyroid nuclei causes a drop in %-values for the other nuclear
components.

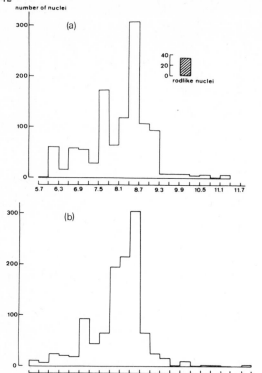

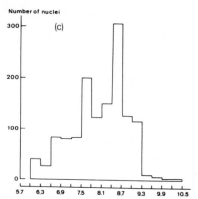

Fig. 3. Histograms of nuclear
diameters of (a) bovine liver
nuclei, (b) bovine hypertrophic
thyroid nuclei, (c) bovine
thyroid (normal) nuclei.
Ordinate: no. of nuclei.
Abscissa: nuclear diameter, μm.
*Nuclear sizes are determined
by means of an eyepiece micro-
meter fitted on a phase-contrast
microscope.*

LIPID COMPOSITION OF HYPERTROPHIC THYROID NUCLEI

The lipids of hypertrophic thyroid nuclei (nuclear fraction c, $2-3 \times 10^9$
nuclei) are extracted, purified and fractionated according to Rouser [7]:
(a) extraction of lipids by means of solvent mixtures with increasing polarity, (b) purification of the lipid extract by either a Folch distribution
or by Sephadex G 25 chromatography separating a total lipid fraction from
gangliosides and residues, (c) a fractionation of the total lipid fraction by
chromatography on silicic acid columns yielding neutral lipids, glycolipids
and total phosphatides. The total phosphatide fraction is also subjected to
two-dimensional thin-layer chromatography, the phosphatide content of each
spot being determined by phosphorus assay. In hypertrophic thyroid nuclei
∿88% of total lipids are phosphatides. This is supported by the value of
35 μg lipid P per mg lipid. This value is of interest for demonstrating
contamination by plasma membranes. Indeed, in membranes ∿90% or more of the
lipids is phosphatide except that plasma membranes contain large amounts of
neutral lipids [8]. For bovine liver plasma membranes a lower ratio of 19 μg
lipid-P/mg lipid was reported by Rouser [9].

It was assumed by Gurr [10] that 1.3×10^{-11}g lipids (phosphatides and cholesterol) on a Langmuir trough acquire a surface of about $800\mu m^2$. If this is related to the surface of the nucleus isolated in citric acid (outer membrane removed), the available lipid would provide 2.1 monolayers. For nuclei (rat liver) isolated in 2.2 M sucrose (double envelope still present) this number would be 5.6 [10]. For hypertrophic thyroid nuclei a similar value is found. However this represents more than the amount of lipid required to build two unit membranes (4 monolayers). It is however possible that the distance between two phosphatide molecules was over-estimated. Also the molecules could be stacked more closely in a membrane structure than in a Langmuir trough. On the other hand the nuclear surface is not necessarily smooth but could display a 'pleated sheet' aspect. Finally it is not excluded that some fragments of the endoplasmic reticulum remained attached to the nuclear envelope during the whole purification procedure. Comparing with other tissues the phosphatide composition seems normal (Table 3) although in our preparations the phosphatidylcholine content is somewhat higher than in the nuclei of other tissues.

Table 2. Percentage composition of bovine hypertrophic thyroid nuclei compared with rat liver nuclei.

Tissue	Protein	DNA	RNA	Phosphatide	Ref.
Hypertrophic thyroid c nuclei	75.5	17.12	4.34	2.34	[4]
Rat liver nuclei	63	26	6.5	4.4	[5]
	72.4	20.0	3.4	4.1	[6]

Finally the presence of cholesterol esters, lipid-bound sialic acid and cerebrosides could not be demonstrated. This is in agreement with experiments that showed increasing amounts of lipid-bound sialic acid in smooth endoplasmic reticulum; Golgi apparatus, secretion vesicles and plasma membranes [12]. It is therefore unlikely that the nuclear envelope, being free of those components. is implicated in their synthesis.

References

1. Widnell, C.C. & Tata, J.R., *Biochem. J. 92* (1964) 313-317.

2. Johnston, I.R., Mathias, A.P., Pennington, F. & Ridge, D., *Biochem J. 190* (1968) 127-135.

3. Hilderson, H.J. & Dierick, W., *Arch. Int. Physiol. Biochim.* (1973) in press.

4. Hilderson, H.J., Lagrou, A. & Dierick, W., *as for ref.* [3].

5. Monneron, A., Blobel, G. & Palade, G.E., *J. Cell Biol. 55* (1972) 104-125.

6. Kay, R.K., Fraser, D. & Johnston, J.R., *Eur. J. Biochem. 30* (1972) 145-154.

Table 3. Percentage composition of the phosphatides of bovine hypertrophic thyroid nuclei in comparison to nuclei from other tissues. PC = phosphatidyl choline; PE = phosphatidyl ethanolamine; PS = phosphatidyl serine; PI = phosphatidyl inositol; Sph = sphingomyelin. 'X' consists mainly of 'lyso' derivatives.

Nuclei	PC	PE	PS + PI	Sph	'X'	Ref.
human brain nuclei	52.9	25.7	14.6	5.2	0.5*	[7]
rat liver nuclei	14.5	7.7	30	5.1		[11]
rat liver nuclei	52.1	25.1	9.7	6.3		[10]
rat kidney nuclei	30.0	18.9	14.9	10.9		[11]
rat uterus nuclei	42.2	7.9	20.0	3.9		[11]
rat ovary nuclei	36.0	9.0	16.2	-		[11]
bovine liver nuclei	53.9	20.6	12.0	2.6	10.8	[9]
bovine thymus nuclei	52.1	19.6	12.2	2.6	13.4	[9]
bovine pancreas nuclei	52.4	23.5	10.6	2.6	11.2	[9]
bovine heart nuclei	45.4	24.6	12.6	5.5	11.7	[9]
bovine brain nuclei	52.9	25.7	14.6	5.2	6.7	[9]
bovine hypertrophic thyroid nuclei	57.06	22.26	12.24	2.16	6.30	

*phosphatidic acid

7. Rouser, G., Kritchevsky, G., Siakotos, A.N. & Yamamoto, A., in *Neuropathology: Methods and Diagnosis* (Tedeschi, C.G., ed.), Little, Brown, Boston (1970) pp. 691-753.

8. Thinès-Sempoux, D., *this volume,* Article 17.

9. Rouser, G., in *Biolological Membranes* (Chapman, D., ed.), Academic Press, London (1968) pp. 5-69.

10. Gurr, M.I., Finean, J.B. & Hawthorne, J.N., *Biochim. Biophys. Acta 70* (1963) 406-416.

11. Biezenski, J.J., Spaet, T.H. & Gorton, A.L., *Biochim. Biophys. Acta 70* (1963) 75-82.

12. Morré, D.J., Yunghans, W.N., Vigil, E.L. & Kennan, T.W., *this volume,* Article 21.

16 SELECTION CRITERIA FOR MARKER MOLECULES IN IDENTIFICATION OF ORGANELLES

Pierre J. Jacques
Laboratoire de Chimie Physiologique
Université Catholique de Louvain
B-3000 Leuven, Belgium

Whatever their nature, enzymic or not, and their constitutive or exogenous character, markers for cell organelles or their substructures must obey several criteria, either as isolated molecules or in relation to the organelles. The marker-molecule must be identifiable in its own right, without ambiguity, amongst other molecular species with analogous properties, detectable in strictly quantitative terms, stable throughout the experimental process, and unaffected by the reagents used for fractionation as well as by natural or experimental cellular alterations.

With respect to cell organelles and their substructures, the marker molecules must specifically be bound to or carried by a single population of structures, be present at fairly comparable levels in each of the individuals composing that population, and keep these two properties throughout the experiment. Besides the foregoing general conditions, others may be imposed by the type or the scope of experimentation. The structure-linked latency of marker enzymes, the availability of techniques for morphological identification of the marker, the ubiquity of the marker molecule in homologous organelles from different organs, species and even biological kingdoms, are amongst the other points that may have to be taken into account.

A. MORPHOLOGICAL *VERSUS* BIOCHEMICAL ANALYSIS OF FRACTIONS

In the pioneer studies, which furnished the basis for fractionation of cell homogenates by physical means, morphological examination was used as the principal guide for the delimitation of fractions and the assessment of their purity. Such a methodology suffers numerous limitations, even nowadays, although the morphological approach has been reinforced by the high resolution power of electron microscopy and the refinements of quantitative morphology. Its major disadvantages are the considerable time needed for quantitative analysis of fractions, the number of profiles which cannot be identified with certainty, and the almost unsurmountable difficulty of quantitating or even merely detecting rare organelles.

For these reasons, workers in the field soon switched to analysis of fractions by means of marker molecules, most often enzymes, corresponding to specific structures. That choice is all the more justifiable in that enzymes can deservedly be considered as the elementary biological units articulating cellular structures and functions [1]. The method proved fruitful to the point that it even permitted the simultaneous discovery of both the existence and the function of relatively rare organelles, e.g. lysosomes [2] and peroxisomes [3].

However, the above considerations do not imply that morphological methods have no important role to play in cell fractionation, nor that enzymes are the only marker molecules of interest.

B. TYPES OF MARKER MOLECULES

There is probably little if any theoretical interest in trying to build up a classification of marker molecules; yet this might be useful merely to illustrate the diversity of solutions for the sometimes arduous problem of searching for or choosing a marker for a given organelle. The ingenuity of the reader will no doubt amplify the following elementary considerations.

In fact, any physical or chemical property of a particle population can be used as a marker, provided that that property obeys the criteria described below. For the sake of brevity, we shall only consider marker molecules; the latter in turn, we reiterate, are by no means limited to enzymic constituents. Many types of non-enzymic constituent can indeed be used: antigenic and non-antigenic macromolecules, micromolecules as trivial as phospholipids [4], cholesterol [5], even minerals [6, 7]. Thus, one can have recourse to the whole arsenal of chemical or physical analytical methods and immunology, in addition to quantitative enzymology.

Whichever their molecular species, marker molecules can also differ by the nature of their association with the particles. Most often, molecular agents belonging to the structural or catalytic equipment of the organelles are used; but this is no necessity. Stored substances, whether endogenous or exogenous in origin, can be good candidates for markers in some cases, as well as compounds in transit through vesicles, especially when their transport is in steady-state and involves cytosis [8].

Amongst stored substances are the contents of secretory granules, yolk platelets, terminal phagosomes, terminal telo- and post-lysosomes ... At the extreme, there is no reason to exclude exogenous compounds applied to cell or to homogenate; e.g. acridine orange seems to bind and display specific fluorescence on certain lysosomes [9, 10].

Rightly or wrongly, one however would expect constitutive molecules to respect marker criteria better than do stored compounds or compounds in transit.

However delicate its application may be, the use of compounds that are in transit may be of unique help, at least in the case of the exoplasmic apparatus. Consider the example of phagosomes, for which no suitable constituent-marker is as yet available. Just as in the morphological approach, one takes an exogenous substance which becomes sequestered within phagosomes through endocytosis [8, 11]. The problem is that in most cells, the phagosomes discharge their contents into lysosomes. The exogenous marker is thus present after a while within both the population of phagosomes and those lysosomes to which it gets access, that is the bulk of lysosomes if the proportion of primary lysosomes is small enough (Fig. 1).

In this simple though frequent case, with the reservations that the marker-criteria must be fulfilled and that lysosomes can be completely separated from phagosomes, the distribution of the latter should be equivalent to that of the exogenous marker after subtraction of the part which coincides with the distribution of genuine lysosomal markers (e.g. constituent acid hydrolases). It seems that such a labelling of phagosomes 'by difference' has been obtained in rat liver, using plasma ribonuclease and lysosomal enzymes as markers [11].

Quite a number of conditions must be fulfilled in such an experiment, but the approach is no less than Utopian when the uptake of the marker is continuous. The problem is considerably more difficult to solve when the uptake is temporary [12, 13]; at least the marker should be supplied long enough as to allow its migration front in the exoplasmic compartment to reach the vesicles located at the distal end of the functional cytotic path.

C. VALIDITY CRITERIA FOR MARKER MOLECULES

There are two hierarchical steps in the validity criteria for marker molecules. Some criteria comprising the first step must be respected in all cases. They are sufficient when the scope of experimentation is limited to identification or purification of particulate populations, to the localization of a compound in a mixture of populations, or when it aims at measuring physical parameters such as sedimentation coefficient, size, and equilibration density, and at establishing the frequency distribution of individual organelles following one of these parameters, in the various analytical or preparative fractionation systems for a given tissue.

The second step consists of additional criteria which must be superimposed on the preceding ones, when the study has a multidisciplinary character, when it aims at analyzing some of the organelle properties which are not revealed in fractionation systems, or when it covers the comparison of homologous particles in various organs, species and even biological kingdoms.

1. General criteria

Several selection criteria are inherent properties of the marker molecule itself; others refer to the relationship between the marker and the population

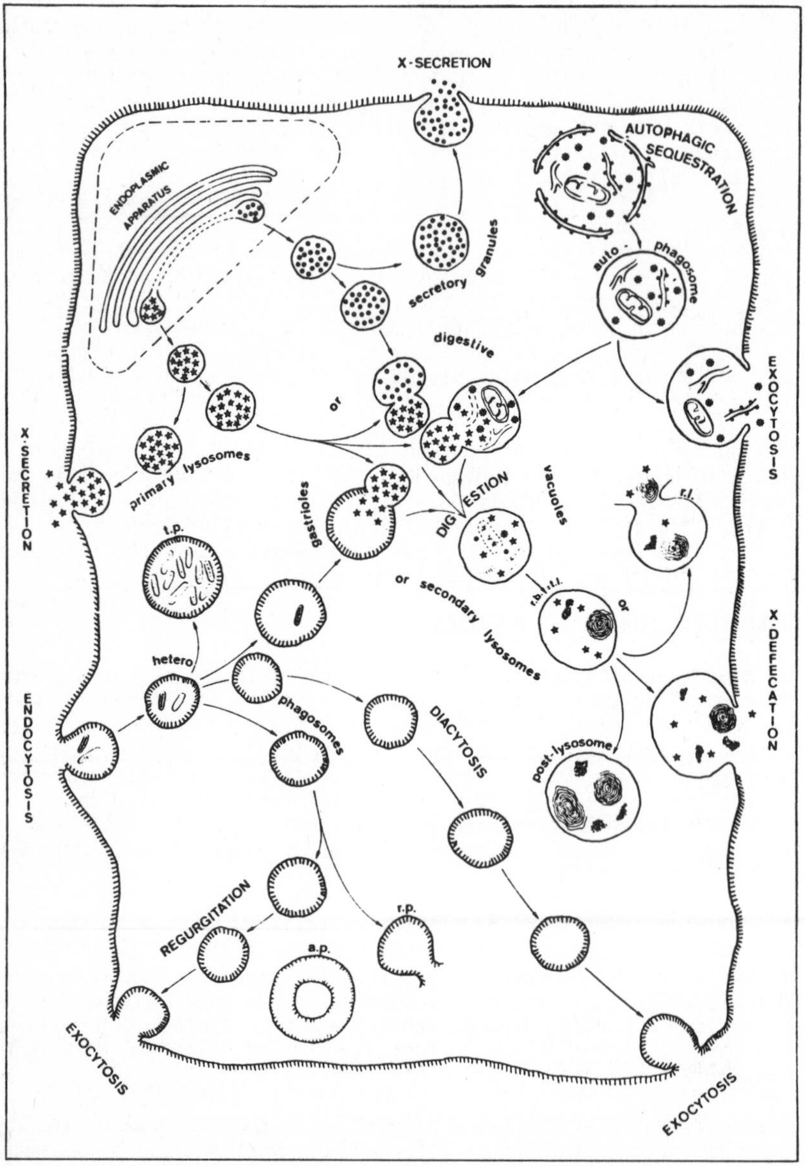

Fig. 1.
*Legend
opposite.*

of organelles which it helps trace.

a. The marker molecule itself

The first set of criteria results from the need to establish the concentration shown by the marker before homogenization, within the corresponding structures.

This requires that the assay method be both specific and quantitative, and that the activity of the marker be stable throughout, from the intact cell until the end of the assay.

In this section, the examples illustrating theoretical considerations will be enzymes; the latter are indeed most frequently chosen as markers, and their relative fragility offers a maximum number of possible complications.

To develop a *specific assay method* allowing quantitation of a single molecular species is not always an easy task. This holds especially true in the case of enzymic markers, which often belong to families of enzymes characterized by a catalytic activity towards analogous if not identical substrates.

The assay of rodent liver lysosomal acid-phosphatase activity particularly well exemplifies the difficulties which can be encountered, and the means through which they can be overcome. At pH 5, which is the optimum for that non-specific phosphatase, the interference from liver alkaline phosphatases is almost negligible. The *choice of pH* for the incubation mixture is sufficient in this respect, in order to ensure assay specificity, because there is little if any overlap between the respective pH-curves of the enzymes concerned; but this is not sufficient to distinguish lysosomal phosphatase from the other two liver 'acid phosphatases', viz. the relatively specific glucose-6-phosphatase of endoplasmic reticulum, and the non-specific acid phosphatase of the cytosol.

Table 1, which describes the respective activities of the three phosphatases on usual substrates, clearly indicates that β-glycerophosphate must be utilized for the assay of lysosomal acid phosphatase, except in preparations of pure lysosomes in which it may be judicious to replace it by phenyl phosphate; the latter is indeed more readily hydrolyzed into products which,

---

Table 1. Activity of liver 'acid phosphatases' at pH 5, towards various substrates.

	Lysosomal acid phosphatase(s)	Glucose-6-phosphate	Hyaloplasmic acid phosphatase
Phenyl phosphate	+++	+++	+++
Glucose-6-phosphate	+	+++	?
α-Glycerophosphate	++	++	++
β-Glycerophosphate	++	0	0

in turn, are more easily measured. Mis-recognition of these implications led several workers to publish totally erroneous conclusions about lysosomes. The *choice of substrate* thus determined the specificity of the assay, as can hold equally in other cases, despite the low specificity of the enzyme species which should be distinguished from one another.

However, lysosomal acid phosphatase might be distinguished from its hyaloplasmic homologue by adequate *choice of inhibitors*, and from glucose-6-phosphatase through *selective inactivation* of the latter by heating at pH 5.

A *strictly quantitative assay* is a crucial need for the purpose of using marker molecules, that is, to establish the frequency distribution curve, in an analytical set of fractions, for the organelles hereby marked, or at least to assess their purity in fractions made for preparative purposes. It is also necessary in order to appreciate the quality of the fractionation procedure itself, through establishing the recovery balance sheet.

Maximal care should thus be exercised during development of the assay procedure. For marker enzymes, the reaction rate must be proportional to enzyme concentration, and the proportionality coefficient must be the same in all the fractions.

A favourable recovery is by no means a guarantee that the assay meets these requirements. Indeed, an important assay defect in a fraction containing only a small percentage of the marker molecule will not visibly alter the recovery; besides, assay defects in richer fractions sometimes compensate one other, resulting in a seemingly acceptable recovery figure.

The *marker molecule must be stable* to the extent of retaining, throughout homogenization and subsequent operations until the end of the assay, the concentration or activity which it had in the cell. Many a factor can impair that stability.

Firstly, the physiological intracellular degradation of a potential marker, which sometimes proceeds after homogenization.

Secondly, the various reagents and physical treatments inflicted on marker molecules during the overall *post-mortem* processing. Amongst possibly interfering reagents are the media for tissue homogenization, homogenate fractionation, fraction dilution and marker determination. Amongst physical treatments, attention should be paid to the considerable lowering of temperature from *in vivo* values to those of fractionation process and, in some cases, of the freeze-drying of the tissue before homogenization and the subsequent storage of homogenate and fractions.

Thirdly, marker molecules can, as a result of homogenization or assay conditions, be exposed to effectors with which they had no contact *in vivo* because of inter- or intracellular compartmentation. Thus, cytosol marker-vacuoles are suddenly exposed to the numerous effectors present at the start in the blood plasma impregnating the tissue and in the cytosol of other cell types from the same organ, or that are leaking from damaged particles. Also, enzymes showing structural or dynamic latency in the intact organelle are assayed under conditions which disrupt the particles; as a result, they are exposed, at time of assay, to all the potential effectors present in the fraction.

A fourth type of 'destabilizing' mechanism will be discussed below, in Section *c.*

b. *Relationship between marker molecule and corresponding organelle*

Two major selection criteria have already been proposed for the relationship between a valid marker and its corresponding organelle [14].

The first one, that of *single location,* stipulates that the marker be (essentially) localized in a single population of particles.

The notion of single population as presented by de Duve [15] seems perfectly clear for structures such as peroxisomes whose population, as far as one can presently judge, does not represent a sub-set inside a wider functional population. In the case of other organelles however, this notion may suffer from two ambiguities. Take, as an example, the valid markers that constitutive lysosomal enzymes represent. On the one hand, they are present in both primary and secondary lysosomes, that is in two populations which are different in several respects; for instance, the former are secretory granules waiting to discharge their contents outside the cell or into other vacuoles belonging to the exoplasmic apparatus, whereas the latter are digestive vacuoles into which substances are poured by auto- and heterophagosomes (Fig. 1).

On the other hand, these markers are not present in the phagosomes, post-lysosomes and secretory granules as distinct from primary lysosomes, although these particles form, together with lysosomes proper, the coherent functional set named 'exoplasmic apparatus'.

One would thus be tempted to say that, in this example, it is the markers themselves (acid hydrolases) which, in fact, define the population of corresponding organelles (lysosomes proper); actually, lysosomes were discovered in this very manner, and this explains why de Duve presented his two criteria as *real postulates* to be honoured when ascertaining the localization of enzymes.

The second criterion, that of *biochemical homogeneity*, stipulates that the marker concentration be (essentially) the same in all the organelles forming the population. Indeed, once the criterion of single location has been duly met, the finality in using marker molecules reduces itself to establishing an unequivocal relationship between their concentration in a fraction, and the quantity of corresponding organelles in the same fraction, in such a manner that the latter can be deduced from the former and reciprocally.

In general, the 'quantity' of organelles refers either to their frequency or to their global mass or volume relative to that of the fraction. It must be realized that the homogeneity criterion as it was formulated by de Duve makes the marker concentration in the fraction serve as an index of the relative mass or volume occupied, in the fraction, by the corresponding organelle, *inasmuch as* within each individual organelle, whatever its size, the concentration of the marker is supposedly constant. But one could conceive of another situation, in which the absolute amount and not the concentration of the marker would be the same inside each organelle; then, the marker concentration in the fraction would reflect the frequency instead of the relative mass or volume of the overall granule population. Lysosomes as labelled by acid hydrolases probably correspond to the first case, whereas nuclei with labelled DNA molecules would rather fall in the second category. Obviously, the foregoing distinction vanishes when all the organelles have the same volume. In our opinion, the criterion of biochemical homogeneity should be reformulated as follows: that either the absolute amount or the concentration be the same for the marker in all the individual organelles of the population.

These two criteria refer to the situation which prevails in the intact cell. They would lack practical interest if they were not reinforced by a third criterion, viz. *stable location*. Indeed, several famous artifacts of cell fractionation have arisen from re-distribution of compounds during homogenization and subsequent operations. Examples include the transfer of cytochrome *c* from mitochondria to microsomal entities in hypertonic sucrose, the transfer to cytosol of enzymes originating from nuclei suspended in aqueous media devoid of citrate, or from organelles with osmotic behaviour exposed to hypotonic solutions or to media whose solutes can readily penetrate into the granules.

Finally, the above three criteria can be extended to the substructural entities of the organelles themselves, e.g. the membrane, the nucleoid and the sap of peroxisomes, or the outer and inner membranes, the interstitial fluid and the matrix of mitochondria.

Thus, within the set of markers for a given organelle, one may have to make a further selection; e.g. both catalase and urate oxidase are good markers for peroxisomes in rodent liver, but of these two enzymes, catalase must be chosen as marker for peroxisomal sap, and urate oxidase for the nucleoid.

c. Residual problems

The general criteria described above refer to the ideal situation prevailing in an homogenate of a single cellular population from a normal animal. Their direct application to homogenates of heterogenous tissues, their extrapolation from one tissue to another one in the same animal or to the homologous tissue in another species, as well as the sensitivity of some markers to pharmacological treatment of the animal, raise corresponding problems, lack of recognition of which has already led to serious mistakes.

The *assay conditions* suitable for a marker enzyme in a given tissue commonly differ appreciably from those for the homologous enzyme in other tissues or other species.

The *pharmacological treatment* of the animal can make it imperative to choose another marker for a given organelle, although it was perfectly adequate in the normal animal. Thus, prolonged treatment of rats with dextran profoundly alters both the hepatic level of β-glycerophosphatase and its distribution after differential pelleting, whereas it leaves unchanged the behaviour of other lysosomal markers [16]; in such situations, β-glycerophosphatase should preferably not be used as marker.

However, a change in tissue level of an enzyme does not necessarily invalidate it as a marker. Indeed, suramin treatment of rats causes the liver levels of cathepsin D and β-glycerophosphatase to respectively increase and decrease by a factor of from 2 to 3; but the distribution of these two enzymes after differential pelleting is normal and similar to that of lysosomal markers whose tissue levels are insensitive to treatment with the trypanocidal compound [17].

We should also mention the peculiar difficulty associated with the choice of marker molecules in the case of *composite functional apparatuses*. Perhaps the most familiar example would be that entity named vacuome or vacuolar apparatus [18, 19], whose various structures (phagosomes, lysosomes, post-lysosomes, secretory granules, Golgi apparatus, endoplasmic reticulum and nuclear envelope) have this in common, that their contents are separated from the cytoplasmic or the nuclear matrices, by a membrane which is thought to derive functionally or at least phylogenetically from the plasma membrane. Within this vast apparatus, secondary lysosomes (digestive vacuoles), for instance, are at a cross-roads on the traffic map for membrane flow (Fig. 1); their membrane is thus likely to contain constituents originating from the plasma membrane, the Golgi apparatus and the endoplasmic reticulum.*

It is, then, urgent to discover marker constituents common to all parts of the vacuome cytomembranes; but it must be kept in mind that the functional differentiation of the various 'organs' of the vacuome is in fact reflected in the chemical and sometimes physical diversification of the cytomembranes bounding these organs. Up to the present, mainly 'organ-specific' membrane markers have been identified in the vacuome [20].

2. ADDITIONAL CRITERIA

The general criteria of section C-1. are always pre-requisites for the validity of marker constituents, but they are not always sufficient; additional criteria must sometimes be introduced. A few examples may make help delineate this notion.

A fascinating step in the development of the peroxisome concept was the recognition of the phylogenetic relationship linking animal 'microbodies' to plant 'glyoxysomes'. Within each of these two biological kingdoms, and even between different tissues within a single species, the enzymic equipment of the organelle shows considerable variability; only catalase, probably because it is involved in the detoxication of the hydrogen peroxide liberated by all the other peroxisomal enzymes, was invariably found in all microbodies and glyoxysomes [20]. That *biological ubiquity* of catalase renders it a first-choice molecule within the battery of peroxisome markers; however, no sooner was this discovered, than it was challenged by an exception [22].

All markers for a given organelle are not necessarily suitable when, for instance, the sensitivity of the organelle towards *in vitro* treatments is being studied, either for its own sake or for comparative purposes. Thus, it seems possible to distinguish sharply in an ordinary homogenate, on the basis of their differential sensitivity to digitonin, structures as varied as lysosomes, peroxisomes, mitochondrial matrix or external membrane, endoplasmic reticulum vesicles and ill-defined structures deriving from the plasma membrane (a_2 sub-fraction of microsomes) [23].

In the collection of markers for these various structures, there is a second determinant: only enzymes showing *latency* as long as the organelle is intact, or those endowed with *non-adsorbability* onto the constituents of the homogenate after granule lysis, are suitable as markers in such experiments.

---

* *Articles 21 (Morré et al.), 22 (Glaumann et al.) and 23 (Touster) also bear on this aspect. The present article is essentially a summary and expansion of points made by Dr. Jacques during the Symposium and the subsequent Lysosomes Course. - Editor.*

A need for additional criteria also arises in multidisciplinary studies. It frequently happens, not frequently enough dare we say, that biochemical and morphological approaches are applied jointly to the same problem. In such cases, it may be wise to select, from among the (biochemically) valid markers, those which, in addition, lend themselves to *morphological identification*.

References

1. Gaebler, O.H. (ed.), *Enzymes: Units of Biological Structure and Function* Academic Press, New York (1956).

2. de Duve, C., Pressman, B.C., Gianetto, R., Wattiaux, R. & Appelmans, F., *Biochem. J. 60* (1955) 604-617.

3. de Duve, C., Beaufay, H., Jacques, P.J., Rahman-Li, Y., Sellinger, O.Z., Wattiaux, R. & De Coninck, S., *Biochim. Biophys. Acta 40* (1960) 186-187.

4. Hallinan, T., *this volume*, in Art. 39.

5. Thines-Sempoux, D., *this volume*, Art. 17.

6. Beaufay, H., Bendall, D.S., Baudhuin, P. & de Duve, C., *Biochem. J. 73* (1959) 623-628.

7. Pujarniscle, S., *Etude Biochimique des Lutoides. - O.R.S.T.O.M. Mémoire 48*, Paris (1971).

8. Jacques, P.J. in *Pathologic Aspects of Cellular Membranes* (Trump, B.J. & Arstila, A. eds.) Academic Press, New York, *in press*.

9. Dingle, J.T. & Barrett, A.J., *Proc. Roy. Soc., Ser. B, 173* (1969) 85-93.

10. Allison, A.C. & Young, M.R. in *Lysosomes in Biology and Pathology*, Vol. 2 (Dingle, J.T. & Fell, H.B., eds.) North-Holland, Amsterdam (1969) pp. 600-628.

11. Bartholeyns, J., *Les Ribonucléases Hépatiques*. Doctoral thesis, Université Catholique de Louvain (1973).

12. Jacques, P.J., *Epuration Plasmatique de Protéines Étrangères, Leur Capture et Leur Destinée dans l'Appareil Vacuolaire du Foie*. Librairie Universitaire, Leuven (1968).

13. Jacques, P.J., *this volume*, in Art. 39.

14. Creemer, J. & Jacques, P.J., *Exp. Cell Res. 67* (1971) 188-203.

15. de Duve, C., in *Enzyme Cytology* (Roodyn, D.B., ed.) Academic Press, New York (1967) pp. 1-26.

16. Thines-Sempoux, D., *personal communication*.

17. Smeesters, C., *Influence de l'Accumulation de Suramine dans le Foie sur l'Activité des Hydrolases Lysosomiales*. Thesis, Université Catholique de Louvain (1968).

18. de Duve, C. in *Lysosomes in Biology and Pathology*, Vol. 1 (Dingle, J.T.& Fell, H.B., eds.) North-Holland, Amsterdam (1969) pp. 3-40.

19. Jacques, P.J., *Ann. Anesth. Franc., Special 1* (1972) 18-28.

20. Hauser, P., Remacle, J., Trouet, A. & Beaufay, H., *Arch. Intern. Physiol. Biochim. 81* (1973) 186-187.

21. de Duve, C. and other authors, in *The Nature and Function of Peroxisomes: Ann. N.Y. Acad. Sci. 168* (1969) 211-381.

22. Lloyd, D. & Cartledge, T.G., *this volume,* Art. 32.

23. Godin, G., *personal communication.*

17 APPROACHES TO THE COMPARATIVE STUDY OF RAT LIVER CYTOMEMBRANES

Denyse Thinès-Sempoux[*]
Laboratoire de Chimie Physiologique[†]
Université de Louvain, Belgium

Consideration is given to the problem of preparing pure and truly representative samples of the different types of membrane in the cell. Conjoint experimental approaches to the problem are presented. Although inherent difficulties remain, it can be concluded that membranous cholesterol is almost confined to the cell vacuome, viz. plasma membrane, endocytotic vesicles, secondary lysosomes and Golgi apparatus.

Knowledge of the composition and of the dynamic structure of the biomembranes is essential for the understanding of the main physiological features of the cell, as well as of their ontogenesis. - The membranes partition the cell and support the enzymes; they are selective barriers for metabolites and they transport them from one compartment to another; they must either carry or understand signals for recognition and fusion phenomena in which they participate.

With these considerations in mind, we started to investigate the intracellular localization of the membranous cholesterol of rat liver, a chemical constituent which is generally accepted as playing an important structural role in all types of cytomembrane. We now present experimental results which have led to the conclusion that contrary to the idea of the ubiquity of the membranous cholesterol, this lipid belongs almost exclusively to the vacuome of the cell, namely the pericellular membrane (plasma membrane), the endocytotic vesicles, the secondary lysosomes, the Golgi apparatus. But firstly we would like to emphasize the difficulties inherent in this kind of study and how to remedy them.

The comparative study of cytomembranes is possible only with very pure samples which should be representative of the entire population. This rather difficult goal is almost never reached because, even disregarding a general tissue heterogeneity, all the cytomembranes have similar physico-chemical properties and are heterogeneous and dynamic structures, with functional or even morphological interrelationships of at least temporary nature.

[*] *Present address: Laboratoire de Biologie Cellulaire (Director: Prof. A. Claude), Université de Louvain.*
[†] *Director: Prof. C. de Duve.*

Moreover, their characterization must rest on safe biochemical and morphological criteria which should be entirely specific.

We now show how, in our study of the intracellular localization of rat-liver cholesterol, we were able to by-pass these unavoidable difficulties in order to get definitive conclusions. Thus, we combined with the preparative approach the very important analytical one; we modified the nature of the gradients in differential and isopycnic centrifugation experiments; we loaded lysosomes with substances which markedly modify their density; we treated our preparations by chemical agents or mechanically, we studied concomitantly the distributions of as many markers as possible; we examined the preparations in the electron microscope.

TOTAL LIVER

Rat liver contains, per gram, about 220 mg of proteins, 35 mg of phospholipids and 2.5 to 3.0 mg of cholesterol. After differential centrifugation of the liver homogenate, analysis[*] shows about 20% of the protein in each of the three particulate sub-fractions, i.e., N, ML and P, leaving 40% soluble. But the findings for lipid are quite different: the microsomal pellet P is 2.5 times as rich in phospholipids as the fractions N and ML. Moreover the weight ratios of phospholipids to cholesterol are 9, 58 and 11.5 for N, ML and P respectively. This indicates that the nuclear and the microsomal fractions arise from membranes which are much richer in cholesterol than those of the mitochondria.

PLASMA MEMBRANES

The nuclear fraction consists essentially of cells, debris and nuclei; it also contains large sheets of plasma membranes. Pericellular membranes have been purified by many authors from many different tissues, with various techniques but usually from a very low-speed sediment. There is a general acceptance of the conclusion first published by Emmelot *et al.* in 1962 [1], that plasma membranes are truly characterized by a very high level of cholesterol, higher than that of all the other types of cytomembranes.

We ourselves purified rat liver plasma membranes by Neville's method [2] as modified by other authors [3, 4]. The method entailed isolating the plasma membranes from a very low-speed sediment after subjecting the liver cells to osmotic shock by hypotonic $NaHCO_3$ at pH 7.4. It furnished preparations containing up to 2 mg of protein per g of wet liver. As compared to the total liver homogenate, they were enriched 10 times in cholesterol and 20-25 times in two marker enzymes, 5'-nucleotidase and alkaline phosphodiesterase I. Samples thus obtained of purified plasma membranes contain 0.1 mg of cholesterol per mg protein on the average, as discussed by Thinès-Sempoux [5], whether the membranes are obtained from homogenates in hypotonic or in isotonic media.[*]

[*] *Editor's Note: Intended tabular documentation has had to be omitted.*

LYSOSOMAL MEMBRANES

We have purified rat liver lysosomes after having loaded them either with the detergent Triton WR-1339, or with Dextran 500 [6, 7]. We have observed that the insoluble fraction of osmotically disrupted granules has a high cholesterol content, quite similar to that found for purified plasma membranes. The cholesterol to protein weight ratios were in fact 0.13 and 0.10 for these tritosomes and for dextranosomes respectively [5]. Values only half of these have been reported by Henning [8, 9] for the insoluble fraction of tritosomes and by Colbeau *et al.* [10] for the whole Triton WR-1339-loaded granules.

These results do not exclude the possibility that this cholesterol could be absent from the membrane of the granules and rather be a constituent of the insoluble indigestible residue of the lysosomal matrix. Indeed, with regard to its protein content, the insoluble fraction of purified lysosomes must contain material foreign to their membrane.

MICROSOMAL FRACTION

The P fraction was pelleted according to de Duve *et al.* [11] from a post-mitochondrial supernatant by centrifuging it at $\omega = 3 \times 10^{10}$ rad^{2}sec^{-1}. Accordingly the fraction contains up to 75% of the glucose-6-phosphatase, the nucleoside-diphosphatase, the esterase, the glucuronyl transferase, and the NADPH-cytochrome *c* reductase; 60% of the NADH-cytochrome *c* reductase, which has a dual location, namely the endoplasmic reticulum and the mitochondria; and $\sim$ 50% of two plasma-membrane enzymatic markers, 5'-nucleotidase and alkaline phosphodiesterase I. On the other hand, only 4% of the cytochrome *c* oxidase was to be found in it, as well as 6% of the catalase and 10-17% of lysosomal acid hydrolases. This composition shows that our microsomal fractions are very representative but not too much contaminated by mitochondria, by lysosomes or by peroxisomes.

As had been done for the mitochondrial fraction [12], the heterogeneity of the microsomal fraction has been analyzed in density-gradient experiments, both isopycnic and differential [13, 14]. Isopycnic centrifugation experiments have been performed with the aid of the annular rotor designed by Beaufay [15] (see also [16]). Fig. 1 shows typical distributions observed for some constituents after equilibration of the preparation in a sucrose gradient. All the components exhibit broad distribution patterns, overlapping each other. One could only classify them into various groups on the basis of the shapes of their distributions and of their median densities [17].

One of these categories, designated *a*, features the microsomal 5'-nucleotidase, the alkaline phosphodiesterase I and the alkaline phosphatase, and also cholesterol, the distribution of which differs markedly from that of

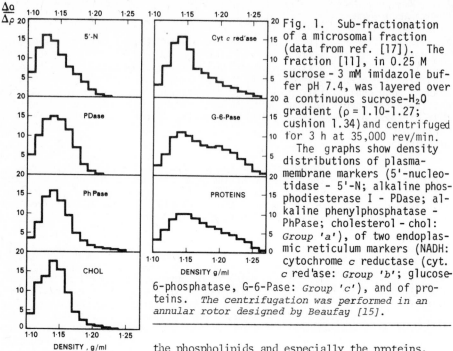

Fig. 1. Sub-fractionation of a microsomal fraction (data from ref. [17]). The fraction [11], in 0.25 M sucrose - 3 mM imidazole buffer pH 7.4, was layered over a continuous sucrose-H_2O gradient ($\rho = 1.10$-1.27; cushion 1.34) and centrifuged for 3 h at 35,000 rev/min.

The graphs show density distributions of plasma-membrane markers (5'-nucleotidase - 5'-N; alkaline phosphodiesterase I - PDase; alkaline phenylphosphatase - PhPase; cholesterol - chol: *Group 'a'*), of two endoplasmic reticulum markers (NADH: cytochrome c reductase (cyt. c red'ase: *Group 'b'*; glucose-6-phosphatase, G-6-Pase: *Group 'c'*), and of proteins. *The centrifugation was performed in an annular rotor designed by Beaufay [15].*

the phospholipids and especially the proteins.

Similar individuality of behaviour of cholesterol and of three plasma-membrane enzymatic markers was confirmed in equilibration density experiments on the P fraction when the gradients contained D_2O instead of H_2O or when they were made from Ficoll in 0.25 or 0.50 M sucrose.

Moreover, differential centrifugation experiments in stabilizing gradients (Fig. 2) show that the sedimentation profiles of the constituents of group a - 5'-nucleotidase, alkaline phosphodiesterase I, alkaline phosphatase and cholesterol - also differ markedly from those of the oxido-reductases (group b) represented on the graph by NADH cytochrome c reductase and cytochrome b_5, and from those of the hydrolases (group c) as exemplified here by glucose-6-phosphatase and nucleoside diphosphatase.

At this stage of the work, it was concluded that, besides fragments of the endoplasmic reticulum, the microsomal fraction contains a minor constituent which was characterized by particles of low density and high sedimentation coefficient but was responsible for its high level of cholesterol and of plasma-membrane markers. This constituent was attributed to pieces of the plasma membranes and/or to structures at least derived from or related to it.

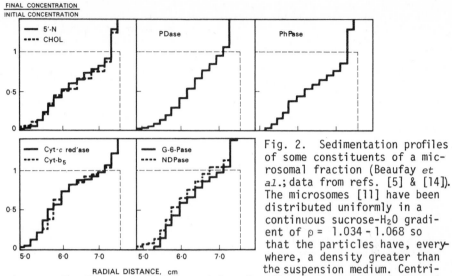

Fig. 2. Sedimentation profiles of some constituents of a microsomal fraction (Beaufay *et al.*; data from refs. [5] & [14]). The microsomes [11] have been distributed uniformly in a continuous sucrose-H_2O gradient of $\rho = 1.034 - 1.068$ so that the particles have, everywhere, a density greater than the suspension medium. Centrifugation was for 30 min at 20,000 rev/min. See Fig. 1 for rotor and abbreviations (*also* cytochrome b_5, cyt. b_5; nucleoside diphosphatase, NDPase). Dashes give the effective initial distribution of microsomes in the gradient.

MICROSOMAL FRACTION TREATED WITH DIGITONIN

This conclusion about the individual nature of group *a* was reinforced by the results of similar isopycnic and differential experiments on the same microsomal fraction in sucrose-H_2O gradients, but now after pre-treatment with small amounts of digitonin [18]. This compound is known to build equimolecular complexes with free cholesterol. Its quantity was chosen so that it did not solubilize microsomal proteins and did not alter the latency of the nucleoside diphosphatase, and it appeared that this amount of digitonin was somehow stoichiometrically equivalent to the cholesterol content of the microsomal fraction.

The density distributions of constituents of digitonin-treated microsomes were compared with those for controls. The digitonin treatment did not at all alter profiles of the constituents of either group *b* (NADH cytochrome *c* reductase)†or group *c* (glucose-6-phosphatase), but increased drastically and specifically the median equilibration density of the cholesterol and of the three enzymes of group *a* (5'-nucleotidase as example).

Sedimentation experiments on similarly treated microsomes showed that the digitonin did not specifically alter microsomal membranes: the sedimen-

† *Here and subsequently, intended Figs. missed the press date for the book.-Ed.*

tation profiles were similar for both control and detergent-treated fractions, whilst showing that constituents of group *a* sediment even faster after digitonin, probably due to their increased density. Electron microscopy (by Dr. M. Wibo) has demonstrated that, with the exception of a few profiles, the appearance of the microsomal vesicles remains unchanged under the influence of digitonin; only a few smooth and large profiles appear discontinuous.

The specificity of the action of the digitonin on all the microsomal constituents of group *a* excludes the possibility that vesicles derived from the endoplasmic reticulum could contain significant amounts of cholesterol.

PLASMA MEMBRANES TREATED WITH DIGITONIN

After the digitonin effect had been observed on microsomes, it became imperative to perform similar experiments on purified plasma membranes. Such preparations [4], equilibrated in sucrose density gradients, formed a fairly narrow band of material with density distribution patterns almost identical and even superimposable for the different constituents. Treatment of the membranes with digitonin in an amount comparable, on a cholesterol basis, to that used with microsomes increased the median equilibration density of all the studied marker constituents to about the same extent as it did in microsome experiments.

Morphological examination of both control and treated preparations showed that the digitonin altered the appearance of the plasma membranes. These appeared more or less discontinuous, as did the few profiles of the digitonin-treated microsomes responsible for the presence of the constituents of group *a* [19]. Some contamination with rough vesicles arising from the endoplasmic reticulum was evident.

MITOCHONDRIAL MEMBRANES

The mitochondrial fraction contains only around 10% of the 5'-nucleotidase and of the alkaline phosphodiesterase I, and as little as 5% of the total liver cholesterol. Outer mitochondrial membranes [20] have now been examined. Only 0.5% of the liver cholesterol is associated with monoamine oxidase, an outer mitochondrial membrane marker. The density distributions have been determined for monoamine oxidase and NADH-cytochrome *c* reductase (markers for the outer membrane), cytochrome oxidase (marker for the inner membrane) and also for 5'-nucleotidase, alkaline phosphodiesterase I and the cholesterol. Both groups of mitochondrial enzymatic markers had sharp distributions, very different from each other; the density equilibration positions for the plasma membrane markers lay in between.

Moreover, treatment of the outer mitochondrial membrane preparations with minute amounts of digitonin, stoichiometrically equivalent to the quantity of the cholesterol present, resulted, as in experiments with microsomes or with purified plasma membranes, in a specific and significant increase of the median equilibration density of the plasma membrane markers and of the cholesterol. This indicates that both the inner and the outer mitochondrial membranes are devoid of cholesterol. Even if one takes account of small shoulders in the cholesterol density distribution at the peak of the outer membrane components in the case of the control, and at the peak of the cytochrome oxidase in the case of the treated preparation, calculation establishes that less than 1% of the total liver cholesterol can be associated with the mitochondrial membranes, which represent approximately 40% of all the liver cytomembranes [21]. These shoulders could be artefacts of concentration or unspecific adsorption. If this cholesterol indeed belongs to the mitochondrial membranes, its very low concentration suggests that it may be concentrated in very specific areas of the mitochondrial membranes, for example, connections of the mitochondria with other organelles such as the Golgi apparatus or the endoplasmic reticulum.

GOLGI APPARATUS

Experiments performed on microsomal fractions have shown that the median equilibration density of the galactosyl transferase, which is a specific marker for the Golgi system, was also increased under the influence of the digitonin treatment, although much less than that of the components of group *a*. Golgi apparatus purified by Wibo [22], with a technique modified from that of Morré *et al.* [24], contain a significant amount of cholesterol, estimated as 5 to 6% of the liver content. This analysis was, however, made on an unfractionated Golgi preparation, so that it cannot at present be taken as proven that all the membranes of the Golgi apparatus contain some cholesterol. It seems to us more probable that this lipid is concentrated in specific areas of the Golgi apparatus (small vesicles, plates, tubules, concentrating vesicles), and it must not be overlooked that the lipoprotein granules of the concentrating vesicles might be responsible for some quantity.

GENERAL DISCUSSION AND CONCLUSIONS

All the foregoing experiments strongly support the hypothesis that, in rat liver, the bulk of the cholesterol belongs primarily to the plasma membrane and to structures which are directly related to it insofar as they derive from it, like the secondary heterolysosomes, or fuse with it and become part of it, like the vesicles which transport the lipoprotein granules. Our conclusion is supported by the work of Werb and Cohn [24], who showed that, in mouse peritoneal macrophages, more than 95% of the total cholesterol is bound to membranes and belongs almost entirely to the plasma membrane and to the lysosomal membrane.

However, if there is indeed similarity in respect of cholesterol amongst the many preparations studied, there remains one difficulty: based on protein and 5'-nucleotidase content, the plasma membrane-related structures of the microsomal fraction must be much richer in cholesterol than the fragments of the pericellular membranes which are purified from a low speed sediment obtained from homogenates in iso- or hypotonic media [5].

Glaumann and Dallner [25] conclude that some cholesterol of the P fraction belongs to the endoplasmic reticulum, as also concluded for 5'-nucleotidase by Widnell [26]. Concerning cholesterol it is argued that the content in the microsomal fraction increases after smooth endoplasmic proliferation in the livers of phenobarbital-injected rats, and that microsomal sub-fractions exhibit inhomogeneous labelling after injection of radioactive precursors to animals [27]*.

Our own conclusion is different and multifarious.
1. Purified plasma membranes are still contaminated by foreign proteins adsorbed onto them (30% of these proteins are washed out by saline solutions).
2. Microsomal fractions contain fragments of the Golgi apparatus which is responsible for 10% of the microsomal cholesterol.
3. Microsomal fractions must also contain endocytotic vesicles which, since they originate from the plasma membrane, must contain cholesterol but might exhibit no 5'-nucleotidase activity, this enzyme being inactivated once it is no longer needed.
4. Within the rat-liver pericellular membrane there are specific areas that are morphologically and functionally different (for example, membranes lining the bile canaliculi play a different role from those lining the sinuses or joining to another hepatocyte). It seems rather implausible that there should be complete chemical homogeneity in spite of these differences.
5. Heterogeneity of plasma-membrane fragments purified according to Neville [2] or other authors has been proved by Evans [28] and by Thinès-Sempoux [5].
6. An increased level of cholesterol in P fractions from phenobarbital-treated rats might reflect merely a difference in the subcellular distribution of the lipid, the total liver content being not reported, or an enhancement of the cholesterol content of the pericellular membrane, the volume and the surface of the centrolobular hepatocytes being increased after phenobarbital [29].
7. The cholesterol of the microsomal subfractions could be in various metabolic states, in that the cholesterol of the Golgi apparatus could be metabolically different from that of the plasma membrane–related structures, or the liver cells, like the macrophages studied by Werb and Cohn [24], could have different compartments for cholesterol exchange, one being more rapid than the other.

In conclusion, although personally convinced that cholesterol is not present as a structural component in all the types of rat-liver cytomembrane, we accept that many problems remain to be elucidated, all related to the dyna-

---

* *Editor's Note: A re-capitulation by H. Glaumann of these and other arguments is to be found in the final article of this book.*

mism of cytomembranes and of cells in general.

Acknowledgements

This paper rests on experimental results obtained in Professor de Duve's laboratory in a collaboration involving A. Amar-Costesec, H. Beaufay, J. Berthet, E. Feytmans, M. Robbi, D. Thinès-Sempoux and M. Wibo.

References

1. Emmelot, P & Bos, C.J., *Biochim. Biophys. Acta 59* (1962) 495-497.

2. Neville, D.M., *J. Biophysic. Biochem. Cytol. 8* (1960) 413-422.

3. Emmelot, P., Bos, C.J., Benedetti, E.L. & Rümke, Ph., *Biochim. Biophys. Acta 90* (1964) 126-145.

4. Song, C.S., Rubin, W., Rifkind, A.B. & Kappas, A., *J. Cell Biol. 41* (1969) 124-132.

5. Thinès-Sempoux, D. in *Lysosomes in Biology and Pathology*, Vol. 3 (J. T. Dingle, ed.) North-Holland, Amsterdam (1972) pp. 278-299.

6. Beaufay, H., in *Lysosomes in Biology and Pathology,* Vol. 2 (J. T. Dingle & H. B. Fell, eds.), North-Holland, Amsterdam (1969) pp. 516-546.

7. Beaufay, H. in *Lysosomes. A Laboratory Handbook.* (J. T. Dingle, ed.) North-Holland, Amsterdam (1973) pp. 1-45.

8. Henning, R., Kaulen, H.D. & Stoffel, W. in Hoppe-Seyler's *Z. Physiol. Chem. 351* (1970) 1191-1199.

9. Henning, R. *this volume,* Article 20.

10. Colbeau, A., Nachbaur, J. & Vignais, P.M., *Biochim. Biophys. Acta 249* (1971) 462-492.

11. de Duve, C., Pressman, B.C., Gianetto, R., Wattiaux, R. & Appelmans, F., *Biochem. J. 60* (1955) 604-617.

12. Beaufay, H., Jacques, P., Baudhuin, O.Z., Sellinger, O.Z., Berthet, J. & de Duve, C., *Biochem. J. 92* (1964) 184-205.

13. Amar-Costesec, A., Thinès-Sempoux, D., Wibo, M., Beaufay, H. & Berthet, J. (*also* Amar-Costesec, A., Beaufay, H., Wibo, M., Thinès-Sempoux, D., Feytmans, E., Robbi, M., & Berthet, J.), *J. Cell Biol.* (1974) *in press.*

14. Beaufay, H., Amar-Costesec, A., Thinès-Sempoux, D., Wibo, M., Robbi, M. & Berthet, J., *J. Cell Biol.* (1974) *in press.*

15. Beaufay, H., *La Centrifugation en Gradient de Densité. Application à l'Etude des Organites Subcellulaires,* Imp. Ceuterick, Louvain (1966).

16. Leighton, F., Poole, B., Beaufay, H., Baudhuin, P., Coffey, J.W., Fowler, S. & de Duve, C., *J. Cell Biol. 37* (1968) 482-513.

17. Amar-Costesec, A., Beaufay, H., Feytmans, E., Thinès-Sempoux, D. & Ber-
    thet, J., *Microsomes and Drug Oxidations* (J.R. Gillette, A.H. Conney,
    G.J. Cosmides, R.W. Estabrook, J.R. Fouts & G.J. Mannering, eds.),
    Academic Press, New York (1969) pp. 41-58.

18. Thinès-Sempoux, D., Amar-Costesec, A., Beaufay, H. & Berthet, J., *J.
    Cell Biol. 43* (1969) 189-192.

19. Wibo, M., Thinès-Sempoux, D. & Amar-Costesec, A., in *Microscopie Elec-
    tronique,* Vol. 3, (Rés. Comm. 7th Congr. Int. Grenoble; P. Evrard, ed.),
    Soc. Franç. Microsc. Electr. Paris (1970) pp. 21-22.

20. Parsons, D.F., Williams, G.R. & Chance, B., *Ann. N.Y. Acad. Sci. 137*
    (1966) 643-666.

21. Weibel, E.R., Stäubli, W., Gnägi, H.R. & Hess, F.A., *J. Cell Biol. 42*
    (1969) 68-91.

22. Wibo, M., *Arch. Int. Physiol. Biochim. 81* (1973) 398-399.

23. Morré, D.J., Hamilton, R.L., Mollenhauer, H.H., Mahley, R.W., Cunning-
    ham, W.P., Cheetham, R.D. & Lequire, V.S., *J. Cell Biol. 44* (1970) 484-
    491.

24. Werb, Z. & Cohn, Z.A., *J. Exptl. Med. 134* (1971) 1545-1569.

25. Glaumann, H. & Dallner, G., *J. Cell Biol. 47* (1970) 34-48.

26. Widnell, C.C., *J. Cell Biol. 52* (1972) 542-558.

27. Glaumann, H. & Dallner, G., *J. Lipid Res. 9* (1968) 720-729.

28. Evans, W.H., *Biochem. J. 166* (1970) 833-842.

29. Wanson, J.C., Mosselmans, R. & Baudhuin, P., *Arch. Int. Physiol.Biochim.
    81* (1973) 397-398.

18 COMPARATIVE STUDY OF THE LABILIZATION OF LYSOSOMES, PEROXISOMES AND MITOCHONDRIA BY INORGANIC MERCURY *in vitro*

H. Roels, J.-P. Buchet[*] and R. Lauwerys
Unité de Toxicologie Industrielle et Médicale
Université Catholique de Louvain
4, Av. Chapelle-aux-Champs
B-1200 Brussels, Belgium

The effects of increasing Hg^{2+} concentration in labilizing lysosomes (β-galactosidase), peroxisomes (catalase), and mitochondria (glutamate dehydrogenase, GDH), were compared simultaneously using the large-granule fraction of rat liver. It has been shown that the peroxisomes are more rapidly labilized than the lysosomes, but the behaviour of the mitochondria is clearly different from that of those two organelles. The response of mitochondria to increasing Hg^{2+} concentration is biphasic: a steep rise in free glutamate dehydrogenase activity was first observed at very low Hg^{2+} concentrations $(0.1-0.3 \times 10^{-4}M)$ followed by a decrease and a second rise in activity. There is evidence that mitochondrial swelling is involved in the first peak of glutamate dehydrogenase activity.

The action of mercurial compounds on membrane functions may reveal biochemical phenomena reflecting the toxic effect of that metal at the cellular or subcellular level. The literature mostly deals with alterations in permeability and transport processes of the cell membrane (e.g. erythrocyte, yeast cell) or the mitochondrial membranes, viz: the action of mercurial compounds [1, 2, 3]. Only a few results have been reported concerning the effect of mercurials on the membrane of other subcellular organelles like lysosomes and peroxisomes. Robinson *et al.* [4] and Coonrod & Paterson [5] have shown that kidney damage induced in rats by mercury is accompanied by the release of lysosomal enzymes into urine. Lauwerys and Buchet [6] suggested that the increased activity of catalase and β-galactosidase in blood plasma of laboratory technicians moderately exposed to mercury vapours might result from mercury-induced changes in the membrane of peroxisomes and lysosomes accompanied by enzyme release. In a pioneer paper on the effect of mercurial compounds on a partially purified lysosomal fraction of mouse liver, Verity and Reith [1] demonstrated *in vitro* the breakdown of structure-linked enzyme latency in the presence of low concentrations of mercury ions. They emphasized the essentiality of thiol groups for lysosomal membrane integrity, and suggested that the formation of mercaptide bonds might damage the lysosomal membrane irreversibly. Lauwerys & Buchet [7], however, demonstrated that the lysosomal labilization *in vitro* by Hg^{2+} (using partially purified lysosomal fractions from rat liver

[*] *Dr. Sc., Chargé de Recherches FNRS*

and kidney) is merely triggered by mercury and is followed by an autocatalytic breakdown of the remaining intact lysosomes.

We have now extended the *in vitro* study of subcellular organelle labilization by $HgCl_2$ to mitochondria and peroxisomes. Our preliminary results
were obtained on the large-granule fraction from rat-liver homogenate (M + L
fraction) in order to compare simultaneously the mercury-induced release of
latent enzymes from the lysosomes, mitochondria, and peroxisomes, viz. β-galactosidase (E.C. 3.2.1.23), L-glutamate dehydrogenase (GDH) (E.C. 1.4.1.2),
and catalase (E.C. 1.11.1.6) respectively.

MATERIALS AND METHODS

Preparation of M + L fraction and its pre-incubation with Hg^{2+}

Adult male Sprague-Dawley rats fasted overnight and weighing 250-350 g were
used. Homogenization of the liver and differential pelleting of the hepatic
homogenate were performed according to de Duve *et al.* [8] in 0.25 M sucrose
using a Sorvall RC-2B refrigerated centrifuge with the angle-head rotor SS-34.
The nuclear fraction, spun down at 11,000 × *g*-min, was washed once and then
the combined supernatants were spun at 630,000 × *g*-min. The resulting pellet
(M + L fraction) was washed once and re-spun at 630,000 × *g*-min. Finally the
pellet was re-dispersed in 0.25 M sucrose so that 10 ml corresponded to 1 g
of fresh liver. The protein concentration was measured by the method of
Lowry *et al.* [9].

A series of $HgCl_2$ solutions ranging in concentration from 2×10^{-5} to
10^{-3} M was prepared in 0.25 M sucrose. An appropriate volume of the 10%
M + L fraction was added to an identical volume of 0.25 M sucrose or $HgCl_2$
solution and pre-incubated for 10 min at 37°; for the β-galactosidase assay,
20 μl of the M + L fraction was taken, for GDH and catalase, 0.1 ml. The
[Hg^{2+}] values in the text and in Figs 1, 2, 4 and 5 are those set in the pre-
incubation.

Enzyme assays

The β-galactosidase assay was slightly modified from that of Lauwerys and
Buchet [7] using 4-methylumbelliferyl-β-D-galactopyranoside monohydrate (Koch-
Light Laboratories, Colnbrook, U.K.) as substrate. A fresh 1.2 mM stock solution of this substrate in a 2:1 (v/v) mixture of 0.125 M acetate buffer (pH 5)
and 1 M sucrose was prepared, and the following substrate mixtures were obtained by adding 3 vol of this substrate stock solution to 1 vol of either
water alone, or water containing 4.8 mM L-cysteine (activator), or water containing 4.8 mM L-cysteine and 0.48% Triton X-100, or water containing 0.48%
Triton X-100 only. At the end of the pre-incubation period 0.2 ml substrate
mixture was added and the incubation continued for 10 min at 37° (final concentration: substrate = 0.75 mM, L-cysteine = 1 mM). The enzymic reaction was

stopped by the addition of 3 ml 0.5 M carbonate buffer (pH 10.7); the amount of released 4-methylumbelliferone was determined fluorimetrically [10].

GDH was assayed according to a modification of the method of Beaufay *et al.* [11], using triethanolamine-HCl buffer instead of glycylglycine-HCl buffer and 10 × more EDTA for the preparation of the substrate stock solution: 0.2 M triethanolamine-HCl buffer (pH 7.7), 0.1 M disodium EDTA (pH 7.7), 0.3 M nicotinamide, 0.2 M sodium glutamate (pH 7.7), 1 M sucrose, and 4 mM KCN (freshly prepared) were mixed in a volume ratio of 2:2:2:1.5:2. To 7 vol of this substrate stock solution 1 vol of either water (solution A) or 1% Triton X-100 in water (solution B) was added. The final substrate mixtures were obtained by mixing 1 vol of water or water containing 14 mM NAD (Boehringer), with 9 vol of solution A or B. These substrate mixtures were kept at 37°. At the end of the pre-incubation period (at 9.5 min) 0.2 ml of a 20 mM EDTA solution in 0.25 M sucrose (activator) was added, followed 30 sec later by 2 ml substrate mixture, and another 30 sec later the production of $NADH_2$ was recorded spectrophotometrically at 340 nm over a period of 3 min (final concentrations: L-glutamate = 14 mM, NAD = 1.4 mM).

For catalase assay [12] the substrate mixture containing 250 ml of solution I, 30.3 mg L-cysteine (activator), and 40 µl 30% H_2O_2, was used for the assay of free activity (solution I: 100 ml of 0.2 M imidazole-HCl buffer pH 7, 250 ml of 1 M sucrose, 650 ml water, and 1 g bovine serum albumin). A similar substrate mixture was prepared for the assay of total activity except that 1 ml of water of solution I was replaced by 1 ml Triton X-100; another mixture was prepared to assay the total activity in the absence of cysteine. All the substrate mixtures were kept at 0°. At the end of the pre-incubation period 5 ml of ice-cold solution I was added and these diluted suspensions were kept in an ice bath at 0°. The enzymic incubation was started by adding 5 ml substrate mixture to 0.1 ml of the diluted suspension and was continued for 10 min at 0° (final concentration: H_2O_2 = 36 µM, L-cysteine = 1 mM). The reaction was stopped with 3 ml of a mixture (2 N) of $TiSO_4$ and H_2SO_4 and the yellow colour was allowed to develop at 20° for 30 min. The absorbance was measured at 410 nm in 1 cm cuvettes.

The addition of an activator (L-cysteine or EDTA) was needed in order to overcome the inhibitory action of mercury ions on the enzymatic activities that must be measured. The 'free' activity as measured in tubes containing no Triton X-100 is expressed as a percentage of the total activity measured in its presence (0.1%).

Determination of mercury

In a separate experiment 0.5 ml portions of the $HgCl_2$ solutions were pre-incubated for 10 min at 37° with 0.5 ml of the 10% M + L fraction. Then 2 ml 20% trichloroacetic acid was added and the tubes kept at 0° for 30 min. The precipitate was spun down (15 min at 2600 rev/min in a Sorvall GLC-1) and the supernatant decanted, the pellet washed once with 2.5 ml water and re-spun. The combined supernatants with the free mercury (5.5 ml) and the final pellet with the bound mercury (solubilized in 2 ml1 N NaOH) were analyzed for mercury [13] with a Perkin-Elmer Model 305 atomic absorption spectrophotometer. The sum of free and bound mercury equals the amount added (see Fig. 5).

Mitochondrial swelling

Mitochondrial volume changes were measured at room temperature (22°) by following for 5 min the absorbance of suspensions at 520 nm [14] with a Zeiss spectrophotometer PMQ II. The basic test system consisted of 4.7 ml 0.25 M sucrose containing 0.02 M Tris-HCl buffer (pH 7.4) to which 0.2 ml of that buffered sucrose solution without or with $HgCl_2$ was added; then 0.1 ml of an M + L fraction prepared in 0.25 M sucrose (no EDTA was used) was added in order to give an initial absorbance of about 0.5. The final mercury concentration ranged from 0.2 to 15×10^{-5} M. The decrements of absorbance were recalculated as percentage swelling [14].

RESULTS AND DISCUSSION

Fig. 1A shows the mercury-induced loss in structure-linked latency of β-galactosidase in an M + L fraction. The percentage labilization of lysosomes as a function of increasing Hg^{2+} concentration is similar to that reported for partially purified lysosomal fractions (ref. [7], Fig. 1). Apparently, four stages can be distinguished in this labilization curve: an initial phase of mercury binding (0 to 0.6×10^{-4} M Hg^{2+}), an autocatalytic phase of lysosomal rupture (0.6 to 2×10^{-4} M), an equilibrium phase with the maximal degree of labilization at 2.5×10^{-4}, and an inhibition phase (2.5 to 5×10^{-4} M) which is usually more evident at lower protein concentration (see ref. [7], Fig. 1). The degree of labilization of the lysosomes increased from 10% (without Hg^{2+}) to a maximal value of about 80% at 2.5×10^{-4} M, but the main rise (from 15 to 70%) fell between 0.6 and 1.5×10^{-4} M. The maximal degree of labilization is probably the result of an equilibrium between the lytic activity of the lysosomal enzymes (liberated by the trigger action of Hg^{2+}) and the inhibition of the same enzymes by mercury ions. In the latent phase the added mercury ions must be entirely bound to proteins and organelle membranes, since hardly any free mercury could be found; moreover, the total β-galactosidase activity remained uninhibited at this very low Hg^{2+} concentration even when no cysteine was added with the substrate. When the amount of free mercury increased progressively the total β-galactosidase activity dropped very sharply to zero, if no activator was present. However, by adding cysteine together with the substrate more than 85% of the total β-galactosidase activity remained measurable even when the Hg^{2+} concentration increased up to 5×10^{-4} M.

The mercury-induced labilization of peroxisomes present in the same M + L fraction showed a classical pattern (Fig. 1B). A lower Hg^{2+} concen-

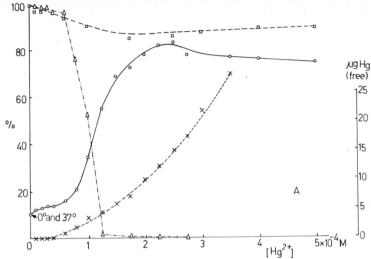

Fig. 1.   Effect of increasing Hg^{2+} concentration on the labilization of
rat liver lysosomes (A), peroxisomes (B) and mitochondria (C) *in vitro*.
The M+L fraction contained 3.29 mg protein /ml.   The amount of free mercury (×---×) present in the pre-incubation system is expressed in μg Hg.
- A *(above)* and B *(below, left)*:   free activity of β-galactosidase (A) or
  catalase (B)   as % of total activity (o—o), and total activity in the
  presence (□---□) and absence (△-·-△) of cysteine, 10^{-3} M.
- C *(below, right)*: free activity of GDH as % of total activity (o—o),
  and total activity in the presence of EDTA, 10^{-2} M (□---□).
The 'floating' axis line below (centre) refers to both B and C.

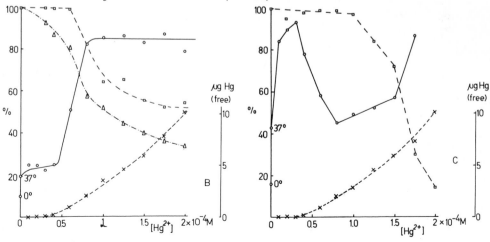

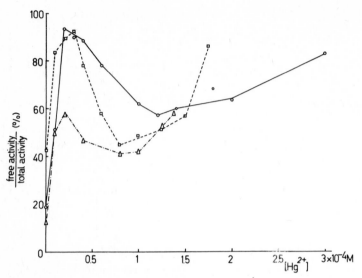

Fig. 2 *(above)*. Effect of increasing Hg^{2+} concentration on liver mitochondria in three M + L preparations containing 2.57 (△), 3.29 (□) and 7.27 (○) mg protein/ml.

Fig. 3 *(below)*. Hg^{2+}-induced swelling of liver mitochondria in a M + L fraction.

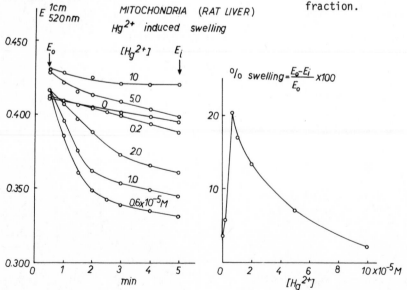

tration than with the lysosomes (0.4 instead of 0.6 × 10⁻⁴ M) was required to start the release of catalase, and maximal labilization was already reached at 0.8 ×10⁻⁴ M (more than 80% in this experiment; 100% in another one). The mercury ions have only a moderate inhibitory effect on catalase, since still 35% of the total activity remained at 2 × 10⁻⁴ M Hg²⁺ without addition of an activator. With cysteine the activity of catalase increased only by ∿20%.

The behaviour of the mitochondria in the same M + L fraction upon increasing Hg²⁺ concentration was clearly different from that of the lysosomes and the peroxisomes: the curve representing %-free GDH is biphasic (Fig. 1C). A very steep increase in free GDH activity to >90% of the total activity was first observed on raising the Hg²⁺ concentration only to 0.1 - 0.3 × 10⁻⁴ M. A further increase to 0.8 × 10⁻⁴ M was accompanied by a decrease in free GDH activity to values similar to those observed without Hg²⁺. At higher Hg²⁺ concentrations the free GDH activity rose again (>80% of the total at 1.75 × 10⁻⁴ M). Up to 10⁻⁴ M the total GDH activity remained near 100% with EDTA as activator, but at higher concentrations it decreased progressively (at 1.75 × 10⁻⁴ M only about one-third of the total GDH remained enzymatically active). The first peak of free GDH activity occurred entirely at mercury levels where hardly any free mercury is present. The position of the second GDH maximum depends highly on the protein concentration of the M + L fraction, whereas the maximum of the first peak still appears at ∿0.2 - 0.3 × 10⁻⁴ M (Fig. 2). There is some evidence that mitochondrial swelling might be linked with the first peak, since the maximal swelling was obtained at 0.1 × 10⁻⁴ M (Fig. 3). There are previous reports [14-16] of mercury-induced swelling of mitochondria.

It is notable that in the absence of Hg²⁺ a pre-incubation period of 10 min at 37° did not alter the degree of labilization of the lysosomes (i.e. 12%) compared to that observed at 0°, whereas peroxisomes and mitochondria showed marked thermal labilization.

From the comparative results in Fig. 4 for subcellular organelle labilization it is evident that apart from the mercury-induced swelling of mitochondria, peroxisomes are more readily labilized than lysosomes. At 0.6 × 10⁻⁴ M Hg²⁺, where hardly any labilization of lysosomes occurred, already 50% labilization of the peroxisomes was obtained and this became very high with the onset of the autocatalytic phase in lysosomal labilization. The first GDH maximum occurs when hardly any free mercury is present and when lysosome labilization has not yet started. Hitherto there is no conclusive evidence whether the second maximum in GDH activity is chiefly due to an effect of mercury or to the attack of lysosomal enzymes released during the autocatalytic phase. The fact that the amount of bound mercury increased progressively with increasing mercury concentration (up to 2.5 × 10⁻⁴ M) (Fig. 5) does not disagree with the assumption that the more Hg²⁺ is added the more organelles are disrupted. Above 2.5 × 10⁻⁴ M (equilibrium and inhibition phase in the labilization of the lysosomes) the binding of Hg²⁺ tends to reach a maximum, probably indicating saturation of the available mercury-binding sites.

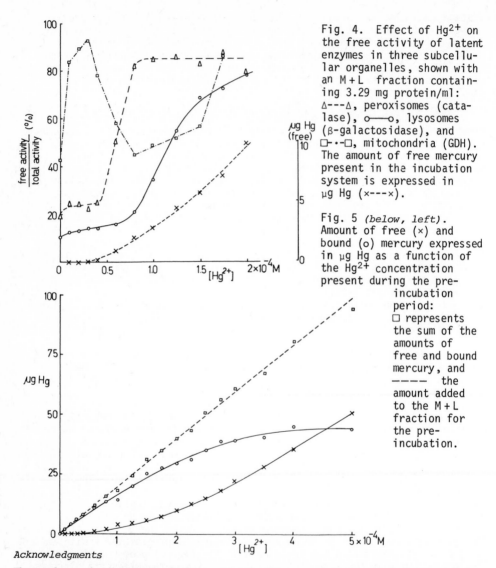

Fig. 4. Effect of Hg^{2+} on the free activity of latent enzymes in three subcellular organelles, shown with an M+L fraction containing 3.29 mg protein/ml: Δ---Δ, peroxisomes (catalase), o——o, lysosomes (β-galactosidase), and □·-·□, mitochondria (GDH). The amount of free mercury present in the incubation system is expressed in μg Hg (×---×).

Fig. 5 *(below, left)*. Amount of free (×) and bound (o) mercury expressed in μg Hg as a function of the Hg^{2+} concentration present during the pre-incubation period: □ represents the sum of the amounts of free and bound mercury, and ———— the amount added to the M+L fraction for the pre-incubation.

Acknowledgments

The authors thank Miss N. Girboux, Mr. H. Bauer, and Mr. A. Vranckx for their skilful technical assistance. Mr. T. Seminck is gratefully acknowledged for the mercury determination.

References

1. Verity, A. & Reith, A., *Biochem. J. 105* (1967) 685-690.

2. Selwyn, M.J., *Biochem. J. 130* (1972) 65-67*P*.

3. Scott, K.M., Knight, V.A., Settlemire, C.T. & Brierley, G.P., *Biochemistry [U.S.A.] 9* (1970) 714-724.

4. Robinson, D., Price, R.G. & Dance, N., *Biochem. J. 102* (1967) 533-538.

5. Coonrod, D. & Paterson, P.Y., *J. Lab. Clin. Med. 73* (1969) 6-16.

6. Lauwerys, R. & Buchet, J.-P., *Arch. Environ. Health 27* (1973) 65-68.

7. Lauwerys, R. & Buchet, J.-P., *Eur. J. Biochem. 26* (1972) 535-542.

8. de Duve, C., Pressman, B.C., Gianetto, R., Wattiaux, R. & Appelmans, F., *Biochem. J. 60* (1955) 606-617.

9. Lowry, O.H., Rosebrough, N.J., Farr, A.L. & Randall, R.J., *J. Biol. Chem. 193* (1951) 265-275.

10. Barrett, A.J., in *Lysosomes, a Laboratory Handbook* (Dingle, J.T., ed.), North-Holland, Amsterdam (1972) pp. 47-135.

11. Beaufay, H., Bendall, D.S., Baudhuin, P. & de Duve, C., *Biochem. J. 73* (1959) 623-628.

12. Baudhuin, P., Beaufay, H., Rahman-Li, Y., Sellinger, O.Z., Wattiaux, R., Jacques, P. & de Duve, C., *Biochem. J. 92* (1964) 179-184.

13. Hatch, W.R. & Ott, W.L., *Anal. Chem. 40* (1968) 2085-2087.

14. Arcos, J.C., Griffith, G.W. & Cunningham, R.W., *J. Biophysic. Biochem. Cytol. 7* (1960) 49-60.

15. Tapley, D.F., *J. Biol. Chem. 222* (1956) 325-339.

16. Gritzka, T.L. & Trump, B.F., *Amer. J. Pathol. 52* (1968) 1225-1277.

19 SEPARATION STUDIES ON DISRUPTED LYSOSOMES

M. Dobrota and R.H. Hinton
Wolfson Bioanalytical Centre
University of Surrey
Guildford GU2 5XH, U.K.

Normal liver lysosomes from untreated rats have been purified using a two-step zonal centrifugation method. The same method is used to separate fragments of partially disrupted lysosomes. From the density at which these fragments equilibrate they appear to be membranous. With whole lysosomes, all the acid hydrolases studied are located in the same peak, whilst with disrupted lysosomes a complex, heterogeneous pattern is obtained.

Since lysosomes and rough endoplasmic reticulum vesicles are similar in size and in their isopycnic densities, they are difficult to separate by centrifugal methods. The problems of purifying normal lysosomes have previously been discussed by Burge & Hinton [1]. The method described in the present paper is comparable, in the purification achieved, with other centrifugal methods for separating normal lysosomes [2, 3], although the purification is less than that reported for free-flow electrophoresis [4], or for lysosomes loaded with Triton WR-1339 [5] or iron-sorbitol-citrate complex [6] (see also B. Arborgh *et al.*, this volume, in final article).

Although the lysosomal membrane and other fragments have been investigated elsewhere [7], most of the work has been done on lysosomes loaded with Triton WR-1339 or dextran. These lysosomes represent a narrow spectrum of the lysosome population, namely secondary lysosomes; but even more seriously their structure is likely to have been affected by the very agents which were used to achieve the high purity. Because of these drawbacks we have chosen to work with normal lysosomes, in spite of the difficulties in purifying them.

MATERIALS AND METHODS

Materials

Gradients were prepared from a stock solution (2 M) of Mineral Water sucrose (Tate & Lyle Ltd., London) which was treated with activated charcoal (Sigma Chem. Co.), 25 g per kg sucrose, in order to remove any protein and ribonuclease activity. All gradient solutions were buffered with 5 mM Tris pH 7.4.

Sample preparation

A fraction enriched in lysosomes was separated from a crude mitochondrial and lysosomal fraction of rat liver homogenate by rate sedimentation in an HS zonal rotor as described by Burge & Hinton [1]. The sample was derived from approximately 10 g of rat liver. If more than this quantity is used the lysosomal region may become heavily contaminated with rough endoplasmic reticulum which has tumbled into it due to overloading of the sample band.

a) *For isopycnic banding of lysosomes*

Rather than pelleting and resuspending the lysosomes as in earlier experiments [1] the whole lysosome-rich region from the HS rotor (approx. 160 ml) was loaded directly onto a B-XIV zonal rotor for the equilibrium spin. In the HS zonal run the lysosome-rich region is found between the microsomes and the mitochondria (tubes 10-20 if 20 ml fractions are collected) and appears as a long trough in the 650 nm trace. Alternatively it can be picked out visually. The microsomal region is usually amber in colour and slightly turbid, while the mitochondria are very turbid. Between these two regions there are usually eight tubes which are almost perfectly clear and these contain most of the lysosomes. The sample obtained by pooling these tubes was then loaded onto the B-XIV zonal rotor. The refractive index of the sample should be checked before loading, to ensure that the sample is less dense than the 0.8 M sucrose at the start of the gradient. The time between recovering the fractions from the HS rotor and loading them onto the B-XIV should be as short as possible.

b) *For isopycnic banding of lysosomal fragments*

The lysosomes are broken by vigorous re-homogenization. The pooled lysosomal region (prepared as above) was split into three approximately equal parts for easier handling. Each part (about 50 ml) was treated with a Polytron PT 35 OD homogenizer (Kinematica, Switzerland) for 1 min at maximum speed (about 13,000 rev/min). To keep the sample effectively cooled a measuring cylinder or large test tube, immersed in ice, was used as the homogenizing vessel. The samples were kept in ice during the whole procedure and were finally pooled. Again the sample should be loaded onto the B-XIV rotor as quickly as possible.

Gradient preparation

The B-XIV rotor was filled with 400 ml of linear (with volume) sucrose gradient 2 M, and with a cushion of 2 M sucrose. The gradient was made with an apparatus described previously [8]. The mixing vessel contained 285 ml of 0.8 M sucrose. The gradient was pumped from the mixing vessel to the rotor at 40 ml/min, while the 2 M sucrose was added to the mixing vessel at 20 ml/min. When the volume in the mixing vessel became too low (10-20 ml) to mix efficiently, the pump was stopped. **The** rest of the rotor was then filled with 2 M sucrose.

During loading the gradient was cooled to 4° by passing through a stainless steel cooling coil (immersed in ice/salt mixture) immediately before entry into the rotor. In our laboratory the gradient shape is usually checked by a continuous-flow recording refractometer (Hilger & Watts; no longer in production).

Centrifugation

X A B-XIV titanium zonal rotor (M.S.E. Ltd.) was operated in a Super Speed 65 centrifuge. The loading speed of the rotor was about 2,000 rev/min. Approximately 150 ml of sample were loaded manually into the rotor using 50 ml disposable syringes. Great care has to be taken to load the sample slowly and evenly in order to avoid any cross-leakage. When the sample was completely loaded it was followed by 50 ml of overlay, the density of which must be lower than that of the sample. The rotor was then accelerated to 47,000 rev/min and left at maximum speed for 2 h 50 min. After decelerating the rotor to unloading speed (2,000 rev/min) the gradient was unloaded with continuous monitoring of the refractive index and extinction at 280 nm, the latter with an SP500 (Pye Unicam) spectrophotometer. Fractions of 20 ml were collected and immediately put in ice.

Re-centrifugation of the lysosomal 'membrane' region

After centrifugation of disrupted lysosomes a peak of acid phosphatase was obtained at a density of 1.175 g/ml. In the attempt to clarify the nature of this material a sample from this region was subjected to further centrifugation. This was done in a 3 × 23 ml swing-out rotor run in a SS 65 centrifuge (M.S.E. Ltd.), the sample constituents being floated or sedimented to equilibrium on a 1 → 2 M sucrose gradient.

A 4 ml sample from the peak tube (usually No. 21) was adjusted with sucrose to give a density of 1.26 g/ml and was then layered under a linear sucrose gradient (1 → 2 M) in one tube of the swing-out rotor. A similar sample was diluted with distilled water to a density of about 1.12 g/ml and was layered on top of a similar gradient. These samples were then spun in the swing-out rotor at 30,000 rev/min for $3\frac{1}{2}$ h. At the end of the spin the tubes were unloaded and 1 ml fractions were collected. These were later assayed for acid phosphatase and protein.

Assay methods

Catalase was chosen as the peroxisomal marker in preference to uricase since it is much easier to assay, this being done by the AutoAnalyzer method of Leighton *et al.* [9]. Details of other enzyme and protein assays are given elsewhere [10, 11]. Acid phosphatase was assayed using sodium β-glycerophosphate as the substrate and in the presence of 0.1% Triton X-100. More recently we have found that acid *p*-nitrophenyl phosphatase gives exactly the same distribution pattern, the distinctive soluble *p*-nitrophenylphosphase being removed during the initial separation in the HS rotor. Acid phosphodiesterase was assayed with *bis*(*p*-nitrophenyl)phosphate.

RESULTS

The enzyme distribution patterns obtained on the isopycnic banding of intact lysosomes (Fig. 1) are as expected from the classical work of de Duve and his co-workers [12]. Acid hydrolases are all located in a well-defined peak banding at a density of 1.21 g/ml. The catalase trace shows that, as expected, peroxisomes are denser than lysosomes. Glucose-6-phosphatase is found at a density of 1.228, much higher than is usually found even for fragments of rough endoplasmic reticulum. Succinate dehydrogenase activity could not be detected. Very little 5'-nucleotidase activity was present in the original sample, but a small peak could be detected at a density of 1.157 g/ml, as expected for plasma membrane fragments. In the lysosome peak both acid phosphatase and acid ribonuclease were purified about 25 times with respect to the homogenate (Table 1).

If the partially disrupted lysosomes are spun to equilibrium, on an identical gradient, a radically different enzyme pattern is obtained. The peak of acid hydrolase enzymes at density 1.21 g/ml (corresponding to whole lysosomes) is much reduced, whilst a new peak of acid phosphatase appears at a density of 1.175 g/ml. For ease of identification this new peak of acid phosphatase will be designated 'Region 3' as in Fig.2. As expected, a broad peak of acid hydrolases, solubilized by the homogenization, is found in the sample region. It is interesting to note that there is little activity of β-galactosidase or acid phosphodiesterase in 'Region 3'. A rather poorly defined peak of acid ribonuclease is found at a density slightly lower than that of the acid phosphatase peak. 5'-Nucleotidase was not measurable. Acid phosphatase was purified about 41-fold in the peak tube (No. 21) of 'Region 3', indicating that the material in this region is richer in acid phosphatase than in the intact lysosome preparation.

For further characterization of the lysosomal fragments a sample from 'Region 3' was floated on a linear sucrose gradient (1→2 M). A sample was also sedimented on an identical gradient. Fig.3a shows the pattern of acid phosphatase after flotation. A peak of activity remains in the sample region (density 1.255 g/ml), whereas a considerable amount has floated to densities of 1.171 and 1.135 g/ml. When sedimented (Fig.3b) the same sample gives a single acid phosphatase peak at a density of ∿1.16 g/ml.

DISCUSSION

The purification of acid hydrolases, when intact lysosomes are isolated by the method now described, is comparable to that obtainable with other centrifugal methods [2, 3]. It is probable that the degree of purification is the

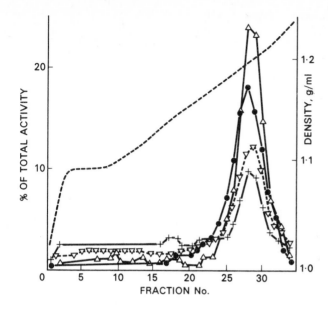

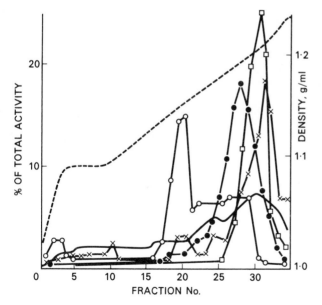

Fig. 1. Isopycnic banding of lysosomes. The B-XIV zonal rotor, filled with 400 ml of sucrose gradient and 150 ml of sample, was spun for 2 h 50 min at 47,000 rev/min.

Upper diagram shows the distribution of lysosomal enzymes, as % in each tube:

●——●, acid phosphatase;
△——△, acid ribonuclease;
▽— —▽, acid phospho-
       diesterase;
+——+, β-galactosidase.

Lower diagram shows the distribution of protein and various other enzymes:

——, protein;
○——○, AMPase (5'-nucleo-
       tidase);
×——×, glucose-6-phospha-
       tase;
□——□, catalase;
also, to facilitate com-
parison,
●——●, acid phosphatase.

Both diagrams show:
———, density.

Table 1.   Enrichment in acid phosphatase
and acid ribonuclease. The values are the
ratios of the specific activity of the
stated fractions or tubes to that of the
whole homogenate.

FRACTION	Acid phos-phatase	Acid ribo-nuclease
NORMAL LYSOSOMES (as in Fig. 1)		
Lysosomal fraction *as loaded → B rotor*	9.71	8.8
Region 4 (tubes 24-29)	17.1	15
Peak tube, no. 28	23	25.7
DISRUPTED LYSOSOMES (as in Fig. 2)		
Lysosomal fraction *as loaded → B rotor*	9.15	8.5
Region 1 (tubes 1-12)	22	13
Region 3 (tubes 18-24)	23	10
Peak tube, no. 8	-	18
Peak tube, no. 21	41	-

most that can be obtained
by centrifugal methods and
that other methods, such
as free-flow electrophoresis [4], must be used if
greater purity is to be
obtained. Our method,
which involves no pelle-
ting and resuspending, besides being convenient also minimizes the risk of
lysosomal breakage. This
is well illustrated in
Fig. 1 which shows little
solubilized enzyme activity in the sample region,
confirming that no ulti-
mate breakage of lysoso-
mes has occurred. Within
limitations the method is
therefore successful for
the isolation of liver
lysosomes from normal
rats.

A particularly important result is the appearance, after the lysosomes
had been partially disrupted with the Polytron homogenizer, of a peak of acid
phosphatase at an equilibrium density (in sucrose) of 1.175 g/ml ('Region
3').   The distribution of other acid hydrolases shows this material to be
quite different in its enzyme composition from intact lysosomes. β-Galactosidase is almost completely absent, which fits well with the observations
of Baccino *et al.* [13] that this enzyme is very readily solubilized yet
shows little tendency to re-adsorb onto particulate material. Acid ribonuclease appears as a poorly defined peak, the main part of which is at a
density slightly lower than the acid phosphatase peak, suggesting that it
may be attached to a lysosomal fragment with slightly different properties.
It would be of interest to ascertain whether the same degree of heterogeneity of truly lysosomal enzymes is obtained if the same method is applied to a
more clearly defined spectrum of the lysosome population, such as Tritosomes
or, better, the iron-sorbitol-citrate loaded lysosomes [6].

From the results presently available it is difficult to reach a conclusion on the nature of the material located in 'Region 3'. However, some
indication is obtainable by closer examination of the results. If acid phosphatase is assayed with and without Triton X-100 the results show that the
enzyme exhibits no latency in 'Region 3'. Therefore the lysosomal membrane

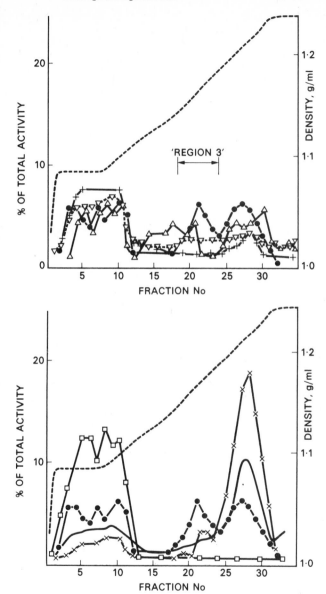

Fig. 2. Isopycnic banding of partially disrupted lysosomes. The conditions and symbols are as in the Legend to Fig. 1. (AMP-ase *(not shown)* was also determined.)

is obviously ruptured; but whether or not it is still attached to the matrix is unclear. The results of the further centrifugation of this region are rather more difficult to interpret. Flotation gives an acid phosphatase pattern which might indicate that some of the enzyme remains attached to high-density material, possibly protein, which remains at the bottom of the tube (density 1.255 g/ml). However, the most likely explanation is that the enzyme is not associated with protein at all but perhaps consists of soluble enzyme released during the preparation of the sample for the flotation run. The occurrence of the enzyme in two other regions, one at 1.171 and the other at 1.135, would indicate that the enzyme is also

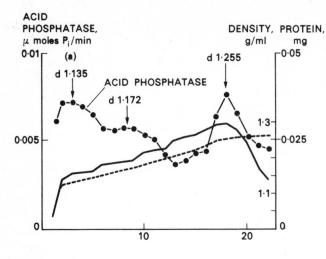

ACID
PHOSPHATASE,
μ moles P$_i$/min

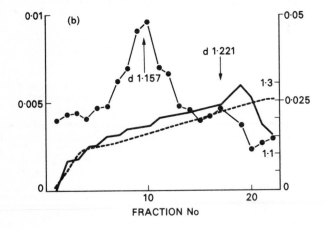

FRACTION No

Fig. 3. Re-centrifugation of 'lysosomal fragments' in a swing-out rotor.

(a) Flotation. A large sample (4 ml) taken from tube no. 21 (peak tube of 'Region 3') of Fig. 2 was adjusted to give a sucrose concentration of 2 M, and layered under a sucrose gradient *(see text)*: centrifugation for 3½ h gave a distribution of acid phosphatase (●──●) and protein (──) as shown.

(b) Sedimentation. The conditions were as in (a) except that the sample was diluted to just below 1 M sucrose and was layered on top of the gradient.

Note that enzyme activity per tube is given in absolute terms (cf. the % distribution basis in Figs. 1 & 2).
The gradients in Fig. 3 are offset due to the sample being at the bottom (Fig. 3a) or at the top (Fig. 3b) of the tube; thus the density of tube 10 in Figs. 3a and 3b is 1.184 and 1.160 g/ml respectively.- As in Figs. 1 & 2,
---- denotes density.

associated with membranous fractions. However, the pattern of the sedimented sample shows only one distinct peak at density 1.16 g/ml. In both cases the peak of protein is at densities greater than 1.23. If acid phosphatase is associated with protein in the flotation run it seems odd that it should not be found in the same region in the sedimentation run. In view of the different patterns obtained with flotation and sedimentation, one could speculate that acid phosphatase, as well as being readily solubilized, is also readily picked up by other particles, again in agreement with the work of Baccino *et al.*[13]. We therefore conclude that the enzyme should not be considered an integral part of the lysosomal membrane until its localization can be determined by other techniques.

Preliminary results on the electron microscopy of fractions separated from disrupted lysosomes have shown that both the region thought to contain intact lysosomes and that containing lysosome 'membranes' were heavily contaminated with rough endoplasmic reticulum. The lysosomes which were visible in the membrane region ('Region 3') all appear to lack the precise outline of a membrane. It was, however, difficult to identify the source of the membrane material present. It is hoped that the application of cytochemical methods [14] to the separated fractions will help elucidate the results.

Acknowledgements

The authors would like to thank Dr. J.T.R. Fitzsimons of the Department of Physiology and Biochemistry, University of Southampton, for carrying out the electron microscopy of the fractions.

References

1. Burge, M.L.E. & Hinton, R.H., in *Separations with Zonal Rotors* (Reid, E., ed.) Wolfson Bioanalytical Centre, University of Surrey, Guildford (1971) pp. S-5.1 - S-5.10.

2. Beaufay, H., in *Lysosomes in Biology and Pathology*, Vol. 2 (Dingle, J.T. and Fell, H.S., eds.) North-Holland, Amsterdam (1969) pp. 515-546.

3. Reid, E., in *Subcellular Components, Preparation and Fractionation,* 2nd edn. (Birnie, G.D., ed.), Butterworth, London (1972) pp. 93-118.

4. Stahn, R., Maier, K.P. & Hannig, K., in *Separation with Zonal Rotors* (Reid, E., ed.), Wolfson Bioanalytical Centre, University of Surrey, Guildford (1971) pp. S-6.1 - S-6.10.

5. Wattiaux, R., Wibo, M. & Baudhuin, P., *Lysosomes* [CIBA Foundation Symp.] (de Reuck, A.V.S. and Cameron, M.P., eds.), Churchill, London (1963) pp. 176-196.

6. Arborgh, B., Ericsson, J.L.E. & Glauman, H., *FEBS Lett. 32* (1973) 190-194.

7. Thinès-Sempoux, D., in *Lysosomes in Biology and Pathology,* Vol. 3 (Dingle, J.T. and Fell, H.S. eds.), North-Holland, Amsterdam and London (1973) pp. 278-299.

8. Hinton, R.H. & Dobrota, M., *Anal. Biochem. 30* (1969) 99-110.

9. Leighton, F., Poole, B., Beaufay, H., Baudhuin, P., Colley, J.W., Fowler, S. & de Duve, C., *J. Cell Biol. 37* (1968) 482-513.

10. Prospero, T.D., Burge, M.L.E., Norris, K., Hinton, R.H.& Reid, E., *Biochem. J. 132* (1973) 449-458.

11. Hinton, R.H. & Norris, K.A., *Anal. Biochem. 48* (1972) 247-258.

12. Beaufay, H., Jacques, P., Baudhuin, P., Sellinger, O.Z., Berthet, J. & de Duve, C., *Biochem. J. 92* (1964) 184-205.

13. Baccino, F.M., Rita, G.A. & Zuretti, M.F., *Biochem. J. 122* (1971) 363-371.

14. El-Aaser, A.A., Fitzsimons, J.T.R., Hinton, R.H., Norris, K.A. & Reid,E., *Histochem. J. 5* (1973) 199-223.

20 THE LYSOSOMAL MEMBRANE - CHARACTERISTIC FEATURES OF COMPOSITION AND FUNCTION

Roland Henning
Institut für Physiologische Chemie der Universitat Köln
D-5 Köln 41, Josef-Stelzmannstr. 52
W. Germany *

The membrane fraction of rat liver lysosomes isolated by the Triton WR-1339 method differs in lipid composition from other intracellular membranes but resembles the plasma membrane. This accords with electron microscopic evidence for pinocytotic uptake of this detergent. A high carbohydrate content is also found. Investigations of mono- and oligosaccharides by the use of carbohydrate-specific agglutinins from plants (lectins) and snails show striking similarities with the plasma membrane when compared with other intracellular membranes. The nature and localization of acidic groups on the lysosomal membrane provides evidence of apparently unique functional properties. The unspecific and reversible binding of lysosomal enzymes having an isoelectric point around pH 5 to lysosomal membranes at pH 4 suggests the presence of strongly anionic groups on the membrane. Sialic acid, being the most probable molecule carrying the anionic charge, could be visualized in the electron microscope by colloidal iron staining and was found to be localized on the inside of the lysosomal envelope. Internal accumulation of membrane-bound sialic acid leads to the development of a Donnan equilibrium. This might serve as an energy-independent means to maintain an intralysosomal acid milieu requisite for the full activity of most lysosomal enzymes.

The lysosomal membrane is one of the most important features of the lysosome concept as worked out by de Duve. The heterogeneity of this class of organelles (primary and secondary lysosomes, heterolysosomes, autolysosomes and others) was the main obstacle to intensive investigation of this membrane. Therefore we took advantage of lysosomes isolated by the Triton WR-1339 method developed by Wattiaux *et al.*[1] which results in pure secondary lysosomes. All evidence available supports a pinocytotic uptake of this non-ionic detergent by rat liver cells [2].

CHARACTERISTIC FEATURES OF THE LYSOSOMAL MEMBRANE LIPIDS

The analysis of various membrane constituents of rat-liver lysosomes filled with Triton WR-1339 (tritosomes) revealed distinct similarities with the plasma membrane in the lipid and the carbohydrate sector. High concentrations of cholesterol and sphingomyelin, predominantly saturated fatty acids in phospho-

* Present address: Rockefeller University, New York, N.Y. 10021, U.S.A.

lipids and the presence of glycolipids are four generally accepted features of the lipid composition of the plasma membranes of animal cells from various sources.

The results of the phospholipid and cholesterol analyses on plasma membranes, tritosomal membranes and lysosomal membranes from normal rat liver are given in Table 1. The isolation of plasma membranes and tritosomal membranes and the criteria of purity have been described elsewhere [3]. Membranes of normal lysosomes have been prepared by osmotic shock and subsequent high-speed centrifugation of lysosomes isolated from rat liver by free-flow-electrophoresis in a modification [5] of the method of Stahn *et al.* [4]. The similarity of the two types of lysosomal membranes with the plasma membrane in respect of their cholesterol and sphingomyelin concentration is clearly shown in Table 1 when compared with mitochondria.

Table 1. Cholesterol and phospholipid composition of plasma membranes [3], tritosomal membranes [3], normal rat liver lysosomal membranes [5] and mitochondria [5]. Values represent % composition, except for the first line which represents the molar ratio of cholesterol to lipid P.

LIPID	Plasma membrane	Tritosomal membrane	Lysosomal membrane	Mitochondria
Cholesterol/phospholipid	0.90	0.52	0.27	0.06
Material at solvent front	2.3 %	6.8 %	4.6 %	
Cardiolipin			4.6 %	11.2 %
Phosphatidyl ethanolamine (PE)	22.4 %	17.9 %	26.0 %	31.0 %
Phosphatidyl inositol	} 17.4 %	8.9 %	{ 9.7 %	} 7.5 %
Phosphatidyl serine			3.0 %	
Phosphatidyl choline (PC)	34.8 %	33.5 %	41.1 %	50.5 %
Sphingomyelin	20.5 %	32.9 %	7.6 %	0
Lyso derivative of PE			2.9 %	0
Lyso derivative of PC	2.6 %	0	2.9 %	0

The gas-liquid chromatographic analysis of the fatty acid composition of the phospholipids of both the plasma and the tritosomal membrane [3] showed in the phospholipids of the latter even less unsaturated fatty acids than in the plasma membrane (for example, 36.6% polyunsaturated fatty acids in the phosphatidyl ethanolamine of the tritosomal membrane and 46.6% in that of plasma membrane). This fact meets the third specific criterion for plasma membrane lipids.

Glycolipids were detected in higher concentrations in the tritosomal membrane than in the plasma membrane. Ceramide monohexoside, dihexoside and trihexoside could be identified in tritosomal membranes by thin-layer chromatography. The trihexoside could not be detected in plasma membranes.

Amongst the gangliosides hematoside was the main component in tritosomal membranes whereas GM_1-ganglioside was predominant in plasma membranes. The fatty acid and sphingosine base composition of these compounds isolated from the tritosomal membrane corresponds well to that of other extraneurally occurring glycolipids. Their concentrations, their carbohydrate composition and sequences are given in Table 2.

Table 2. Glycolipids in tritosomal membranes [6]. Concentrations were determined by the anthrone method according to Razin *et al.* [7] or by the thiobarbituric acid assay (hematoside)[8]. Molar ratios of sugars were obtained by gas-liquid chromatography of the corresponding alditol acetates. Sequences were determined by partial hydrolysis studies.

Glycolipid	Concn. (µg/mg prot.)	Molar sugar ratios	Sugar* sequence
Monohexosyl ceramide	11.8	glc	cer-glc
Dihexosyl ceramide	9.4	glc/gal 1:1	cer-glc-gal
Trihexosyl ceramide	3.2	glc/gal 1:2	cer-glc-gal-gal
Hematoside (*N*-ac-neuraminic acid)	8.4	glc/gal 1:1	cer-glc-gal

Abbreviations in heading to Table 3

CARBOHYDRATES IN TRITOSOMAL MEMBRANES

It has been found that the tritosomal membrane contains higher carbohydrate concentrations than the plasma membrane (neutral sugars, amino-sugars and sialic acid) [3]. Therefore a more detailed examination of the carbohydrate residues present on the plasma and the tritosomal membrane was performed by agglutination studies, using carbohydrate-specific agglutinins from plants (lectins) [9]. These results were compared with the agglutinability of mitochondria and microsomes, and it was found that the plasma and tritosomal membranes exhibit a very similar agglutination pattern when compared with mitochondria and microsomes (Table 3).

To summarize these data, the lipids and the carbohydrates of the tritosomal membrane appear to be unique among other intracellular membranes. Its similarities with the plasma membrane might be explained by the formation of the tritosomal membrane by fusion of endocytotic vesicles derived from the plasma membrane with the primary lysosome during the pinocytosis process. It should be emphasized, however, that a contribution of the primary lysosomal membrane as well as that of other intracellular material taken up by autophagocytosis cannot be judged on the basis of these analyses. It cannot yet be decided which of the compounds described are genuine lysosomal constituents and which of them are incorporated by other membranous material.

Table 3. Agglutination titres of rat liver cell organelles.
Agglutinins were used in 1% solutions in 0.9% NaCl. The specificity of
all reactions was checked by inhibition tests with the appropriate inhibit-
or substances. Cell fractions were washed twice to remove sucrose before
incubation with agglutinins. Titres are given in serial dilution factors
of the agglutinin solutions used.
0 = no agglutination, (++) strong agglutination after treatment with 0.1%
pronase solution for 30 min at 20°, gal = D-galactose, glc = D-glucose,
glc-ʍac = ʍ-acetyl-glucosamine, ʍac-gal =ʍ-acetyl-galactosamine,
man = D-mannose.

AGGLUTININ	Plasma membrane	Tritosomal membrane	Mito-chondria	Micro-somes	SPECIFICITY OF AGGLUTININ
Wheat germ lipase	64	2	0 (++)	0	β-glc-ʍac
Ricinus communis (0.9% NaCl extract)	16	32	0	0	β-D-gal
Concanavalin A	128	16	0 (++)	0 (++)	α-D-glc, α-D-man
Phytohaemagglutinin from *Phaseolus vulgaris*	256	16	0	0	D-gal - β-glc-ʍac-man

FUNCTIONAL ASPECTS

Among the various functional tasks of the lysosomal membrane, the maintenance
of an intralysosomal acid pH is one of the most important ones. Since most
of the hydrolases involved in the intralysosomal digestion exhibit an acid pH
optimum, the intralysosomal pH is expected to be acid. Besides some rough
estimations of the intralysosomal pH reported earlier [10], recently more
precise data indeed indicate an acid pH within lysosomes (tritosomes) [11].
Among the mechanisms which have been discussed to maintain this pH-gradient,
Coffey and de Duve [12] pointed to the possibility of a Donnan-equilibrium
produced by an intra-lysosomal accumulation of non-diffusible anions. The
following experiments were designed to detect the anionic charges on the
tritosomal membrane, and to describe their nature in terms of the chemical
structure and their localization on the membrane.

Isolated tritosomes were incubated under osmotic protection for 10 min
at 37°at various pH values. It can be seen in Fig. 1 that the tritosomal
membrane is sensitive to acid pH, as demonstrated by an increased release of
[3H]-Triton WR 1339 from tritosomes determined by radioactivity measurements

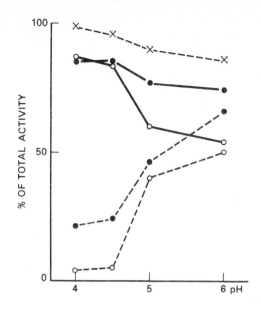

Fig.1. Dependence on pH of the 'free' and non-sedimentable activities of β-glucuronidase and acid phosphatase of tritosomes; and release of [^{3}H] Triton WR-1339 from tritosomes determined in the 12,000 *g* (10 min) supernatant. [15]. Enzyme activities were determined after pre-incubation of 0.3 mg of protein in 1.0 ml 0.037 M acetate buffer (0.25 M sucrose) of desired pH for 10 min at 37°.

———•——— acid phosphatase
———○——— β-glucuronidase
———— 'free' activity
————— non-sedimentable activity

———x—— release of [^{3}H]-Triton WR-1339

on the supernatant after centrifugation (12,000 *g*, 10 min). In addition, the determination of the 'free' activity of β-glucuronidase and acid phosphatase according to Appelmans *et al.* [13] demonstrates also an increasing injury to the membrane with acidification. The non-sedimentable activities, however, decrease with acidification of the medium. This indicates the binding of enzymes to membranes by electrical forces. The negative counterpart is the membrane since both enzymes are positively charged at pH 4, their isoelectric points being 4.8 (acid phosphatase) and 5.8 (β-glucuronidase).

The ionic nature of this type of binding was established by its inhibition by increasing ionic strength. Fig. 2a shows that the enzyme remains bound at low ionic strength and is released with increasing concentrations of various salts. Increasing concentrations of added non-lysosomal proteins also release enzyme bound to the membrane at pH 4, indicating the unspecific nature of this binding.

THE ROLE OF SIALIC ACID

The ionic nature of this binding is demonstrated once more by the nearly complete reversibility, depending only on the pH. This allows one to attack isolated tritosomal membranes with various enzymes, to test the binding capa-

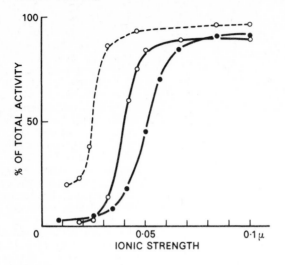

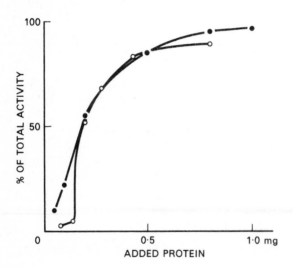

Fig. 2. Influence of ionic strength and non-lysosomal proteins [15] on the binding of β-glucuronidase to tritosomes at pH 4.

(a) Ionic strength:

o---o $MgCl_2$,
o—o $MgSO_4$,
●—●   NaCl

(b) Non-lysosomal proteins (dialyzed albumin, mitochondria)

●—●   dialyzed bovine serum albumin

o—o   rat liver mitochondria

Data represent non-sedimentable activity as % of total.

city of β-glucuronidase and to compare it with that of untreated membranes. Digestion of tritosomal membranes with trypsin, papain, pronase and phospholipase C had no effect (95% reversibility). Extensive treatment with neuraminidase significantly reduced the reversibility of binding at acid pH (66% reversibility). This indicates sialic acid as most probable molecule bearing the anionic charge on the lysosomal membrane.

LOCALIZATION OF SIALIC ACID

The question of the localization of sialic acid was followed up by several experimental approaches. The best results were obtained from electron micrographs stained with colloidal iron at pH 1.8 according to Gasic *et al.* [14]. These pictures clearly show a preferential staining on the inner site of tritosomal membranes [15] (Fig. 3). In no case could an outside localization of the stain be observed. Occasionally the iron was found on both sites, but in these cases tilting of the membrane could not be ruled out. The specificity of the staining method was checked by staining of membranes after extensive neuraminidase treatment. This treatment inhibits the staining almost completely.

In summary, these results demonstrate an apparent accumulation of non-diffusible anions in the form of bound sialic acid within the tritosome, potentially setting up a Donnan-equilibrium. This could serve as a simple means to maintain the intralysosomal acid pH by an increased influx of cations such as K^+ and H^+. Since the K^+ concentration within the cell is about 10^6 times higher than the H^+ concentration, two auxiliary mechanisms should be discussed. To keep out the potassium ion the lysosomal membrane should be equipped either with a selective permeability for the proton or with a K^+/H^+ energy-dependent exchange pump as discussed recently by Mego *et al.* [16].

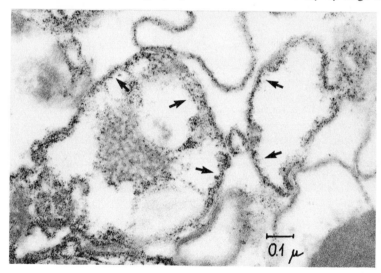

Fig. 3. High-power electron micrograph of isolated tritosomes stained with colloidal iron according to Gasic *et al.* [14]. Glutaraldehyde fixation; section contrasted with hot uranyl acetate in ethanol. *The colloidal iron stain is located predominantly on the inner side of the bounding membrane (ARROWS) and also along fragments within the interior of tritosomes.* Courtesy of Dr. H. Plattner, Inst. for Cell Biology, Univ. of Munich. Mag. × 77,000

References

1. Wattiaux, R., Wibo, M. & Baudhuin, P., in *Ciba Foundation Symposium on Lysosomes* (De Reuck, A.V.S. & Cameron, M.P., eds.), Little Brown, Boston (1963) pp. 176-196.

2. Henning, R. & Plattner, H. (1973) *manuscript submitted for publication.*

3. Henning, R., Kaulen, H.D. & Stoffel, W., *Hoppe-Seyler's Z. Physiol. Chem. 351* (1970) 1191-1199.

4. Stahn, R., Maier, K.-P. & Hannig, K., *J. Cell Biol. 46* (1970) 576-591.

5. Henning, R. & Heidrich, H.G., *Biochim. Biophys. Acta*   (1974) *in press.*

6. Henning, R. & Stoffel, W., *Hoppe-Seyler's Z. Physiol. Chem. 354* (1973) 760-770.

7. Razin, N.S., Lavin, F.B. & Brown, J.R., *J. Biol. Chem. 245* (1955) 789-796.

8. Warren, L., *J. Biol. Chem., 234* (1959) 1971-1975.

9. Henning, R. & Uhlenbruck, G., *Nature New Biol. 242* (1973) 120-122.

10. Mandell, G.L., *Proc. Soc. Exp. Biol. Med. 134* (1970) 447-449.

11. Reinjngoud, D.J. & Tager, J.M., *Biochim. Biophys. Acta 297* (1973) 174-178.

12. Coffey, J.W. & de Duve, C., *J. Biol. Chem. 243,* (1968) 3255-3263.

13. Appelmans, F. & de Duve, C., *Biochem. J. 59* (1955) 426-433.

14. Gasic, G.J., Berwick, L. & Sorrentino, M., *Lab. Invest. 18* (1968) 63-71.

15. Henning, R., Plattner, H. & Stoffel, W., *Biochim. Biophys. Acta 330* (1973) [61-75.

16. Mego, J.L., Farb, R.M. & Barnes, J., *Biochem. J. 128* (1972) 763-769.

EDITOR'S NOTE.- Attention is drawn to the final page ('ADDENDUM') of the paper by D.J. Morré and co-authors, submitted subsequently to Dr. Henning's Symposium presentation as given above. As is mentioned in the final 'NOTES & COMMENTS', Dr. Henning agrees that trihexosyl ceramide could not be used as a lysosomal membrane marker: it has to be regarded as a degradation product of globoside.

21 ISOLATION OF ORGANELLES AND ENDOMEMBRANE COMPONENTS FROM RAT LIVER: BIOCHEMICAL MARKERS AND QUANTITATIVE MORPHOMETRY

D. James Morré, W.N. Yunghans, E.L. Vigil and T.W. Keenan
Departments of Botany & Plant Pathology, Biological Sciences and Animal Sciences
Purdue University
West Lafayette, Indiana 47907
and
Department of Biology
Marquette University
Milwaukee, Wisconsin 53233, U.S.A.

Biochemical, cytochemical, and morphological markers, used together, identify membranous cell components isolated from rat liver. These markers also can provide estimates of the yield and purity of fractions. For rat liver, cell components (and appropriate markers) include: rough e.r.* (membrane-bound ribosomes and glucose-6-phosphatase); G.a. (stacked cisternae and galactosyl transferase); p.m. (junctional complexes and 5'-nucleotidase); microbodies (crystalline cores, association with e.r., and urate oxidase); lysosomes (Gomori staining, single limiting membrane, and acid phosphatase); and nuclear envelope (pores and DNA). Additionally, within the endomembrane system, transitional membrane elements such as G.a. have morphological and biochemical properties intermediate between those of generating elements (such as e.r.) and end products (such as p.m.). Smooth e.r. is identified from a combination of characteristics (continuity with rough e.r., thin membranes, glucose-6-phosphatase, drug-induced mixed function oxidases). Golgi membranes exhibit these same properties but to a lesser degree.

Chemical constituents such as sterols, sphingomyelin, sialic acid, cerebrosides, gangliosides, glycoproteins, ubiquinones, vitamin A, cytochrome b_5, or cytochrome P_{450} are concentrated in components of the endomembrane system, but are not sufficiently restricted to any one component to serve as definitive markers. Common membrane constituents that are shared in varying degrees among different endomembrane components provide evidence for a functionally continuous endomembrane system. Additionally, these enzymes and constituents complicate the use of biochemical markers to analyze Golgi apparatus fractions for contamination by e.r. and/or p.m.

---

* Editor's abbreviations :- e.r., endoplasmic reticulum; p.m., plasma membrane(s); G.a., Golgi apparatus.

Within the endomembrane system the nuclear envelope, e.r., G.a., secretory vesicles, etc. exist as an interconnected, three-dimensional complex. Processes of membrane flow and membrane differentiation account for the origin and biogenesis of G.a. and p.m. through the e.r.- G.a. - vesicle (lysosome)- p.m. export route. Thus, it is not surprising that common membrane constituents are shared by more than one cell component within the endomembrane system. For endomembrane components, estimates of fraction purity from biochemical analyses can be complemented by evidence from quantitative electron microscopic morphometry. Cytochemistry is helpful to identify specific types of membranes. Thus, glutaraldehyde-resistant NADH-ferricyanide oxidoreductase of p.m. and mature secretory vesicles identify p.m. as a contaminant of microsomal fractions.

The problem of endomembrane markers is exemplified by mixed populations of lysosomes where G.a., e.r. and p.m. all are thought to contribute to the origin and biogenesis of the lysosomal membranes. Thus, the transitional nature of the endomembrane components may even preclude distinct markers which would be unique to each cell component within the endomembrane system. Until truly specific markers (if they exist) are established for endomembrane components, combinations of characteristics must be used to evaluate the yield and purity of endomembrane fractions.

Biochemical and morphological markers are utilized to identify components isolated from living cells and to provide estimates of yield and of fraction purity. Reliability and ease of application vary among different markers. Markers for mitochondria and chloroplasts (the oxygen-mediating organelles) are applicable to a wide range of cell types [1]. Those markers for endomembrane components (nuclear envelope, e.r., G.a., vesicles) and end-products of endomembrane function (p.m. and vacuolar apparatus) are not. They tend to appear in more than one cellular fraction and vary among different cell types and tissues.

Here we discuss the availability of marker constituents for endomembrane constituents with emphasis on a single tissue, rat liver. Additionally, we address the problem of why marker constituents may appear in more than one endomembrane component in this tissue. Finally, we offer evidence that quantitative morphometry may provide a suitable alternative or at least supporting data in situations where biochemical markers are of unknown or uncertain validity.

MATERIALS AND METHODS
Isolation of organelles and endomembrane components

The procedures provided outline basic techniques for isolation of endomembrane components from rat liver. Details are provided in the references.

Male Holtzman rats (Holtzman Company, Madison, Wisconsin), 40-60 days old (200-250 g), provided with standard diet and drinking water *ad libitum* are killed by decapitation and drained of blood. The livers ($\sim$ 10 g) are excised and minced with razor blades. All subsequent operations and solutions are at 0° to 4° Centrifugal forces are calculated for the middle of the tube. All solutions are prepared in deionized water.

Golgi apparatus

The procedure [2-7] is shown in Fig. 1. In the first transfer step, the yellow brown portion of the differential pellet (upper $\frac{1}{2}$ to $\frac{2}{3}$) which lies above the red to pink layer (containing whole cells) and dark brown (containing nuclei and mitochondria) layer is suspended in a small amount of the supernatant to yield a final volume of 6 ml per 10 g liver. This is accomplished with a large-bore Pasteur pipette fitted with a rubber bulb. Hand homogenization at this step causes unnecessary disruption of Golgi apparatus structure.

Following the next centrifugation step, the Golgi apparatus collect as a layer at the homogenate/1.2 M sucrose interface and are carefully removed using a Pasteur pipette. If the 1.2 M sucrose layer is turbid with microsomal fragments, special care must be taken to collect the Golgi apparatus from the interface without removing any of the 1.2 M sucrose layer.

Finally the purified Golgi apparatus are suspended in 5 to 10 ml of either 1) clear supernatant from the 100,000 g centrifugation step for preservation of morphology, or 2) distilled water for highest fraction purity, or 3) homogenization medium (enzyme assay buffer) for preservation of enzymatic activities. After concentration by centrifugation at 5,000 g, the yield is 5 to 10 mg of Golgi apparatus protein per 10 g fresh weight of liver [3]. Recovery based on measurements of galactosyl transferase [8] is 40% (30 to 70%).

Endoplasmic reticulum

The procedure [3,9] is shown in Fig. 2. The supernatant from the differential pelleting step (at 5,000 g) of the procedure for isolation of Golgi apparatus is depleted in both G.a. and mitochondria and provides a convenient starting material for isolation of e.r. fragments.

For preparation of e.r., the post-G.a. supernatant is diluted 1:3 to 1:5 with homogenization medium. Centrifugation at 10,000 g for 10 min, or at 8,000 g for 20 min, is useful to remove residual mitochondria (Fig. 2; Sorvall HB-4 rotor). The diluted supernatant is then top-loaded onto a discontinuous sucrose gradient as shown, and centrifuged (Spinco SW-27 rotor). Rough e.r. fragments are recovered from the 1.5 M/2.0 M sucrose interface. Free ribosomes and glycogen enter the 2.0 M sucrose layer and form a pellet at the bottom of the tube. Smooth e.r. fragments and transitional elements consisting of part smooth and part rough e.r. collect at the 1.3 M/1.5 M sucrose interface. The supernatant/1.3 M sucrose interface contains p.m. and

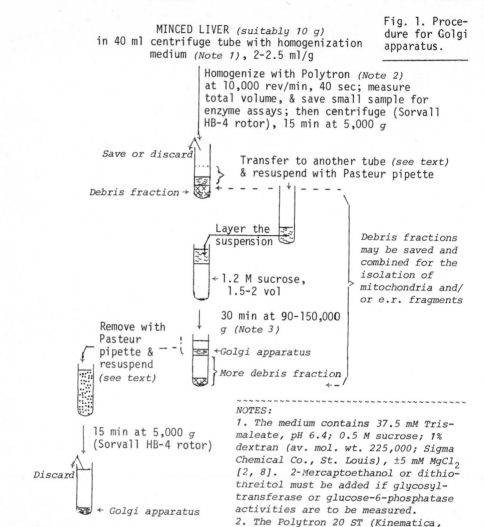

Fig. 1. Procedure for Golgi apparatus.

MINCED LIVER *(suitably 10 g)* in 40 ml centrifuge tube with homogenization medium *(Note 1)*, 2-2.5 ml/g

Homogenize with Polytron *(Note 2)* at 10,000 rev/min, 40 sec; measure total volume, & save small sample for enzyme assays; then centrifuge (Sorvall HB-4 rotor), 15 min at 5,000 *g*

Save or discard

Transfer to another tube *(see text)* & resuspend with Pasteur pipette

Debris fraction →

Layer the suspension

Debris fractions may be saved and combined for the isolation of mitochondria and/ or e.r. fragments

←1.2 M sucrose, 1.5-2 vol

30 min at 90-150,000 *g (Note 3)*

Remove with Pasteur pipette & resuspend *(see text)*

←*Golgi apparatus*

More debris fraction

15 min at 5,000 *g* (Sorvall HB-4 rotor)

Discard

← *Golgi apparatus*

NOTES:
1. *The medium contains 37.5 mM Tris-maleate, pH 6.4; 0.5 M sucrose; 1% dextran (av. mol. wt. 225,000; Sigma Chemical Co., St. Louis), ±5 mM MgCl$_2$ [2, 8]. 2-Mercaptoethanol or dithiothreitol must be added if glycosyltransferase or glucose-6-phosphatase activities are to be measured.*
2. *The Polytron 20 ST (Kinematica, Lucerne, Switzerland) is calibrated by converting flashing light reflected from the rotating, half-blacked shaft into an electrical signal by means of a photocell, and matching the frequency of an audiosignal generator to this signal with an oscilloscope.*
3. *This centrifugation is done with a swinging bucket rotor (Spinco SW-39), suitably at 30,000 rev/min.*

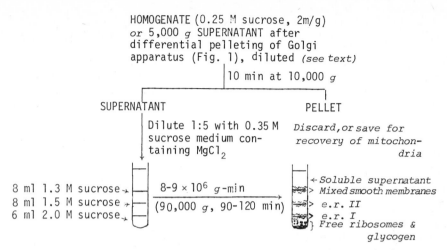

HOMOGENATE (0.25 M sucrose, 2m/g)
or 5,000 *g* SUPERNATANT after
differential pelleting of Golgi
apparatus (Fig. 1), diluted *(see text)*

10 min at 10,000 *g*

SUPERNATANT   PELLET

Dilute 1:5 with 0.35 M   *Discard, or save for*
sucrose medium con-   *recovery of mitochon-*
taining MgCl$_2$ *dria*

8 ml 1.3 M sucrose →   8-9 × 10^6 *g*-min   ← *Soluble supernatant*
8 ml 1.5 M sucrose →   ─────────────────   > *Mixed smooth membranes*
6 ml 2.0 M sucrose →   (90,000 *g*, 90-120 min)   > *e.r. II*
                                                       > *e.r. I*
                                                       } *Free ribosomes &*
                                                           glycogen

Fig. 2.   Procedure for
endoplasmic reticulum.

Each fraction, recovered with a
Pasteur pipette, is diluted 1:1
with 37.5 mM Tris-maleate buffer
containing 4 mM MgCl$_2$ or with
water, and pelleted (SW-27 rotor,
minimum 20 min at 50,000 *g*).

Golgi apparatus fragments and is discarded. Unless all residual mitochondria
are removed by initial differential pelleting, some contaminating mitochondria may appear in all three fractions. Depending on the size range of the
fragments, a variable amount of e.r. is also removed by this initial centrifugation; it may be omitted to increase yield of e.r., but mitochondrial contamination is also increased (compare Tables 3 and 4).

Fractions are recovered from the gradient as indicated in Fig. 2.   The
yield is 10-15 mg of protein per 10 g fresh weight of liver [3].   Recovery of
e.r. based on that   of glucose-6-phosphatase is about 50% [3].

Plasma membrane [10-12]

The procedure is shown in Fig. 3.   The dilution and mixing of the bicarbonate
homogenate are necessary to increase cell breakage and to help disperse the
contents of ruptured cells.

The centrifugation at 3,500 rev/min is primarily a concentrationstep
and the supernatant is left to avoid losing any of the pellet.   The portion
of the pellet rich in p.m. is loose and friable and overlies a reddish-
brown layer of packed nuclei and unbroken cells.   The supernatant solution

Legend to Fig. 3 (opposite)
Fig. 3. Procedure for plasma membrane. Volumes are approximate. Scale: about 4 rat livers (altogether 40 g of liver). To obtain the SUSPENSION IN BICARBONATE, homogenates are prepared by successive Polytron treatments (*Note 1*; 6,000 rev/min, 90 sec), each on a 3-4 g portion of liver in 15 ml 1mM bicarbonate; immediately transfer into 700 ml ice-cold bicarbonate, swirling after each addition.

remaining in the bottle is gently swirled over the pellet to resuspend the friable layer without disrupting the pellet of nuclei and debris (Fig. 3). The supernatant containing the resuspended p.m. is poured through a double layer of cheesecloth premoistened with bicarbonate (to trap fragments of the nuclear pellet) while the main mass of the pellet of nuclei and debris is held back with a rubber policeman. Resuspension of the p.m. is completed using a glass-teflon homogenizer and the volume is made to 300 ml. After the next centrifugation step (Fig. 3), most of the supernatant is aspirated and discarded leaving a few ml covering the loosely packed pellet, which is resuspended by swirling the tubes. This centrifugation and resuspension step using 300 ml is performed a total of three times, followed by further similar steps but with only four and then only two 50 ml tubes. The result is pellets of loosely packed, washed membranes.

The final pellets from the differential centrifugation steps are suspended in bicarbonate and homogenized briefly (Fig. 3). To the suspension in a 40 ml cellulose nitrate tube are added 10 ml of sucrose d = 1.3 g/ml (dropwise with constant mixing). This mixture forms the bottom layer of the sucrose gradient (final density = 1.22 g/ml). On top of the d = 1.22 layer are added 8 ml sucrose d = 1.20, 8 ml sucrose d = 1.18, 4 ml sucrose d = 1.173, 4 ml d = 1.166, and enough sucrose d = 1.16 g/ml to fill the tube [4, 12]. Sucrose solutions contain respectively 81 (d = 1.3), 53.4 (d = 1.2), 48 (d = 1.18), 46 (d = 1.173), 44 (d = 1.166), and 42.6 (d = 1.16) g per 100 ml of solution. The sucrose gradient is centrifuged for 90 min (Spinco SW-27 rotor, 26,000 rev/min, 90,000 g.) Purified p.m. is collected as a thick 'rug' from the d = 1.16/1.166 interface using a Pasteur pipette and removing as little sucrose as possible. The membranes are resuspended in 40 ml of bicarbonate and centrifuged 20 min at 8,000 rev/min (Sorvall SS-34 rotor).

Additional p.m. may be recovered from the pellet of nuclear material obtained from the first differential centrifugation (Fig. 3). The crude nuclei are resuspended with a loose-fitting glass-teflon homogenizer in 40 ml of bicarbonate. They are then centrifuged at 3,000 rev/min for 10 min to a pellet. A loose friable layer of p.m. collects over the lower mass of nuclear material. This friable layer may be combined with the other p.m. material at the density gradient step. Alternatively, the entire resuspended nuclear material, without differential pelleting, may be applied to the sucrose gradient. Here, purified nuclei are obtained as a pellet at the bottom of the sucrose gradient.

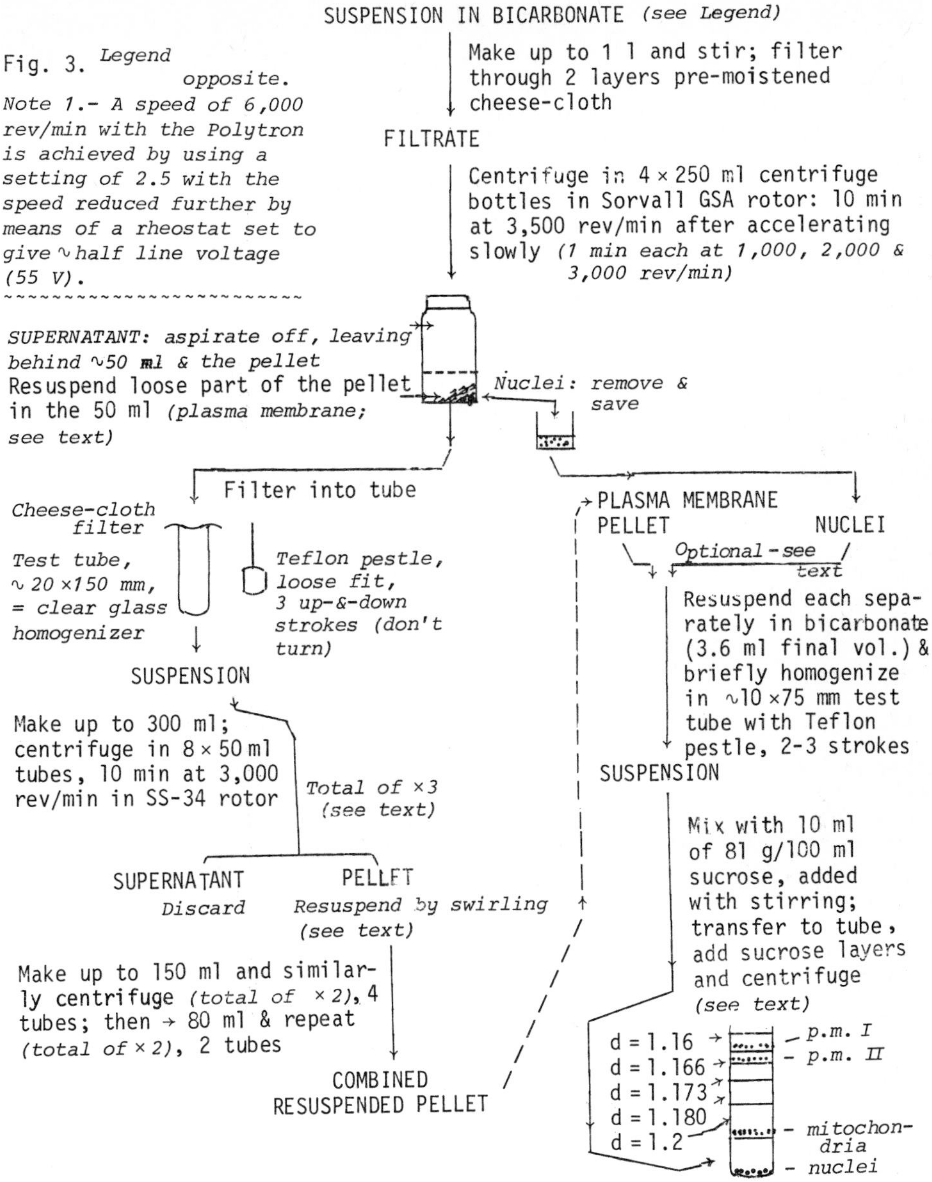

SUSPENSION IN BICARBONATE *(see Legend)*

Fig. 3. *Legend
          opposite.*
*Note 1.- A speed of 6,000
rev/min with the Polytron
is achieved by using a
setting of 2.5 with the
speed reduced further by
means of a rheostat set to
give ∿half line voltage
(55 V).*
~~~~~~~~~~~~~~~~~~~~~~~~~

Make up to 1 l and stir; filter
through 2 layers pre-moistened
cheese-cloth

FILTRATE

Centrifuge in 4 × 250 ml centrifuge
bottles in Sorvall GSA rotor: 10 min
at 3,500 rev/min after accelerating
slowly *(1 min each at 1,000, 2,000 &
3,000 rev/min)*

*SUPERNATANT: aspirate off, leaving
behind ∿50 ml & the pellet*
Resuspend loose part of the pellet
in the 50 ml *(plasma membrane;
see text)*

*Nuclei: remove &
save*

*Cheese-cloth
filter*

*Test tube,
∿ 20 ×150 mm,
= clear glass
homogenizer*

↓ Filter into tube

*Teflon pestle,
loose fit,
3 up-&-down
strokes (don't
turn)*

SUSPENSION

Make up to 300 ml;
centrifuge in 8 × 50 ml
tubes, 10 min at 3,000
rev/min in SS-34 rotor

*Total of ×3
(see text)*

SUPERNATANT   PELLET
Discard   *Resuspend by swirling
               (see text)*

Make up to 150 ml and similarly centrifuge *(total of ×2)*, 4
tubes; then → 80 ml & repeat
(total of ×2), 2 tubes

COMBINED
RESUSPENDED PELLET

→ PLASMA MEMBRANE
  PELLET   NUCLEI
  \   *Optional - see /
   \   text*
Resuspend each sepa-
rately in bicarbonate
(3.6 ml final vol.) &
briefly homogenize
in ∿10 ×75 mm test
tube with Teflon
pestle, 2-3 strokes
SUSPENSION

Mix with 10 ml
of 81 g/100 ml
sucrose, added
with stirring;
transfer to tube,
add sucrose layers
and centrifuge
(see text)

d = 1.16 →   – p.m. I
d = 1.166 →   – p.m. II
d = 1.173
d = 1.180
d = 1.2   – mitochon-
                       dria
                    – nuclei

The yield of p.m. is 1.5-4 mg plasma membrane protein/10 g wet weight of liver [12]. The recovery is 7 to 20% [12] based on a theoretical value of 1.98 mg p.m. protein/g fresh weight of liver from quantitative morphometry [13].

Nuclei, mitochondria and lysosomes [3, 14, 15]

Nuclei may be removed from the plasma membrane gradients as described above or isolated by the procedure of Franke *et al.* [15]. A procedure for isolation of quality rat liver mitochondria based on repeated differential pelleting steps is given by Morré [3]. Lysosomes were prepared as tritosomes from livers of rats injected with Triton WR-1339 [16, 17].

Assay of marker enzymes

Procedures for assay of marker enzymes are as in the references given below. Unless indicated otherwise (Table 7, Figs. 9, 11 & 12), specific activities are μmoles substrate transformed per h per mg protein. Proteins are determined by the Lowry *et al.* [18] procedure.

5'-Nucleotidase (EC 3.1.3.5), [3, 10]. *Glucose-6-phosphatase* (EC 3.1.3.9), [19]. Fractions are isolated and incubated in the presence of 5 mM 2-mercaptoethanol. *Adenosine-5'-triphosphatase* (EC 3.6.1.4), [10]. *Thiamine pyrophosphatase* (EC 3.6.1.), [20,21]. *Succinate dehydrogenase* (EC 1.3.99.1), [22]. Measured as succinate-2(*p*-indophenyl)-3-(*p*-nitrophenyl)-5-phenyltetrazolium (INT) reductase. *Cytochrome oxidase* (EC 1.9.3.1), [23]. *Monoamine oxidase* (EC 1.4.3.4), [24, 25]. *Acid phosphatase* (EC 3.1.3.2), [26].

NADH-cytochrome c *oxidoreductase* (EC 1.6.99.3) and *NADPH-cytochrome* c *oxidoreductase* (EC 1.6.2.3), [27, 28] with either cytochrome *c* or potassium ferricyanide as electron acceptor. Oxidation of NADH is followed at 340 nm when ferricyanide is the acceptor, and reduction of cytochrome *c* (Sigma Type VI) is followed at 550 nm in the other assays. All measurements are recorded with a double-beam spectrophotometer at 30°. Specific activities are calculated using molar extinction coefficients of 6.2 mM^{-1}cm^{-1} for NADH, and 21.1 mM^{-1}cm^{-1} for cytochrome *c*.

Galactosyl transferase (UDP-galactose : *N*-acetylglucosamine galactosyltransferase) (EC 2.4.1.), [8, 29]. Membrane fractions are solubilized in 1% Triton X-100 prior to assay, and the fractions are isolated and incubated in the presence of 2-mercaptoethanol. *Adenyl cyclase*, [30 as modified by 31]. *Urate oxidase* (EC 1.7.3.3), [32]. *UDP-glucuronyl transferase* (EC 2.4.1.17), [33]. *Nucleoside diphosphatases* (ADP or CDP, EC 3.6.1. ; GDP, IDP, or UDP, EC 3.6.1.6), [10]. *Alkaline phosphatase* (EC 3.1.3.1 [10]. *L-gluconolactone oxidase* (EC 1.1.3.8), [34]. *Arylsulphatase c* (EC 3.1.6.1), [35].

Specific activities may vary by a factor of two among different lots of animals during the year. Animal variation within a uniform lot is usually

± 10% or less. Specific activities within a given Table are meant to be comparable. Except for the data of Table 2, comparisons within each Table are based on data from experiments where animal variation was not a significant factor.

Enzyme cytochemistry

NADH-ferricyanide reductase

The procedure is based on studies of Karnovsky and Roots [36] and of Lukaszyk [37] for acetylcholinesterase and α-glycerolphosphate dehydrogenase, respectively. Enzymatic reduction of ferricyanide in the presence of NADH is coupled with removal of copper sulphate from solution to form an electron-dense precipitate of cupric ferrocyanide [$Cu_2Fe(CN)_6 \cdot 7H_2O$] known as Hatchett's brown. To prevent non-specific precipitation of ferricyanide by copper sulphate, the latter is chelated with sodium-potassium tartrate.

   Samples were fixed for 10 min in 0.1% glutaraldehyde at 0° to 4° in 0.05 M potassium phosphate, pH 7.2, containing 0.2 M sucrose. Blocks of tissue or portions of pellets less than 1 mm diam were selected and washed for 1-3 h at 0° to 4° in 0.05 M potassium phosphate, pH 7.2, containing 0.2 M sucrose.

   The incubation medium was freshly prepared for each experiment. The separate components were added dropwise with stirring in the order listed in Table 1. Control solutions were prepared minus NADH or minus ferricyanide or plus 0.1 ml of 1 mM p-hydroxymercuribenzoate. Incubations were for 40 min at 30°. The samples were washed 1 h in the phosphate-buffered sucrose solution at 0 to 4°. They were then fixed for 1 h in 0.05 M potassium phosphate, pH 7.2, containing 1% glutaraldehyde and 2% formaldehyde [38] followed by 2% osmium tetroxide, or were fixed in 2% osmium tetroxide directly. All fixatives were in 0.05 M potassium phosphate, pH 7.2 containing 0.2 M sucrose. The samples were dehydrated through an acetone series and embedded in Epon [39]. Thin sections were viewed directly using a Philips EM/200 without lead or uranium section staining.

5'-Nucleotidase and glucose-6-phosphatase

The procedure was that of Wachstein and Meisel [40] as modified by Schin and Clever [41].

Quantitative electron microscope morphometry

Quantitative estimations of endomembrane components were made on prints enlarged to 8 × 10 inches (20 × 25 cm) at magnifications of 25,000 according to the method of Loud [42] (see also ref. [43] and refs. cited therein). A transparency having a grid of parallel lines spaced 1 cm apart was laid over the electron micrograph. An estimate of the amount of a specific cell component in each of the isolated fractions was obtained from the number of intersec-

Table 1. Complete reaction mixture for histochemical demonstration of NADH-ferricyanide reductase.

Volume added, ml	Stock solution	Final concentration
2.0	30 mM sodium-potassium tartrate	6 mM
1.0	30 mM copper sulphate	3 mM
4.25	0.2 M potassium phosphate, pH 7.2	85 mM
0.20	100% dimethyl sulphoxide	2%
0.50	10 mM potassium ferricyanide	0.5 mM
0.15	5 mM NADH	0.075 mM
1.0	2 M sucrose	0.2 M
0.9	deionized water	

tions of the lines of the grid and the profiles or edges of membranes in thin sections of pellets. Results were expressed as the ratio of the number of membrane intersections of the cell component in question to the number of total membrane intersections ± standard deviation.

Morphometric methods can also be used to obtain approximations of the relative compositions of preparations of membrane vesicles or tubules negatively stained with phosphotungstic acid where multiple layers of membranes (such as intact mitochondria) are not encountered. Here, estimates may be obtained using a transparency having dots spaced 1 cm apart. Approximations of the composition of the preparations are derived from the number of dots which are superimposed over profiles of membranes on the electron micrographs [44].

Since identification of membranes for morphometric analyses is based on morphology, one must be able to clearly recognize all cell components to score them correctly. Those structures not clearly identified are scored as unidentified membrane fragments. Thus, morphometric analyses are useful only on preparations such as those illustrated in Figs. 4-7 where morphological details are exceptionally well preserved.

RESULTS

Estimates of fraction purity based on marker enzymes

Table 2 summarizes results based on use of several standard marker enzymes to estimate the relative purity of e.r., G.a. and p.m. fractions. Included are succinate-INT-reductase and monoamine oxidase to estimate the contributions by inner and outer mitochondrial membranes respectively. Contamination by e.r. (both rough and smooth), p.m. and G.a., and/or enrichment of G.a. and p.m. fractions, is estimated from glucose-6-phosphatase activity. P.m. contamination and/or enrichment is estimated from 5'-nucleotidase activity, and G.a. from galactosyl transferase activity.

For the purpose of estimating contamination, each of the reference fractions is assumed to be pure. Thus, for 5'-nucleotidase, p.m. has a specific activity of 77.5 μmoles AMP hydrolyzed/h/mg protein. From this, we estimate the contamination of highly purified G.a. by p.m. to be 2.5/77.5 = 0.03 or 3% on a protein basis. Similar calculations for each of the three fractions are summarized in the lower half of Table 2. They show that each of the fractions has a purity approaching 90%. Since low levels of some marker enzymes may be indigenous to each of the fractions, these assays may over-estimate contamination and provide under-estimates of fraction purity.

Estimates of fraction purity from quantitative electron microscope morphometry

With fractions where the morphology of the membranes is well preserved and where the various membranous components can be identified (Figs. 4-7), independent estimates of fraction purity are obtained from electron microscope morphometry (Table 3). For e.r. and p.m. fractions, the results are in close agreement with the biochemical analyses of Table 2. For G.a., morphometric estimates of e.r. and p.m. contamination are less than biochemical estimates although the percentage of G.a. remains about the same. Many membranes are unidentified (up to 10%) in morphometric evaluations of Golgi apparatus fractions either from thin sections (Table 3) or from negatively stained preparations [44]. Some of these unidentified membranes may be Golgi apparatus fragments while others may be contaminating cell components. For p.m. fractions, the morphometric estimates of e.r. contamination are lower than the estimates from biochemical analyses (Tables 2 & 3). This may result from mis-identification of membrane fragments, but ribosomes are retained on e.r. during isolation of p.m., and the preparations analyzed consisted mainly of large, continuous membrane sheets (Fig. 6). We do not exclude the possibilities that small vesicles of smooth e.r. adhere to the inner surface of the p.m. and are mis-identified, or that low levels of glucose-6-phosphatase activity are indigenous to rat liver p.m. (see, however, ref. [45]).

Accurate morphological assessment of a fraction depends on a number of conditions [46, 47], one of which is random sampling. With crude fractions, pellets are often stratified. A collection technique by Millipore filtration was developed to provide random sampling, yet avoid the necessity of sampling at different levels [48, 49]. For our purified fractions, some stratification is still evident although not significant (Table 4). We prefer to make biochemical and morphometric analyses on the same pellet rather than risk loss of small vesicles, ribosomes, or changes in morphology which might come from Millipore filtration. Whilst the technique offers definite advantages especially for highly heterogeneous fractions, for most of the fractions we analyze the conditions of random sampling are also met by sampling at different levels through the pellet (Table 4).

Chemical constituents as markers for endomembrane components

Plasma membranes are known to be enriched in the phospholipid *sphingomyelin* (Table 5), *sterols* (Table 6), *sialic acid* (Table 6) and various *glycolipids*

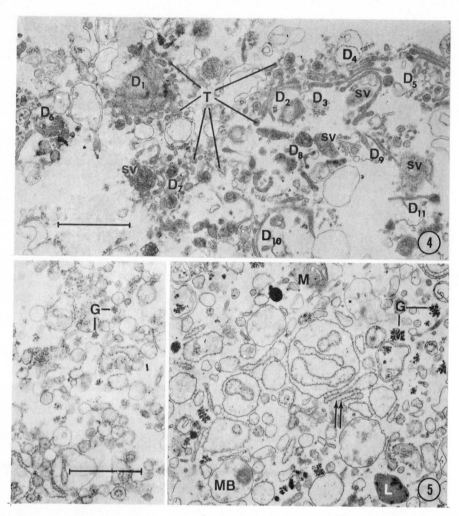

Figs. 4-7. Electron micrographs of purified cell fractions from liver. Bar = 1μm

Fig. 4. Golgi apparatus. *These preparations consist mostly of intact dic-tyosomes (D) with attached secretory vesicles (sv). The plate-like portions of cisternae are seen in face view (D_1) or sectioned tangentially (D_2 to D_{11}). The system of peripheral tubules (T), when sectioned tangentially, accounts for the many, small vesicular profiles of the preparation.*

Table 2. Estimates of fraction purity of isolated cell fractions based on analyses of marker enzyme activities. The ENZYME ACTIVITY units are μmoles product/h/mg protein. In calculating % *CONTAMINATION*, each of the reference fractions is assumed to be of absolute purity. Estimates of urate oxidase (microbodies) and acid phosphatase (lysosomes) show that these cell components account for less than 1% of the total protein of the rough e.r., G.a. and p.m. fractions [6, 9, and unpublished results].

ENZYME ACTIVITY, *units*	mitochondria	*PURIFIED FRACTIONS* rough e.r.	G.a.	p.m.
Succinate-INT-reductase	25	0.2	0.2	1.3
Monoamine oxidase	33	-	0.5	-
Glucose-6-phosphatase	-	34.0	3.0	1.3
Galactosyl transferase	-	0.003	0.22	trace
5'-Nucleotidase (AMP)	-	1.9	2.5	77.5
CONTAMINATION, %				
Mitochondria -------------------------------		1	1	5
Rough endoplasmic reticulum ------------		-	9	4
Golgi apparatus -----------------------		1	-	0
Plasma membrane -----------------------		3	3	-
TOTAL ----------------------------------		5	13	9
FRACTION PURITY, %		95	87	91

Fig. 5 *(opposite).* Rough endoplasmic reticulum. Both fractions were collected from the 1.5 M sucrose/2.0 M sucrose interface of a preparative sucrose gradient for e.r. (Fig. 2). The preparation on the *left* is a 1 : 10 homogenate centrifuged at 10,000 rev/min for 20 min (Sorvall HB-4 rotor, 8,000 *g*) to remove residual mitochondria prior to gradient centrifugation. *The fraction is typical rough microsomes consisting of small membrane vesicles with attached ribosomes and 1 to 2% mitochondrial contamination (Tables 2 & 3). Larger e.r. fragments are removed by the 8,000 g centrifugation and the yield is reduced accordingly.*

The preparation on the *right* is from the 5,000 *g* supernatant of the differential pelleting step for isolation of Golgi apparatus. The supernatant is applied to the gradient without first removing residual mitochondria and large e.r. fragments. *Fractions prepared in this manner contain 5 to 15% mitochondria (M) as well as about 1% each of lysosomes (L) and microbodies (MB) (Table 4). However, the e.r. is isolated as large cisternal fragments, some in stacked arrays* (double arrows). *Rosettes of glycogen (G) contaminate both fractions.*

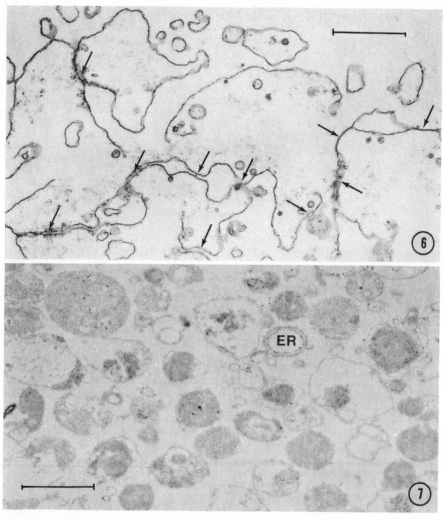

Fig.6. Plasma membrane. *Junctional complexes (ARROWS) identify the long,
undulating profiles as p.m. Many of the small vesicles are derived from p.m.
but some may be fragments of smooth e.r. which adhere or attach to the p.m. sheets.*

Fig. 7. Lysosomes. The lysosomes were prepared as tritosomes from livers
of rats injected with Triton WR-1139 according to Trouet [17] and Leighton
et al. [16]. *Small fragments of endoplasmic reticulum (ER) are the only con-*

[CONTINUED OPPOSITE

(Fig. 8) and *glycoproteins*. As is illustrated by Tables 5 & 6 and Fig. 8, these constituents are not confined to the p.m., but are present in e.r., G.a. and p.m. The amount present on a protein basis is e.r. < G.a. < p.m. [6, 50-53]. For these constituents to be present in G.a. fractions as the result of p.m. contamination, this would have to amount to ∿50% instead of 1-3% as shown by enzymatic (Table 2) and morphometric (Table 3) criteria. Therefore we conclude that sphingomyelin, sterols, protein-bound sialic acid and gangliosides, even though concentrated in p.m., cannot be used as p.m. markers [e.g. 54] since they are present throughout the endomembrane system (e.r. < G.a. < p.m.).

A similar situation is encountered with chemical constituents concentrated in e.r., such as cytochrome b_5, cytochrome P_{450} + P_{420} (Table 7) and phosphatidyl choline (Table 5). *Phosphatidyl choline* (lecithin) is present in all cellular membranes even though it is present in greatest proportion in total phospholipid fractions from e.r. [50, 53]. The *microsomal cytochromes* (b_5, P_{450}) occur in G.a. at levels ∿30% of that in e.r. fractions [7, 55, 56; Table 7]. *Iron* (haem + non-haem) is present in both G.a. and e.r. fractions in about equal amounts on a protein basis (Table 7). G.a. preparations contain only small amounts of RNA [53]. *RNA* determinations provide an approximation of ribosomal contamination but do not provide a valid estimate of contamination even by rough e.r. membranes since ribosomes may be lost from the membranes during preparation of the fractions. Other chemical constituents such as *ubiquinone* [57], *vitamin A* [58] and *vitamin K* [59] are concentrated in G.a. membranes but not specifically localized there.

Table 3. Purity of cell fractions by quantitative morphometry.

CELL COMPONENT	*PROFILES per 100 profiles* ± *standard deviation*		
	ndoplasmic reticulum	Golgi apparatus	lasma membrane
endoplasmic reticulum	95.6 ± 2.6	3.2 ± 1.4	1.3 ± 1.2
Golgi apparatus	0.3 ± 0.2	85.0 ± 3.8	0.1 ± 0.1
plasma membrane	0.7 ± 0.7	1.1 ± 0.9	91.1 ± 6.5
mitochondria	1.2 ± 1.0	0.8 ± 0.4	4.8 ± 4.7
lysosomes	0.5 ± 0.3	0.6 ± 0.4	*None detected*
microbodies	0.2 ± 0.2	0.3 ± 0.3	*None detected*
unidentified	1.5 ± 0.6	9.0 ± 1.3	2.7 ± 1.9

Legend to Fig. 7, CONTINUED
 taminants. Remnants of the original lysosomes and small ferritin-like particles are seen within the swollen lysosomes filled with Triton WR-1339.

Table 4. Distribution of cell components, comparing top, middle and bottom of pellets of rough endoplasmic reticulum (rough e.r.), Golgi apparatus and plasma membranes (p.m.) of rat liver as estimated by quantitative morphometry. *The e.r. was from the 5,000 g supernatant of the differential pelleting step for isolation of G.a. without first removing residual mitochondria.* The deviations given are among 6 micrographs yielding ∿500 intersections with membrane profiles each averaging 3 preparations of comparable purity. This method of data presentation was chosen to show that variation within a given region of the pellet exceeded the variation between regions.

CELL COMPONENT	*PROFILES per 100 profiles ± standard deviation*								
	endoplasmic reticulum			Golgi apparatus			plasma membrane		
	TOP	*MIDDLE*	*BOTTOM*	*TOP*	*MIDDLE*	*BOTTOM*	*TOP*	*MIDDLE*	*BOTTOM*
e.r.	85 ± 5	86 ± 6	87 ± 2	5 ± 3	5 ± 3	6 ± 2	0.5[*]	2 ± 1	3 ± 3
G.a.	1 ± 1	1 ± 1	1 ± 1	86 ± 4	88 ± 5	81 ± 4	0 ± 0	0.5[*]	0.5[*]
p.m.	1 ± 1	1 ± 1	0 ± 0	1 ± 1	0 ± 0	3 ± 3	91 ± 5	92 ± 5	89 ± 7
mitochondria	7 ± 2	4 ± 2	9 ± 3	0 ± 0	0.5[*]	1.5 ± 1.5	4 ± 4	2 ± 2	6 ± 6
lysosomes	1 ± 1	1 ± 1	1 ± 1	0.5[*]	0.5[*]	0.5[*]	0 ± 0	1 ± 0	0 ± 0
microbodies	0.5[*]	0.5[*]	0.5[*]	<0.5	<0.5	<0.5	0 ± 0	0 ± 0	0 ± 0
unidentified	3 ± 1	3 ± 2	2 ± 1	7 ± 1	6 ± 2	8 ± 4	5 ± 4	3 ± 2	2 ± 2

[*] 0.5 ± 0.5

Table 5. Phospholipid composition of endomembranes isolated from rat liver *as % of total lipid P recovered*. Data from Keenan & Morré [50] or (nuclear envelope) Kleinig [60].

PHOSPHOLIPID	Nuclear envelope	Rough endoplasmic reticulum	Golgi apparatus	Plasma membrane
Phosphatidylcholine	54	61	45	40
Phosphatidylethanolamine	22	19	17	18
Sphingomyelin	5	4	12	19
Phosphatidylinositol	8	9	9	7
Phosphatidylserine	7	3	4	4
Others[*]	4	5	12	12

[*] *composed predominantly of lysophosphatides and phosphatidic acid*

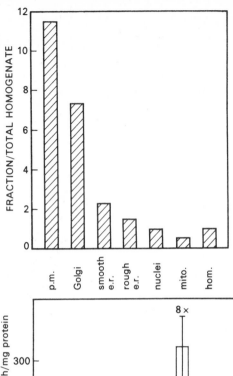

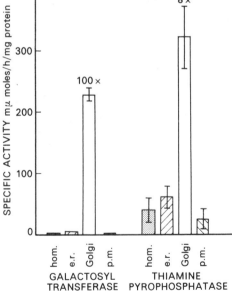

Fig. 8. Distribution of ganglio-sides among purified endomembrane fractions from rat liver (data from Keenan *et al.* [51]). Results are presented to show relative en-richment, the total homogenate (hom) being given a value of 1, = 3.4 n-moles sialic acid/mg protein. *Abbre-viations as on p. 195; also* mito = mitochondria; Golgi = Golgi appara-tus *[Editor's solecism; also on Figs. 9 & 12].*

Fig. 9. Specific activities ± mean absolute deviations, in a comparison of the distribution among cell fractions of galactosyl transferase and thiamine pyrophos-phatase, their enrichment in Golgi apparatus relative to total homo-genate (hom) being 100-fold and only 8-fold respectively. *Enzymes in liver that hydrolyze thiamine pyrophosphate are present in e.r.* [21] *and also in p.m.* From ref.[21].

(Text, CONTINUED)

Enzyme activities as markers for endomembrane constituents

Galactosyl transferase (UDP-galact-ose : *N*-acetylglucosamine galact-osyltransferase) may be a valid marker for Golgi apparatus since this activity is extremely low in e.r. fractions and cannot be de-tected in p.m. (Fig. 9). Up to 70% of the total galactosyltrans-ferase of the homogenate is recov-ered in the G.a. fractions, and we use this activity to estimate yield and recovery of Golgi appa-ratus [8]. However, not all gly-cosyltransferases concentrated in G.a. fractions constitute abso-lute markers.

The *glycosphingolipid trans-*

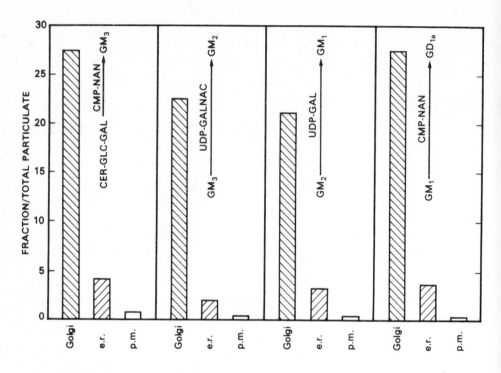

Fig. 10.   Distribution of several sugar-nucleotide : glycolipid transferase activities among purified endomembrane fractions from rat liver. *Abbreviations besides* CMP *and* CDP: CER, ceramide; CMH, glucosyl ceramide; CDH, lactosyl ceramide; GM_3, *N*-acetylneuraminylgalactosylglucosylceramide; GM_2, *N*-acetylgalactosaminyl-(*N*-acetylneuraminyl)-galactosylglucosylceramide; GM_1, galactosyl-*N*-acetylgalactosaminyl-(*N*-acetylneuraminyl)-galactosylglucosylceramide; GD_{1a}, *N*-acetylneuraminylgalactosyl-*N*-acetylgalactosaminyl-(*N*-acetylneuraminyl)-galactosylglucosylceramide; NAN, *N*-acetylneuraminic acid; GAL, galactose; GLC, glucose; GALNAC, *N*-acetylgalactosamine.   *G.a. fractions contained all glycosyl transferases which catalyze in vitro biosynthesis of* GM_3, GM_2, GM_1, *and* GD_{1a} *from their respective precursors. Relative to total particulate fractions, these transferases were enriched 22 to 27 times in Golgi apparatus. Rough e.r. fractions also contained these glycolipid glycosyl transferases with specific activities 2 to 4 times those of total particulate fractions; p.m. fractions had negligible activities. The combined G.a. and e.r. fractions accounted for more than 80% of the total homogenate glycolipid glycosyl transferase activities.* Data from Keenan *et al.* [61].

Table 6. Cholesterol and sialic acid contents of endomembranes isolated from rat liver [6,12,50].

FRACTION	Cholesterol mg/100 mg of lipid	Sialic acid μmoles/100 mg of protein
e.r.	3.7	0.7
G.a.	7.6	2.1
p.m.	13.1	3.3

Table 7. Summary of e.r. activities or constituents of G.a.-rich fractions [56]

CONSTITUENT	Specific activity or Amount /mg protein	RATIO, G.a. /e.r.
Glucose-6-phosphatase	1.3 μmoles/h	0.13
NADH-cyt. c reductase	6.3 μmoles/h	0.14
NADPH-cyt. c reductase	0.6 μmoles/h	0.12
Cytochrome b_5	0.18 mμmoles	0.30
Cyt. $P_{450} + P_{420}$	0.21 mμmoles	0.33
Total iron	36.4 mμatoms	1.03
UDP-glucuronyl transferase (bilirubin)	1.0 mμmoles/h	0.16
Arylsulphatase c	18.0 mμmoles/h	0.23
L-gulonolactone oxidase	87.0 mμmoles/h	0.22

ferases appear to be present in both Golgi apparatus and e.r. [61; Fig. 10], which together account for >80% of the activity of the total homogenates [61]. In this regard, balance sheets [46, 47] provide information critical in deciding the assignment of a given enzyme activity to a subcellular component.

Thiamine pyrophosphatase may be a G.a. marker for some tissues [62] but not for rat liver. This activity has been demonstrated to be present in rat-liver e.r. by both biochemical [9, 21; Fig. 9] and cytochemical [21] criteria [*see also* 63, 64]. Biochemical determinations also indicate that purified p.m. preparations have low levels of this activity [18; Fig. 9].

Adenyl cyclase, often assumed to be restricted to p.m. [*see, however,* 65] is also present in other membranes in rat liver (Figs. 11 & 12). Both the glucagon- and the fluoride-stimulated components are present in e.r. fractions at nearly the same levels as in p.m. G.a. fractions show an intermediate level of glucagon-stimulated adenyl cyclase but little or no fluoride-stimulated activity.

5'-Nucleotidase activity is endogenous to G.a. membranes. This is shown in Fig. 13 from a preparation of secretory vesicles isolated according to the procedure of Merritt and Morré [66]. The section is from the 'heavy' end of the pellet so that p.m. fragments and a swollen mitochondrion are included. The secretory vesicles, which are isolated by release from intact and purified Golgi apparatus, are identified from their contained lipoprotein particles. As shown in the micrograph (Fig. 13), some of the membranes of the secretory vesicles show 5'-nucleotidase activities comparable to that of the p.m. fragments. Not all the vesicles show the activity, nor is the acti-

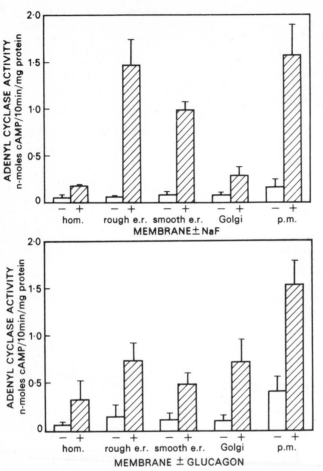

Fig. 11. Adenyl cyclase activities of membrane fractions isolated from rat liver and assayed in the absence (-) or presence (+) of sodium fluoride. *Abbreviations:* hom, *total homogenate; otherwise as on first text page.* The fluoride was 10 mM. Activity is expressed as n-moles cyclic AMP formed/10 min/mg protein ± mean absolute deviation. The radioactivity of AT(α^{32})P was 1-2 mCi/m-mole or ~500,000 cts./min AT^{32}P in 3.0 mM ATP. The incubations were at 37°. From W.N. Yunghans & D.J. Morré [in preparation].

Fig. 12. Adenyl cyclase activity of membranes isolated from liver and assayed in the absence (-) or presence (+) of glucagon (1µg/ml) without sodium fluoride. Other conditions as in Fig. 11. From W.N. Yunghans & D.J. Morré [in preparation].

vity restricted to either mature or immature vesicles. As reported by Farquhar *et al.* [67], the 5'-nucleotidase activity of isolated Golgi apparatus is most evident in the secretory vesicles. The activity is not due to lysosomal acid phosphatases since it is not inhibited by tartrate (Table 8). Based on cytochemical studies, Widnell [68] concludes that 5'-nucleotidase is present in e.r. Interestingly, the 5'-nucleotidase of tritosomes was not inhibited by tartrate, suggesting an endogenous 5'-nucleotidase associated with the tritosome membrane (Table 8). In contrast, the glucose-6-phosphatase activity of tritosomes was markedly inhibited by tartrate (Table 8).

Table 8. Summary of treatments to distinguish among 5'-nucleotidase, glucose-6-phosphatase, and lysosomal acid phosphatase activities of endomembrane fractions. Units of specific activity are μmoles P_i released/h/mg protein [69]. Averages are from two preparations. Glucose-6-phosphatase assays were at pH 6.5 except for those in the column headed *pH 5.0*. In the final column the pH was first adjusted to pH 5.0, and then assays were at pH **6.5**.

| FRACTION | 5'-NUCLEOTIDASE | | GLUCOSE-6-PHOSPHATASE | | | |
	CONTROL	+10 mM tartrate [70]	CONTROL	+20 mM tartrate	pH 5.0	pH 5.0 → pH 6.5
Total homogenate	2.2	2.0	3.4	3.4	3.8	0.8
rough e.r.	-	-	18.0	17.5	20.0	4.5
smooth e.r.	-	-	21.4	21.4	36.4	6.1
G.a.	3.4	3.1	2.6	2.5	2.7	2.3
tritosomes	16.8	16.0	15.5	3.7	31.0	7.8
p.m.	50.0	46.0	1.4	1.3	1.6	0.1

Monovalent ion-stimulated ATPases remain as candidates for p.m.-specific markers for rat liver. We have been unable to consistently demonstrate this activity in Golgi apparatus (Table 9; J.H. Elder, unpublished results; ref. [71]) but do find low levels of activity in e.r. fractions (Table 9). Previously, Ernster and Jones [72] reported monovalent ion-stimulated ATPases in microsomes. The extent to which the ion-stimulated ATPase activities of e.r. fractions may be ascribed to contamination by p.m. fragments remains undetermined.

Glucose-6-phosphatase is one of the best marker enzymes for e.r. of rat liver because of its concentration in e.r. and the ease by which it is estimated [73]. A major difficulty arises in using glucose-6-phosphatase activity to assess e.r. contamination of Golgi apparatus fractions since glucose-6-phosphatase activity may be endogenous to G.a. membranes [56]. Purified Golgi apparatus were unstacked and fragmented by the procedure of Ovtracht *et al.* [44], and fractions consisting of secretory vesicles, boulevard périphérique (smooth e.r. of the Golgi apparatus zone) and saccules (plate-like portions of cisternae) were prepared (Fig. 14); rough e.r. contaminants were collected at the bottom of the gradient. Under these conditions, the specific activity of the purified Golgi apparatus membranes for hydrolysis of glucose-6-phosphate was still about 50% of that of the intact apparatus. Thus far, we have been unable to localize the glucose-6-phosphatase of Golgi apparatus cytochemically.

The glucose-6-phosphatase activity of isolated Golgi apparatus is not inhibited by tartrate, which shows that it is not due to lysosomal phosphatase; nor is the activity markedly different at pH 5.0 (the optimum for lysosomal hydrolases) as compared with pH 6.5 (Table 8). At the suggestion of Dr. P.J. Jacques (Louvain), the pH of the membrane suspensions was adjusted to 5.0, a treatment which inactivates glucose-6-phosphatase, and the activity was again measured at pH 6.5. Under these conditions, the glucose 6-phosphatase activity of the p.m., e.r., and total homogenate fractions was reduced by 70% or

Table 9. Monovalent and divalent ion-stimulated ATPase activities of endomembrane components isolated from rat liver. Data of J.H. Elder [unpublished]. Assay conditions were as described by Emmelot *et al.* [10]. Specific activity units as in heading to Table 8. Results are from duplicate determinations from each of two different membrane preparations. *The rough and smooth e.r. fractions correspond to the endoplasmic reticulum-I and endoplasmic reticulum-II of Morré [3]; the latter fraction contains the smooth e.r. of the hepatocyte, but rough e.r., G.a. fragments and p.m. fragments are also present.*

FRACTION	SPECIFIC ACTIVITY AFTER ADDITION OF ION			
	Mg^{2+}	Ca^{2+}, Mg^{2+}	K^+, Mg^{2+}	Na^+, K^+, Mg^{2+}
Total homogenate	3.9 ± 0.1	4.1 ± 0.3	4.0 ± 0.1	4.2 ± 0.1
mitochondria	15.4 ± 1.1	16.2 ± 2.7	15.6 ± 2.4	13.6 ± 0.5
rough e.r.	8.0 ± 0.4	6.5 ± 0.1	8.4 ± 0.4	9.1 ± 0.1
smooth e.r.	7.2.± 0.3	6.6 ± 0.2	7.2 ± 0.1	9.4 ± 0.3
G.a.	7.5 ± 0.3	6.9 ± 0.2	7.5 ± 0.2	6.2 ± 0.5
p.m.	38.9 ± 0.5	33.8 ± 0.6	41.2 ± 0.4	46.0 ± 1.5

Fig. 13 *(opposite).* Cytochemical demonstration of 5'-nucleotidase in a preparation of secretory vesicles (SV) isolated from Golgi apparatus according to the procedure of Merritt & Morré [66](unpublished electron micrograph from a study by W.D. Merritt and D.J. Morré). The adaptation of the Gomori [24] procedure for cytochemical localization of phosphatases was that of Wachstein & Meisel [40] as modified by Schin & Clever [41]. The preparation was photographed near the 'heavy' end of the pellet so that contaminating mitochondrial (M) and plasma membrane fragments (PM) were included for reference. Secretory vesicles were identified by the osmiophilic low density lipoprotein particles within the vesicle interiors [66]. Bar = 1 µm. *Limiting membranes of some but not all of the secretory vesicles (SV) exhibited 5'-nucleotidase activity comparable to that of the plasma membrane. These findings confirm other studies which show that 5'-nucleotidase is a constituent endogenous to secretory vesicles of rat liver Golgi apparatus [67].*

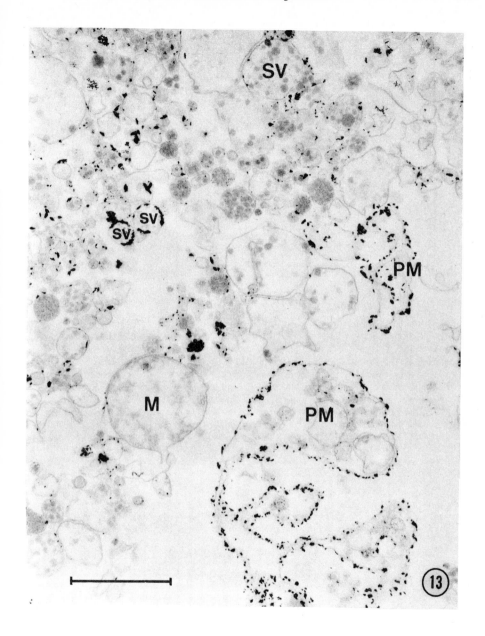

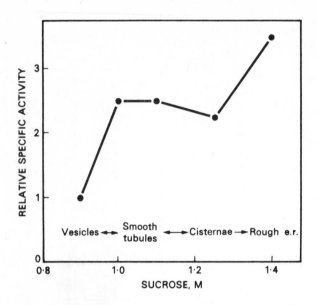

Fig. 14. Distribution of glucose-6-phosphatase among Golgi apparatus sub-
fractions from rat liver isolated by sucrose gradient centrifugation of unstacked and fragmented dictyosomes according to the procedure of Ovtracht
et al. [44]. From Morré *et al.* [56].
Fractions were obtained which were enriched in secretory vesicles, periphe-
ral tubules (boulevard périphérique) and plate-like central portions of
cisternae. Contaminating rough e.r. fragments were separated from G.a.
cisternae, yet the G.a. membranes contained glucose-6-phosphatase at levels
approximately 50% those of carefully isolated fractions of intact Golgi
apparatus [44]. Thus, the possibility remains that a significant part of
the glucose-6-phosphatase activity of G.a. fractions is activity intrinsic
to some component of the Golgi apparatus. Certainly a portion, at least
50%, is due to contamination with e.r.

more. About 50% of the glucose-6-phosphatase activity of the tritosome fraction was inhibited by this treatment, but the activity of G.a. fractions
was merely reduced to a basal level of about 2 μmoles/h/mg protein.

The findings of Table 8 agree with the unstacking experiments of Fig.
14 in that perhaps no more than 50% of the glucose-6-phosphatase activity of
isolated Golgi apparatus is due to contamination by e.r. The origin of the
basal activity remains unclear, except that it is apparently not due to a
lysosomal acid phosphatase. The basal activity is not stimulated by assay at

Table 10. Nicotinamide adenine dinucleotide dehydrogenases of Golgi apparatus and rough endoplasmic reticulum of rat liver. (Unpublished data of C.M. Huang, Purdue University.)

ACTIVITY	SPECIFIC ACTIVITY		
	rough e.r.	G.a.	G.a. rough e.r.
NADH-cytochrome c oxidoreductase, μmoles cyt. c reduced/min/mg protein	1.26	0.30	0.24
NADH-ferricyanide oxidoreductase, μmoles NADH oxidized/min/mg protein	3.47	1.44	0.41
NADPH-cytochrome c oxidoreductase, μmoles cyt. c reduced/min/mg protein	0.101	0.014	0.14
Glucose-6-phosphatase, μmoles P_i liberated/h/mg protein	33.8	4.0	0.12

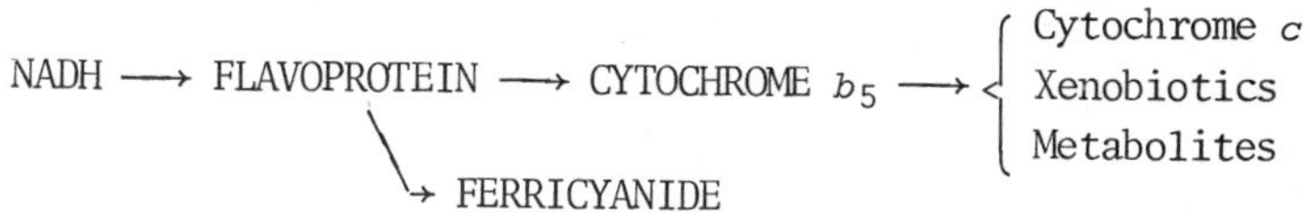

$$\text{NADH} \longrightarrow \text{FLAVOPROTEIN} \longrightarrow \text{CYTOCHROME } b_5 \longrightarrow \begin{cases} \text{Cytochrome } c \\ \text{Xenobiotics} \\ \text{Metabolites} \end{cases}$$
$$\searrow \text{FERRICYANIDE}$$

Fig. 15. Schematic representation of reactions catalyzed by the microsomal NADH-cytochrome c (ferricyanide) oxidoreductases.

pH 5.0, is not inhibited by tartrate, and does not show structure-linked latency (Table 8, and unpublished results).

Antimycin a-*insensitive NADH-cytochrome* c *(ferricyanide) oxidoreductase* is a second enzyme which has been used as a marker for e.r. [74]. As shown in Table 10, this activity is also present in Golgi apparatus especially when electrons are transferred to ferricyanide to bypass the cytochrome b_5 of the membrane (Fig. 15). Studies of Huang *et al.* ([75], & in preparation) show similar properties for this activity in both the e.r. and G.a. fractions. Recent findings show this activity to be present in p.m. as well [76]. The p.m. activity is resistant to glutaraldehyde and differs in K_m from the activity of e.r. and G.a. Cytochemically, NADH-ferricyanide oxidoreductase has been demonstrated throughout the endomembrane system, with a glutaraldehyde-resistant component in p.m. and mature secretory vesicles of the Golgi apparatus (Figs. 16-20).

Glutaraldehyde-resistant NADH-ferricyanide oxidoreductase: a cytochemical marker for plasma membrane, used for quantitative morphometry

Under normal conditions of incubation for cytochemistry (Fig. 16), grains of copper ferrocyanide indicating NADH-ferricyanide oxidoreductase are found over e.r., outer and inner mitochondrial membrane, and Golgi apparatus. Surprisingly, however, the most intense reactivity is shown by membranes of mature secretory vesicles of the Golgi apparatus (Fig. 16) and by p.m. (Fig. 17).

If the time of fixation with 0.1% glutaraldehyde is increased from 10 min to 30 min, the reaction over the mature secretory vesicles is retained but that over e.r. is diminished (Fig. 17). By increasing the conditions of fixation and the concentration of substrate, the reaction of p.m. and mature secretory vesicles is intensified (Fig. 18).

Biochemical studies [76] confirm the existence of a NADH-ferricyanide oxidoreductase associated with p.m. which is resistant to glutaraldehyde. It has a K_m of about 100 µM with respect to ferricyanide, which contrasts with the K_m for e.r. and G.a. of about 23 µM.

In preliminary studies, the glutaraldehyde-resistant NADH-ferricyanide oxidoreductase has been used to help assess p.m. contamination of e.r. and G.a. fractions. Under conditions of 30 min fixation and 5 mM ferricyanide, p.m. fragments are recognized in isolated e.r. fractions as membranes containing dark deposits of copper ferricyanide (Fig. 19). The large vesicle illus-

---

Figs. 16-20. Electron micrographs illustrating the cytochemical localization of NADH-ferricyanide oxidoreductase activities of endomembranes in liver (Figs. 16, 17 & 18) and cell fractions (Figs. 19 & 20). Experimental conditions are summarized in Table 1 and detailed in the text.

Fig. 16. Liver fixed for 10 min with 0.1% glutaraldehyde and incubated for NADH-ferricyanide oxidoreductase with a ferricyanide concentration of 0.5 mM. Bar = 0.5 µm.
Electron-dense grains of cupric ferrocyanide (Hatchett's brown) show reductase activity (single arrows) in outer membranes of mitochondria (M), rough e.r. (RER) and Golgi apparatus (GA). The double arrows mark lipoprotein particles of secretory vesicles of the Golgi apparatus which are characteristically osmiophilic and should not be confused with the cytochemical reaction product which is especially marked over membranes of mature secretory vesicles (SV$_2$).

Fig. 17. Liver fixed for 30 min with 0.1% glutaraldehyde and incubated as for Fig. 16. Bar = 0.5 µm.
The appearance is similar to that of Fig. 16 except that with the longer fixation the reaction product over mitochondrial (M) outer membranes and rough endoplasmic reticulum (RER) is reduced. Reaction product over mature secretory vesicles (SV) of the Golgi apparatus and elements of smooth e.r. (SER) near the p.m. (PM) still show heavy deposits of reaction product. Lipoprotein particles (arrows) with a diameter larger than those of mature secretory vesicles of the Golgi apparatus identify the elements labelled SER as derived from or associated with smooth e.r. [79, 109].

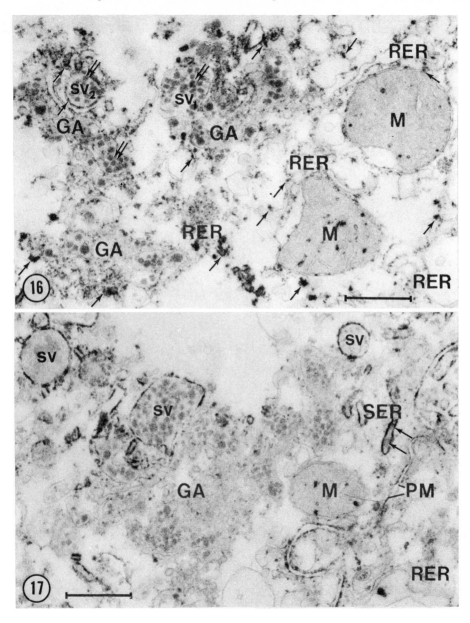

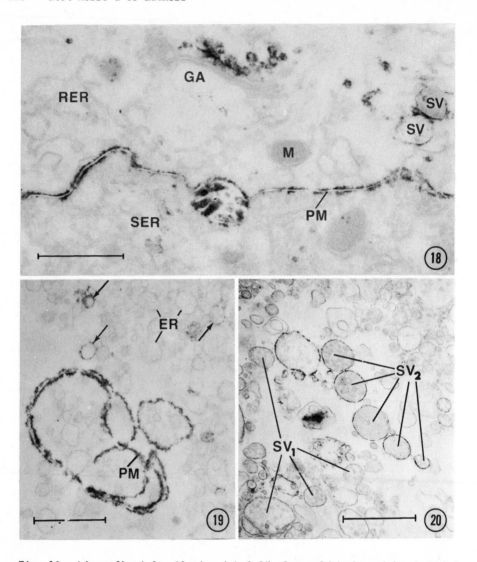

Fig. 18. Liver fixed for 30 min with 0.1% glutaraldehyde and incubated for NADH-ferricyanide oxidoreductase with a ferricyanide concentration of 5 mM to intensify reactivity of p.m. (PM) and mature membranes and secretory vesicles (SV) of the Golgi apparatus (GA). Bar = 0.5 µm.   *continued*

trated would be recognized as p.m. even without the cytochemical reaction but the small vesicles (*arrows*) would not. The cytochemical method is also potentially useful for differentiating between mature and immature secretory vesicles in preparations of isolated secretory vesicles from rat liver (Fig. 20).

Morphometric data based on information from the glutaraldehyde-resistant NADH-ferricyanide oxidoreductase are summarized in Table 11. The findings show that p.m. contamination of rough e.r. fractions is nearly that estimated from evaluations of 5'-nucleotidase, suggesting that p.m. fragments are a major contributor to the 5'-nucleotidase activity of isolated e.r. fractions. In isolated G.a preparations, the NADH-ferricyanide oxidoreductase resistant to glutaraldehyde is concentrated in membranes of mature secretory vesicles. The contamination by p.m. is much less than estimated from 5'-nucleotidase activity because 5'-nucleotidase is present as an indigenous constituent of the secretory vesicle membrane as well.

SUMMARY OF FRACTION PURITY BASED ON MARKER ENZYMES AND QUANTITATIVE MORPHOMETRY

In Table 12 the findings discussed in this report are summarized by correcting the values for each of the isolated fractions in the light of results from morphometry and from the treatments set down in Table 8. There is no need to revise the values for rough e.r. or p.m. since morphometric (Table 3)

LEGEND to Fig. 18, continued

Rough e.r. (RER) and membranes of mitochondria (M) are unreactive under these conditions of fixation. Portions of smooth e.r. (SER) near the p.m. or associated with the Golgi apparatus are reactive. Note in this and the previous Figure that the plasma membranes show a reactivity similar to that of mature secretory vesicles of the Golgi apparatus.

Fig. 19. Portion of a fraction of rough e.r. fixed and incubated for NADH-ferricyanide oxidoreductase as in Fig. 18. Bar = 1 µm.
A large p.m. (PM) fragment and several smaller vesicles (arrows) show the glutaraldehyde-resistant enzyme while e.r. (ER) vesicles do not. This difference in cytochemical reactivity between p.m. and e.r. provides a tentative basis for assessing contamination of e.r. and Golgi apparatus fractions by fragments of p.m. as shown in Table 11.

Fig. 20. Portion of a secretory vesicle fraction isolated by the procedure of Merritt & Morré [66] and fixed and incubated for NADH-ferricyanide oxidoreductase as in Fig. 18. Bar = 1 µm.
Differences in cytochemical reactivity of the limiting membranes of the vesicles permit positive identification of membranes of immature (SV_1) and mature (SV_2) secretory vesicles. Other criteria for distinguishing among the two classes of secretory vesicles, based on the appearance of the vesicle contents, are given by Merritt & Morré [66].

Table 11. Purity of cell fractions from rat liver by quantitative morphometry. Identification of plasma membrane (p.m.) was based on cytochemistry
using the glutaraldehyde-resistant NADH-ferricyanide oxidoreductase. Averages
are of four preparations for each cell fraction.

*Although based on preparations different from those of Tables 3 and 4, the
findings show that small p.m. fragments in e.r. fractions were being scored
as unidentified fragments or as e.r., while small p.m. fragments were generally absent from Golgi apparatus fractions. Correct identification of p.m.
fragments in p.m. preparations is substantiated.*

	PROFILES/100 profiles ± S.D.		
CELL COMPONENT	e.r.	G.a.	p.m.
e.r.	92 ± 1	1 ± 1	1 ± 1
G.a.	0 ± 0	90 ± 4	0 ± 0
p.m.	2 ± 1	1 ± 1	92 ± 4
mitochondria	3 ± 3	2 ± 1	3 ± 3
lysosomes	0.6 ± 0.3	0.3 ± 0.2	0 ± 0
microbodies	trace	0.3 ± 0.3	0 ± 0
unidentified	3 ± 1	4 ± 1	4 ± 3

and marker enzyme (Table 2) assays agree. The major changes occur for Golgi apparatus. No more than 50% of the glucose-6-phos-phatase in the G.a. is due to e.r. contamination (Table 8; Fig. 9). To assess p.m. contamination, cytochemical evidence (Table 11) is used rather than total 5'-nucleotidase activity. At least a portion of the latter is associated with membranes of secretory vesicles (Fig. 13) and is not the result of lysosomal acid phos-phatases (Table 8) or of direct contamination by p.m. Also, G.a. fractions lack significant monovalent ion-stimulated ATPase activity (Table 9), whereas this activity is concentrated in the p.m. [10]. With these revisions, the purity of the isolated G.a. fractions is found to be better than 90% (Table 12).

DISCUSSION

What constitutes a marker enzyme ?

According to the criteria of de Duve [46], a marker enzyme must be located in a single population of cell components and must be distributed homogeneous-ly within that population. As was evident from a discussion led by Dr.P.Jac-ques *[see Addendum]* other criteria such as ubiquity in nature, stability, and ease of visualization by biochemical and cytochemical procedures would be useful additional criteria. By these criteria, few, if any, endomembrane components provide valid markers.

Chemical constituents as markers

None of the common membrane constituents thus far examined emerge as specific markers. These include sphingomyelin, sterols, protein-bound sialic acid, neutral glycolipids, and gangliosides which are concentrated in the p.m. but distributed in lesser quantities throughout the endomembrane system. Micro-

Table 12. Estimates of fraction purity of isolated cell fractions from rat liver, based on analyses of marker enzyme activities and quantitative electron-microscope morphometry in conjunction with cytochemistry. Data from Tables 1, 2, 8 & 11 were integrated to arrive at the values given.

	PURIFIED FRACTIONS		
% CONTAMINATION BY:	e.r.	G.a.	p.m.
mitochondria	1	1	5
e.r.	-	4	4
G.a.	1	-	0
p.m.	*1*	*1*	-
% TOTAL CONTAMINATION	3	6	9
% FRACTION PURITY	97	94	91

somal cytochromes are present in both e.r. and G.a. [2, 55, 56]. Phosphatidyl choline, although perhaps synthesized exclusively in e.r. [77-80], is widely distributed among endomembranes [50]. *(See also Addendum to this article.)*

Enzymatic activities as markers

Enzymatic markers for mitochondria and chloroplasts (oxygen-mediating organelles) are available and applicable to a wide range of cell types [1; see however, D. Lloyd & T.G. Cartledge's article, ref. 81]. It is clear that most markers for endomembrane components (nuclear envelope, e.r., G.a. vesicles, vacuolar apparatus and p.m.) are insufficient and must be established for each tissue and cell type. Additionally, we have shown that for a single tissue, liver, many of the currently accepted marker enzymes for p.m. and e.r. are found in more than one cell component and are present in the G.a. A similar conclusion may be reached for cell fractions from bovine liver [63], bovine and rat mammary gland [82], and guinea pig pancreas [83].

Galactosyl transferase as an example of a marker enzyme ?

Galactosyl transferase, measured by the transfer of galactose from UDPgalactose to *N*-acetylglucosamine with the formation of *N*-acetyl-lactosamine, emerges as one enzyme activity which, for rat liver, may fulfil the criteria of de Duve [46]. The activity seems to be located exclusively in the Golgi apparatus from both biochemical [8, 21, 63, 84-87] and autoradiographic [84, 88, 89] investigations. Up to 70% of the activity is recovered in isolated G.a. preparations [8].

The criterion of homogeneity is more difficult to assess. Merritt and Morré [66] have shown that the activity, while concentrated in the secretory vesicles, is also present in purified cisternal fractions. Data of Ovtracht et al. [44] show the activity to be low or absent in the boulevard périphérique (smooth tubules corresponding to the smooth e.r. of the Golgi apparatus zone). There is at present no procedure for estimating galactosyl transferase cytochemically to show homogeneous distribution among all Golgi apparatus of liver.

Galactosyl transferase is present in Golgi apparatus of rat liver [8, 84, 85-87], and rat [86] and bovine [90] mammary gland, bovine liver [63], a mucopolysaccharide-secreting gland of the snail [91], rat intestine [92],and rat spermatids [93]. It is absent from the Golgi apparatus of plants when N-acetylglucosamine is the acceptor [94]. Therefore, UDP-galactose: N-acetyl-glucosamine galactosyltransferase is not a universal marker for Golgi apparatus of all cells.

Additionally, the activity is somewhat artifactual, which may contribute to its utility as a marker; N-acetyl-lactosamine is not thought to be a product of its normal activity. The activity of the enzyme is greatest in Golgi apparatus of mammary gland [86] where in the presence of specifier protein (α-lactalbumin) the enzyme catalyzes the transfer of galactose from UDP-galactose to glucose with the formation of lactose, the predominant carbohydrate constituent of rat and bovine milks. Liver Golgi apparatus will catalyze lactose formation in the presence of α-lactalbumin, whereas in the absence of α-lactalbumin and N-acetylglucosamine the galactosyl transferase of mammary gland Golgi apparatus will synthesize N-acetyl-lactosamine; the latter activity is inhibited in both liver and mammary gland by α-lactalbumin [86]. The same enzyme is reported to transfer galactose from UDPgalactose to glycoprotein [95] and this latter activity is not influenced by α-lactalbumin.

The endomembrane concept: a functional integration of endoplasmic reticulum, Golgi apparatus and plasma membrane (Fig. 21)

Concepts of membrane flow [96] and membrane differentiation [97, 98] have been combined to explain the formation of eukaryotic endomembranes along a sequence of cell components in subcellular developmental pathways [79]. Membrane differentiation is the gradual conversion of membranes from one type into another and is documented by comparisons of enzymatic activities, lipid composition and progressive modification of the proteins and lipids of membranes along the e.r.−Golgi apparatus - secretory vesicle - p.m. or e.r.-vesicle - p.m. export routes. The biochemical studies show the transitional nature of G.a. membranes which were first revealed by morphological studies [6, 97, 98]. Membrane dimensions and staining characteristics change progressively from e.r.-like to p.m.-like across the stacked cisternae from the forming to the maturing face of the Golgi apparatus. Membrane flow is the physical transfer of membrane from one cell component to another.

Nuclear envelope, e.r., and Golgi apparatus form part of a 3-dimensional interconnected system in eukaryotic cells. This has been verified by analyses of thin sections by electron microscopy for a variety of cell types [79, 99] including liver [100 - 103], and has recently been demonstrated by the elegant studies of Rambourg and coworkers at Saclay [104, 105] and by Favard, Carasso, Ovtracht, Poux, and others at Ivry-sur-Seine [106 - 108] using thick sections and high-voltage electron microscopy.

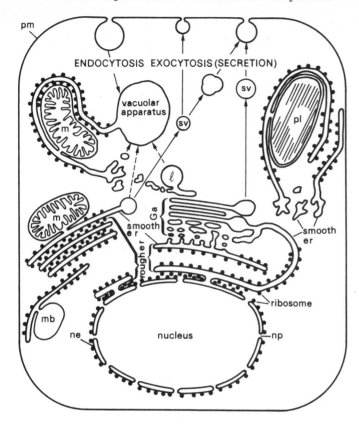

Fig. 21. Diagrammatic representation of the endomembrane system as an interconnected, three-dimensional complex. GA = Golgi apparatus; RER = rough endoplasmic reticulum; SER = smooth endoplasmic reticulum; PM = plasma membrane; SV = secretory vesicle; L = lysosome; MB = microbody; NE = nuclear envelope; NP = nuclear pore; M = mitochondrion; Pl = plastid. *Examples of electron micrographs illustrating each of the inter-endomembrane connections and/or functional associations are provided by Morré & Mollenhauer [79]. Because certain endomembrane components are transitional in nature, membrane constituents may be shared among two to several different components within the endomembrane system.*

Because all endomembranes are functionally interconnected, the presence of glucose-6-phosphatase and NADH-ferricyanide oxidoreductase as well as microsomal cytochromes in Golgi apparatus derives from the functional interrelationships of endomembrane components. If, as we have suggested, G.a. membranes are derived from e.r. membranes, at least some (if not all) e.r. constituents and activities should be present in Golgi apparatus. To explain the gradual loss of these activities in the G.a. we invoke a process of membrane differentiation one aspect of which includes a process whereby the enzyme activities are inactivated or excluded from the membranes at the Golgi apparatus. For the microsomal electron transport enzymes, the sequence (suggested by data of Table 7) is that the enzyme activities are lost first, followed by the loss of cytochromes, and finally only the iron remains in the G.a. [56]. In fact, the cytochemical evidence summarized here suggests the conversion of the NADH-oxidoreductases into glutaraldehyde-resistant forms which appear in the secretory vesicles of the G.a. and subsequently in the p.m. This conversion may take place at the G.a. (see Figs. 16-19).

In liver, a second export route also exists for discharge of lipo-proteins into the circulation [78, 96, 109]. This route involves the direct contribution of e.r.-derived vesicles to the p.m. These findings suggest that direct transfer of membrane from the e.r. to p.m. might also occur. As shown in Figs. 16-18, smooth elements of e.r. at or near the cell surface show the glutaraldehyde-resistant NADH-ferricyanide oxidoreductase activity in cytochemical studies. This means that membrane differentiation of a form similar to that of the Golgi apparatus is also taking place along this path-way. Estimates from isotope incorporation studies suggest that the contribu-tions of the two pathways (e.r.→ G.a. → p.m. and e.r. → p.m.) must be nearly equal in magnitude in liver to account for the labelling of p.m. proteins in short-term labelling and turnover studies [96]. The existence of this second export route helps to explain why adenyl cyclase activity is high in e.r. and low in G.a. Perhaps this activity is being transported from its site of synthesis and assembly into the membrane at the e.r. directly to the p.m. without first passing through the G.a. A similar situation might explain data for monovalent ion-stimulated ATPases (Table 9), if these activities are endogenous to e.r.

Nuclear envelope

Electron microscope studies show that the outer membrane of the nuclear enve-lope is in direct connection with the e.r. [110,111] and a part of the endo-membrane system. Its role in membrane biogenesis, however, is still uncertain.

The nuclear envelope has NADH- [15, 78] and NADPH- [15] cytochrome c oxidoreductase activities, also present in e.r. Franke *et al.* [15] were unable to demonstrate glucose-6-phosphatase in nuclear membranes of liver, whereas Kashnig and Kasper [110] report glucose-6-phosphatase to be present. NAD$^+$ pyrophosphorylase (ATP : NMN adenyltransferase), a nuclear enzyme, is probably confined to the nuclear contents and not a constituent of the nuclear envelope [112]. Nuclear envelope contained Mg^{2+}-ATPase activity but no Na$^+$, K$^+$-stimulated ATPase could be demonstrated [15,110]. We are unaware of any specific enzymatic markers for membranes of the nuclear envelope.

Whether or not DNA is a marker for nuclear membranes appears to depend partly on the method of preparation. Some workers [15, 112, 113] report a small amount of DNA in preparations of nuclear envelope, whilst others [110] do not. The lack of suitable enzymatic or chemical markers for nuclear enve-lope has prompted reliance on the nuclear pore structures as a means of identifying nuclear envelopes and nuclear membrane fragments in isolated preparations [60, 112-119].

The lysosome problem

The origin of lysosomes is far from being solved; but evidence suggests that e.r., G.a. and p.m. all may contribute to the lysosomal membranes found in a mixed population of lysosomes. Primary lysosomes seem to be formed through a cooperative activity of e.r. and G.a. in a specialized region at the mature face of the Golgi apparatus known as GERL in Novikoff's terminology [1]. Once formed, the primary lysosomes are thought to fuse with auto-phagic and heterophagic vacuoles derived from e.r. and p.m., respectively, to give rise to secondary or derived lysosomes [121].

Lysosomes are definitely a part of the endomembrane system, being at times continuous with both e.r. and G.a. and capable of fusing with p.m. or vesicles derived from p.m. Therefore, the situation with regard to lysosomes is likely to be analogous to that encountered with G.a. In Golgi apparatus, a portion of the membrane is derived from e.r. and destined to become p.m. It therefore contains markers for both e.r. and p.m. With lysosomes, the membranes are derived from e.r. and/or G.a. membranes [62, 120, 122] with additional contributions directly from p.m. [121].

The use of marker enzymes or chemical constituents to evaluate e.r., p.m., or G.a. contamination of lysosomal membranes must be with caution. The presence of 5'-nucleotidase (Table 8), sphingomyelin [123] or plant lectin-binding proteins [124] in a lysosomal preparation does not mean that the preparation is necessarily contaminated with p.m. The constituents could arise from contributions to the lysosomal membrane from p.m. or G.a. membranes.

Even though acid phosphatase activity is present in isolated Golgi apparatus, it remains the classic lysosomal marker [46, 47, 62, 120, 122]. Yet, this is a marker for the lysosomal contents and not for the membranes. As with nuclear envelope, there is yet no known marker for the lysosomal membrane which is specific to lysosomal membranes.

Interim solutions to the marker problem for transitional membrane elements

Transitional membrane elements such as Golgi apparatus and primary lysosomes which have properties intermediate between those of generating elements and end products of the endomembrane system offer special problems in determining yields and purity of fractions based on marker criteria. Some markers are more specific than others; galactosyl transferase emerges as a reasonably good marker for G.a. of rat liver, but not all glycosyl transferases appear to be equally well localized. Thiamine pyrophosphatase, on the other hand, is a good cytochemical marker for Golgi apparatus [62] but, for liver, it is not specific enough to be a biochemical marker [21].

As suggested by results for thiamine pyrophosphatase, cytochemistry can introduce an element of specificity not attainable by other methods through sensitivity to glutaraldehyde fixation. This possibility is demonstrated for the NADH-ferricyanide oxidoreductases of the endomembrane system. This activity is concentrated in e.r. in unfixed membranes, with lesser activity in G.a. and p.m. However, upon fixation with glutaraldehyde, the activity falls or disappears in the e.r. and G.a. but remains in p.m. and membranes of mature secretory vesicles of the G.a.   This reaction combined with electron microscope morphometry provides estimates of p.m. contamination of e.r. and G.a. fractions in a manner which relates morphological appearance and biochemical reactivity. Even here the problem is still complicated, since membranes of mature secretory vesicles of the G.a. and segments of smooth e.r. adjacent to p.m. show reactivity (Fig. 17). These membranes are p.m.-like and probably contribute membranes to the p.m. Problems relating to heterogeneity of e.r. are discussed by Eriksson *et al.* [126] and Glaumann and Dallner [127].

Thus, the solution to the marker problem for transitional elements of the endomembrane system is not readily resolved. So far, the approach we have taken is to combine biochemistry, cytochemistry and morphometry in as many ways as possible. With e.r. and p.m. fractions, the morphometric and biochemical evaluations establish a fraction purity in excess of 90%. With G.a., the problem remains complicated. A purity of 85% is found by either biochemical analysis or morphometry alone. By combining the two approaches to evaluate contamination, the fraction purity of the Golgi apparatus can be shown to exceed 90%.

See also ADDENDUM following the refs.

Acknowledgements

We thank John Elder, Kathy Frantz and Sue Hocker, Purdue University, for use of unpublished information. The work was supported by grants NIH CA 13145 & HD 06624 and NSF GB 23183 & GB 25110. T.W.K. is supported by Public Health Service Career Development Award GM 70596 from the National Institute of General Medical Science. Purdue University AES Journal Paper No. 5269.

References

1. Racker, E., in *Structure and Function of Membranes of Mitochondria and Chloroplasts*, (Racker, E., ed.), Reinhold, New York (1969) pp. 127-145.

2. Morré, D.J., *Methods Enzymol. 22* (1971) 130-148.

3. Morré, D.J., in *Molecular Techniques and Approaches in Developmental Biology* (Chrispeels, M.J., ed.), Wiley-Interscience, New York (1973) pp. 1-27.

4. Morré, D.J., Cheetham, R.D., Nyquist, S.E. & Ovtracht, L., *Preparative Biochem. 2* (1972) 61-69.

5. Morré, D.J., Hamilton, R.L., Mollenhauer, H.H., Mahley, R.W., Cunningham, W.P., Cheetham, R.D. & LeQuire, V.S., *J. Cell Biol. 44* (1970) 484-490.

6. Morré, D.J., Keenan, T.W. & Mollenhauer, H.H., in *Adv. Cytopharmacol.* (Clementi, F. & Cecarelli, B., eds.) *[Proc. 1st Int. Symp. Cell Biol. & Cytopharmacol., Venice, 1969],* Raven Press, New York, *1* (1971) 159-182.

7. Glaumann, H. & Ericsson, J.L.E., *J. Cell Biol. 47* (1970) 555-567.

8. Morré, D.J., Merlin, S.M. & Keenan, T.W., *Biochem. Biophys. Res. Commun. 37* (1969) 813-819.
[-500.

9. Cheetham, R.D., Morré, D.J. & Yunghans, W.N., *J. Cell Biol. 44* (1970) 491

10. Emmelot, P., Bos, C.J., Benedetti, E.L. & Rümke, Ph., *Biochim. Biophys. Acta 90* (1964) 126-145.

11. Neville, D.M., *J. Biophys. Biochem. Cytol. 8* (1960) 413-422.

12. Yunghans, W.N. & Morré, D.J., *Preparative Biochem. 3* (1973) 301-312.

13. Lauter, C.J., Solyom, A. & Trams, E.G., *Biochim. Biophys. Acta 266* (1972) 511-523.

14. Beaufay, H., in *Lysosomes,* Vol. 2 (Dingle, J.T. & Fell, H.B., eds.), North-Holland, Amsterdam (1969) pp. 515-546.

15. Franke, W.W., Deumling, B., Ermen, B., Jarasch, E.-D., & Kleinig, H., *J. Cell Biol. 46* (1970) 379-395.

16. Leighton, F., Poole, B., Beaufay, H., Baudhuin, P., Coffey, J.W., Fowler, S. & de Duve, C., *J. Cell Biol. 37* (1968) 482-513.

17. Trouet, A., *Arch. Intern. Physiol. Biochem. 72* (1964) 698-699.

21. Cheetham, R.D., Morré, D.J., Pannek, C. & Friend, D.S., *J. Cell Biol. 49* (1971) 899-905.

18. Lowry, O.H., Rosebrough, N.J., Farr, A.L. & Randall, R.J., *J. Biol. Chem. 193* (1951) 265-275.

19. Nordlie, R.C. & Arion, W.J., *Methods Enzymol. 9* (1966) 619-625.

20. Allen, J.M. & Slater, J.J., *J. Histochem. Cytochem. 9* (1961) 418-423.

22. Pennington, R.J., *Biochem. J. 80* (1961) 649-654.

23. Sun, F.F. & Crane, F.L., *Biochim. Biophys. Acta 172* (1969) 417-428.

24. Schnaitman, C.A., Erwin, V.G. & Greenawalt, J.W., *J. Cell Biol. 32* (1967) 719

25. Schnaitman, C.A. & Greenawalt, J.W., *J. Cell Biol. 38* (1968) 158-183. [-735.

26. Ostrowski, W. & Tsugita, A., *Arch. Biochem. Biophys. 94* (1961) 68-78.

27. Donaldson, R.P., Tolbert, N.E. & Scharrenberger, C., *Arch. Biochem. Biophys. 152* (1972) 199-215.

28. Mahler, H.R., *Methods Enzymol. 2* (1955) 688-693.

29. Palmiter, R.D., *Biochim. Biophys. Acta 178* (1969) 35-46.

30. Krishna, G., Weiss, B., and Brodie, B.B., *J. Pharmacol. Exp. Therap. 163* (1968) 379-385.

31. Rodbell, M., *J. Biol. Chem. 242* (1967) 5744-5750.

32. Henry, R.J., Sobel, C. & Kim, J., *Amer. J. Clin. Path. 28* (1957) 152-160.

33. Nyquist, S.E. & Morré, D.J., *J. Cell. Physiol.* 78 (1971) 9-12.

34. Bublitz, C., *Biochim. Biophys. Acta* 48 (1961) 61-70.

35. Dodgson, K.S. & Thomas, J., *Biochem. J.* 53 (1953) 452-457.

36. Karnovsky, M.J. & Roots, L., *J. Histochem. Cytochem.* 12 (1964) 219-221.

37. Lukaszyk, A., *Folia Histochem. Cytochem.* (1971) 167-186.

38. Karnovsky, M.M., *J. Cell Biol.* 23 (1964) 213-247.

39. Spurr, A.R., *J. Ultrastruct. Res.* 26 (1969) 31-43.

40. Wachstein, M. & Meisel, E., *J. Histochem. Cytochem.* 4 (1956) 499.

41. Schin, K.S. & Clever, U., *Z. Zellforsch. Mikroskop. Anat.* 86 (1967) 262-279.

42. Loud, A.V., *J. Cell Biol.* 15 (1962) 481-487.

43. Weibel, E.R., *Intern. Rev. Cytol.* 26 (1969) 235-302; Weibel, E.R. & Bolender, R.P., in *Principles and Techniques of Electron Microscopy* (Hayat, M.A., ed.), Van Nostrand Reinhold, New York (1973) pp. 237-296.

44. Ovtracht, L., Morré, D.J., Cheetham, R.D. & Mollenhauer, H.H., *J. Microsc., in press.*

45. Barclay, M., Barclay, R.K., Essner, E.S., Skipski, V.P. & Terebus-Kekish, O., *Science* 156 (1967) 665-667.

46. de Duve, C., *Harvey Lectures* 59 (1963-1964) 49-81; *J. Theoret. Biol.* 6 (1964) 33-59.

47. de Duve, C., in *Enzyme Cytology* (Roodyn, D.B., ed.), Academic Press, New York (1967), pp. 1-26.

48. Baudhuin, P. & Berthet, J., *J. Cell Biol.* 35 (1967) 631-648.

49. Baudhuin, P., Evrard, P. & Berthet, J., *J. Cell Biol.* 32 (1967) 181-191.

50. Keenan, T.W. & Morré, D.J., *Biochemistry* 9 (1970) 19-25.

51. Keenan, T.W., Morré, D.J. & Huang, C.M., *FEBS Lett.* 24 (1972) 204-208.

52. Larsen, C., Dallner, G. & Ernster, L., *Biochim. Biophys. Res. Commun.* 49 (1972) 1300-1306.

53. Yunghans, W.N., Keenan, T.W. & Morré, D.J., *Exptl. Mol. Path.* 12 (1970) 36-45.

54. Keenan, T.W., Morré, D.J. & Huang, C.M., *Biochem. Biophys. Res. Commun.* 47 (1972) 1277-1283.

55. Fleischer, S., Fleischer, B., Azzi, A. & Chance, B., *Biochim. Biophys. Acta* 225 (1971) 194-200.

56. Morré, D.J., Franke, W.W., Deumling, B., Nyquist, S.E. & Ovtracht, L., in *Biomembranes,* Vol. 2 (Manson, L.A., ed.) [Proc. Symp. Membranes and Coordination of Cellular Activities, Gatlingburg] Plenum, New York (1971) 95-104.

57. Nyquist, S.E., Barr, R. & Morré, D.J., *Biochim. Biophys. Acta 208* (1970) 532-539.

58. Nyquist, S.E., Crane, F.L. & Morré, D.J., *Science 173* (1971) 939-941.

59. Nyquist, S.E., Matschiner, J.T. & Morré, D.J., *Biochim. Biophys. Acta 244* (1971) 645-649.

60. Kleinig, H., *J. Cell Biol. 46* (1970) 396-402.

61. Keenan, T.W., Morré, D.J. & Basu, S., *J. Biol. Chem., in press.*

62. Novikoff, A.B., Essner, E., Goldfischer, S. & Heus, M., in *Interpretation of Ultrastructure* (Harris, R.J.C., ed.), Academic Press, New York (1962) pp. 149-192.

63. Fleischer, B., Fleischer, S. & Ozawa, H., *J. Cell Biol. 43* (1969) 59-79.

64. Leelavathi, D.E., Estes, L.W., Feingold, D.S. & Lombardi, B., *Biochim. Biophys. Acta 211* (1970) 124-138.

65. Sutherland, E.W., *Science 177* (1972) 401-408.

66. Merritt, W.D. & Morré, D.J., *Biochim. Biophys. Acta 304* (1973) 397-407.

67. Farquhar, M.G., Bergeron, J.J.M. & Palade, G.E., *J. Cell Biol. 55* (1972) 72a.

68. Widnell, C.J., *J. Cell Biol. 52* (1972) 542-558.

69. Fiske, C.H. & SubbaRow, Y., *J. Biol. Chem. 66* (1925) 375-400.

70. Mitchell, R.H. & Hawthorne, J.N., *Biochem. Biophys. Res. Commun. 21* (1965) 333-338.

71. Middleton, A.E., Cheetham, R., Gerber, D. & Morré, D.J., *Proc. Ind. Acad. Sci. 78* (1969) 183-188.

72. Ernster, L. & Jones, L.C., *J. Cell Biol. 15* (1962) 563-578.

73. Leskes, A., Siekevitz, P. & Palade, G.E., *J. Cell Biol. 49* (1971) 264-302.

74. Ernster, L., Siekevitz, P. & Palade, G.E., *J. Cell Biol. 15* (1962) 541-562.

75. Morré, D.J., Huang, C.M., Keenan, T.W. & Vigil, E.L., *J. Cell Biol. 55* (1972) 1812. [353a.

76. Vigil, E.L., Morré, D.J., Frantz, C. & Huang, C.M., *J. Cell Biol. 59* (1973)

77. Jungawala, F.B. & Dawson, R.M.C., *Eur. J. Biochem. 12* (1970) 399-402.

78. Morré, D.J., Keenan, T.W. & Huang, C.M., *Adv. Cytopharmacol.*, Vol. 2 [Proc. 2nd Adv. Study Inst. Cytopharmacol., Venice, 1973], Raven Press, New York, *in press.*

79. Morré, D.J. & Mollenhauer, H.H. in *Dynamics of Plant Ultrastructure* (Robards, A.W., ed.), McGraw-Hill, London, *in press.*

80. Stein, O. & Stein, Y., *J. Cell Biol. 40* (1969) 461-483.

81. Lloyd, D. & Cartledge, T.G., *this volume,* Art. 32.

82. Keenan, T.W., Morré, D.J. & Huang, C.-M., in *Lacation: A Comprehensive Treatise* (Larson, B.L. & Smith, V.R., eds.), Academic Press, New York,

83. Meldolesi, J., Jamieson, J.D. & Palade, G.E., *J. Cell Biol. 49* (1971) 109-158.

84. Cook, G.M.W., in *Lysosomes in Biology and Pathology,* Vol. 3 (Dingle, J.T., ed.), North Holland, Amsterdam (1973), pp. 237-277.

85. Fleischer, B. & Fleischer, S., *Biochim. Biophys. Acta 219* (1970) 301-319.

86. Keenan, T.W., Morré, D.J. & Cheetham, R.D., *Nature 228* (1970) 1105-1106.

87. Schachter, H., Jabbal, I., Hudgin, R.L. & Pinteric, L., *J. Biol. Chem. 245* (1970) 1090-1100.

88. Droz, B., *Compt. Rend. Acad. Sci., Paris 262D* (1966) 1766-1768.

89. Neutra, M. & LeBlond, C.P., *J. Cell Biol. 30* (1966) 137-150.

90. Keenan, T.W., Huang, C.M. & Morré, D.J., *J. Dairy Sci. 55* (1972) 1577-1585.

91. Ovtracht, L., Morré, D.J. & Merlin, L.M., *J. Micros. 8* (1969) 989-1002.

92. Mahley, R.W., Bennett, B.D., Morré D.J., Gray, M.E., Thistlethwaite, W. & LeQuire, V.S., *Lab. Invest. 25* (1971) 435-444.

93. Cunningham, W.P., Nyquist, S.N. & Mollenhauer, H.H., *J. Cell Biol. 51* (1971) 273-285.

94. Morré, D.J., Lembi, C.A. & Van Der Woude, W.J. in *Prakticum der Zytologie* (Jacobi, G., ed.), Georg Thieme Verlag, Stuttgart, *in press.*

95. Schanbacher, F.L. & Ebner, K.E., *J. Biol. Chem. 245* (1970) 5057-5061.

96. Franke, W.W., Morré, D.J., Deumling, B., Cheetham, R.D., Kartenbeck, J., Jarasch, E.-D. & Zentgraf, H.-W., *Z. Naturforsch. 26b* (1971) 1031-1039.

97. Grove, S.N., Bracker, C.E. & Morré, D.J., *Science 161* (1968) 171-173.

98. Morré, D.J., Mollenhauer, H.H. & Bracker, C.E., in *Results and Problems in Cell Differentiation. Origin and Continuity of Cell Organelles,* Vol. 2 (Reinert, T. & Ursprung, H., eds.), Springer-Verlag, Berlin (1971) 82-126.

99. Bracker, C.E., Grove, S.N., Heintz, C.E. & Morré, D.J., *Cytobiol. 4* (1971) 1-8.

100. Claude, A., *J. Cell Biol. 47* (1970) 746-766.

101. Morré, D.J., Merritt, W.D. & Lembi, C.A., *Protoplasma 73* (1971) 43-49.

102. Morré, D.J., Vigil, E.L. & Keenan, T.W., *Proc. Electron Micros. Soc. America 21* (1973) 690-691.

103. Ovtracht, L., Morré, D.J. & Nyquist, S., *Micros. Electronique 1970,* Vol. 3 *[Proc. 7th Intern. Congr. Electron Microscopy]* (1970) pp. 81-82.

104. Rambourg, A., *Compt. Rend. Acad. Sci. (Paris) 269D* (1969) 2125-2127.

105. Rambourg, A. & Chrétien, M., *Compt. Rend. Acad. Sci. (Paris) 270D* (1970) 981-983.

106. Carasso, N., Ovtracht, L. & Favard, P., *Compt. Rend. Acad. Sci. (Paris) 273D* (1971) 876-879.

107. Favard, P. & Carasso, N., *J. Microscopy 97* (1973) 59-81.

108. Favard, P., Ovtracht, L. & Carasso, N., *J. Microscopie (Paris) 12* (1971) [301-316.

109. Morré, D.J. & Van Der Woude, W.J., *Devel. Biol. Suppl. 5* (1974) *in press.*

110. Kashnig, D.M. & Kasper, C.B., *J. Biol. Chem. 244* (1969) 3786-3792.

111. Watson, M.L., *J. Biophys. Biochem. Cytol. 1* (1955) 257-270.

112. Zentgraf, H., Deumling, B., Jarasch, E.D. & Franke, W.W., *J. Biol. Chem. 246* (1971) 2986-2995.

113. Agutter, P.S., *Biochim. Biophys. Acta 255* (1972) 397-401.

114. Berezney, R., Funk, L. & Crane, F.L., *Biochim. Biophys. Acta 203* (1970) 531-546.

115. Franke, W.W., *J. Cell Biol. 31* (1966) 619-623.

116. Franke, W.W., *Z. Zellforsch. Mikroskop. Anat. 80* (1967) 585-593.

117. Gurr, M.I., Finean, J.B. & Hawthorn, J.N., *Biochim. Biophys. Acta 70* (1963) 406-416.

118. Kleinig, H., Zentgraf, H., Comes, P. & Stadler, J., *J. Biol. Chem. 246* (1971) 2996-3000.

119. Monneron, A., Blobel, G. & Palade, G.E., *J. Cell Biol. 55* (1972) 104-125.

120. Novikoff, A.B., in *The Neuron* (Hayden, H. ed.), Elsevier, Amsterdam (1967) pp. 255-319.

121. de Duve, C. & Wattiaux, R., *Ann. Rev. Physiol. 28* (1966) 435-492.

122. Novikoff, A.B., Essner, E. & Quintana, N., *Fed. Proc. 23* (1964) 1010-1022.

123. Thinès-Sempoux, D., in *Lysosomes in Biology and Pathology,* Vol. 3 (Dingle, J.T., ed.), North-Holland, Amsterdam (1973) pp. 278-299.

124. Henning, R. & Uhlenbruck, G., *Nature New Biol. 242* (1973) 120-122.

125. Gomori, G., *Microscopic Histochemistry* (1952) University Press, Chicago.

126. Eriksson, L., Svensson, H., Bergstrand, A. & Dallner, G., in *Role of Membranes in Secretory Processes,* North-Holland, Amsterdam (1972) 3-23.

127. Glaumann, H. & Dallner, G., *J. Cell Biol. 47* (1970) 34-48.

128. Beaufay, H., in *Lysosomes in Biology and Medicine,* Vol. 2 (Dingle, J.T. & Fell, H.B., eds.), North-Holland, Amsterdam (1969), pp. 516-546 (see 525).

129. Frohwein, Y.Z. & Gatt, S., *Biochemistry 6* (1967) 2775-2782.

ADDENDUM
arising from R. Henning's presentation
(this volume)

Dr. Henning reported that, for rat liver, trihexosylceramide (galactosylgalactosyl-ceramide) was a unique constituent of the membrane of tritosomes. The trihexosyl-ceramide could not be found in e.r. or p.m. fractions in his studies, and we have found none in our G.a. preparations even though both the di- and monohexosylceram-ides are present (mono- > di-, in contrast to p.m. where di- and mono- are more near-ly equal).

The possibility that the biosynthe-tic enzymes for the conversion of di- to trihexosylceramide might be present in tritosome membranes was discussed at the Symposium. The localization of this enz-ymatic activity in the lysosome membrane would be of interest not only in estab-lishing a biosynthetic activity associ-ated with this membrane but also in pro-viding a potentially useful marker enz-yme for this membrane. Alternatively, the trihexosylceramide found in trito-somes might arise as a degradation pro-duct of globoside (*N*-acetylgalactosamin-yl-galactosylgalactosylglucosylceramide) since lysosomes have been shown to con-tain *N*-acetylhexosaminidase active against globoside [129].

A routine assay of glycosylsphin-golipid glycosyltransferase activity failed to demonstrate significant levels of the UDP-galactose : lactosylceramide galactosyl transferase in a tritosome fraction (Table A1). The activities of all fractions were low (near the limits of detection), as we have found previous-ly with liver, and the recovery of particulate activity was only 25% with glucosylceramide (CMH) as acceptor and 31% with lactosylceramide (CDH) as acceptor (subtracting activity with no acceptor; all particulate fractions summed).

References are on previous page.

Table A1. Distribution of UDP-gal-actose : glucosylceramide galactosyl transferase and UDP-galactose : lac-tosylceramide galactosyl transferase amongst cell fractions of rat liver. *Rats were injected with Triton WR 1339 and, after 4 days, the livers were fractionated for preparation of tritosomes [128]. The complete reac-tion mixture contained, in final vols. of 0.1 ml: Bicine-HCl pH 6.8, 15 μmoles; MgCl$_2$, 0.5 μmoles; Cut-scums-Triton X-100 (2:1, w/w), 0.6 mg; UDP-galactose (6.6 × 10^5 counts /min/μmole; glucosylceramide (CMH) or lactosylceramide (CDH), 0.1 μmole; and enzyme protein suspended in 0.32 M sucrose containing 14 mM 2-mercap-toethanol. Incubation was for 3 h at 37°. Reactions were terminated by addition of 0.1 ml methanol, and incorporation of radioactivity into glycolipids was determined by des-cending paper chromatography and scintillation counting [61].*

FRACTION	ACCEP-TOR	SPECIFIC ACTIVITY*	TOTAL ACTIVITY†
Total	CMH	17.0	15,147
particulate	CDH	10.2	9,089
	None	1.7	1,616
Low-speed	CMH	3.4	1,215
pellet	CDH	4.5	1,608
	None	1.1	393
Tritosomes	CMH	26.1	60
	CDH	8.7	20
	None	8.7	20
Mixed mito.-	CMH	2.4	81
e.r. fraction	CDH	6.0	203
from trito-	*None*	0.0	0
some gradient			
Microsomes	CMH	14.7	4,542
	CDH	9.8	3,028
	None	6.9	2,132

* *μμmoles/h/mg protein*
† *s.a. × total protein recovered*

22 CHARACTERIZATION OF LIVER MICROSOMAL SUB-FRACTIONS AND THE GOLGI APPARATUS WITH RESPECT TO CHEMICAL COMPOSITION AND INTRACELLULAR TRANSPORT OF MACROMOLECULES

Hans Glaumann, Helena Persson, Anders Bergstrand and Jan L.E. Ericsson
Department of Pathology at Sabbatsberg Hospital,
Karolinska Institutet, Stockholm, Sweden

Liver microsomes have been sub-fractionated on sucrose gradients into rough microsomes and smooth microsomes [1], and the Golgi apparatus isolated [2,3] (see also D.J. Morré and co-authors, this volume). The fractions have been characterized chemically (protein, PLP*, cholesterol and TG*), enzymically (phosphatases, electron transport enzymes, glycosyl transferases, etc.) and ultrastructurally. We have studied the synthesis and intracellular migration of albumin and VLDL* after labelling with radioactive leucine and glycerol, the VLDL precursors being isolated from the fraction at d <1.007 g/ml. The chemical compositions of the VLDL or their precursors originating from rough and smooth microsomes and from the Golgi apparatus were compared with that of serum VLDL. Based on biochemical and electron-microscopic analysis it is concluded that precursors of VLDL can be found in the rough and smooth e.r.* and that the formation of VLDL takes place in a multi-step manner. Consequently, the lipids (mainly TG) are added to the apoprotein while the growing nascent macromolecule travels through the e.r.

It is now well established that the e.r.* is the principal site of lipid synthesis in the liver cell cf.[4]. Certain lipids - especially TG - are thought to be conjugated with an apolipoprotein moiety within the channels of the e.r. to form secretory particles [5,6]. It has also been proposed that the particles presumed to be composed of VLDL are channeled through the e.r. and, at least in part, through the Golgi apparatus before being released to the extracellular space at the sinusoids [7-10]. Although smooth-surfaced e.r. has been implicated in the transport and possible assembly of the VLDL in morphological studies [10], it has not been clear whether or not other ultrastructural specializations of the e.r. are also involved. A comparison of lipoprotein particles isolated from the Golgi apparatus with serum VLDL showed that they had similar chemical composition and immunochemical properties [11].

It has recently become possible to isolate not only fractions of rough and smooth-surfaced microsomes from liver, but also reasonably pure Golgi preparations [2,3,12]. An attempt was therefore made to follow the appearance of labelled TG from one subcellular fraction to another after pulse labelling with [³H-]glycerol, in order to explore the site of synthesis and the intra-

* *ABBREVIATIONS: VLDL, very low density lipoprotein; e.r., endoplasmic reticulum; TG, triglycerides; NL, neutral lipids; PLP, phospholipids*

cellular pathway in the liver cell under functional conditions *in vivo*. Furthermore, the chemical composition of the VLDL or their precursors was determined in smooth- and rough-surfaced microsomes and the Golgi apparatus to evaluate the possibility of a stepwise synthesis of the VLDL. According to such a theory of a multi-site synthesis of VLDL, the attachment of lipids to the apoprotein would start on the rough-surfaced part of the e.r.; the formation of the VLDL particles would be completed while the growing lipoprotein molecule travels through the e.r. and the Golgi apparatus, indicating an active participation of all these membranes in the formation of the VLDL. Moreover, in order to corroborate the biochemical data and further elucidate the sites of synthesis and routes of migration of VLDL in relation to structural specializations and sub-units of the e.r. and Golgi zones, the appearance and distribution of presumed VLDL particles was studied in the electron microscope both in liver tissue (*'in situ'*) and in isolated 'Golgi fractions'.

MATERIALS AND METHODS

Separation of rough and smooth microsomes was performed as described earlier [1]. A Golgi-rich fraction was isolated according to Morré *et al.* [12]* with some modifications as now summarized. Non-starved male rats give better recoveries, in our experience. The livers are minced in 0.5 M sucrose containing 5 mM $MgCl_2$, 1% dextran and 0.05 M Tris-maleate buffer, pH 6.8. The Mg^{2+} addition minimizes microsomal contamination, since it causes aggregation of both rough and smooth microsomes, thereby increasing their sedimentation rates and densities [13]. Consequently, the microsomes will sediment to a pellet through the 1.25 M sucrose layer, whereas the Golgi elements remain at the interface. Homogenization can be performed either with a Polytron 20ST homogenizer for 40-60 sec, at 5,000-10,000 rev/min, or — with better reproducibility in our hands — with a Teflon-glass homogenizer at low speed (about 100 rev/min) with only two complete strokes. In order to minimize possible adverse effects of lysosomal hydrolases, the entire fractionation should be carried out as quickly as possible. The pH of the medium is less critical, because liver homogenate in sucrose exhibits a pH close to neutral or somewhat on the acid side, provided that the procedures are carried out within a temperature range of 3-5° and care is taken to avoid a low protein concentration. Commercial dextran should be further purified by dialysis against doubly distilled water to remove cations, which may drastically and irreversibly change the physical propoerties of several subcellular organelles by initiating clumping and aggregation of cytoplasmic particles. Care must also be taken not to overload the 1.25 M sucrose gradient with material.

    [^{3}H-]Glycerol (20 µCi/100 g) was injected into a branch of the superior mesenteric vein. To dissociate adsorbed material from the membranes, all fractions were washed twice with Tris buffer (pH 8.0) in KCl-EDTA and sonicated [14]. The lipids of the different fractions were extracted with chloroform-ethanol, and neutral lipids (NL) were separated from phospholipids (phos-

* *This paper and Glaumann & Ericsson's paper [3] should be consulted for details*

phatides, PLP) by silicic acid chromatography of the washed chloroform phase [1]. The microsomal and Golgi pellets were frozen, thawed, and sonicated, and the VLDL particles were separated from the microsomes and Golgi elements by flotation at 100,000 g for 16 h through a NaCl solution of density 1.007 g/ml at 12°. Lower temperatures gave much lower recoveries, because the VLDL particles then adhered to the membranous elements. For studies on the distribution of VLDL particles *'in situ'*, liver tissue from rats of variable age and nutritional status (unstarved and starved for up to 20 h) was fixed by immersion in collidine-buffered OsO_4 and embedded in Epon or Maraglas [3]. Pellets intended for fine-structural analyses were fixed in glutaraldehyde and post-fixed in OsO_4 as described previously [3].

RESULTS AND COMMENTS

The Golgi-rich fraction was estimated to consist of at least 75% Golgi apparatus-derived material as indicated by the activities of marker enzymes (5'-nucleotidase, glucose-6-phosphatase, cytochrome oxidase and galactosyl transferase). After removing non-membranous material (adsorbed and intra-cisternal) by washing procedures and sonication, the chemical composition was determined (Table 1). The rough microsomal membranes contain about 26% lipid, as compared to 34% and 44% for smooth and Golgi membranes, respectively. When related to PLP, the cholesterol content increased from rough microsomal to smooth microsomal membranes and to the Golgi membranes. Much higher values have been reported for plasma membranes and lysosomes [15]. Consequently, the Golgi membranes are intermediate in lipid composition between the e.r. and lysosomal membranes and the plasma membrane [cf. 16]. Although it is not generally accepted that the e.r. membranes contain cholesterol [17] - a possibility that has been extensively debated (see *NOTES & COMMENTS*, Art. 39; also Art. 17) - the results support the view that the Golgi membranes are transitional in nature. Recent studies by Franke *et al.* [18] on the turn-

Table. 1. Distribution of protein, PLP, TG and cholesterol in membranous elements of liver microsomes and Golgi apparatus. 'Membranous elements' signifies Tris-washed, sonicated and KCl-EDTA washed membranes [14]. *Abbreviations are given at foot of first page.*

	mg per g of liver				$\frac{Cholesterol}{PLP}$	$\frac{TG}{PLP}$	$\frac{PLP}{Protein}$
	Protein	PLP	Cholesterol	TG			
Total microsomes	12.1	3.2	0.25	0.15	0.078	0.047	0.26
Rough "	6.0	1.4	0.096	0.059	0.068	0.042	0.23
Smooth "	3.6	1.1	0.099	0.058	0.09	0.053	0.30
Golgi apparatus	0.4	0.14	0.021	0.015	0.15	0.10	0.35

over of membrane proteins in rat liver have added evidence to earlier, mainly morphological, findings of a transfer of membrane material from e.r. to the lysosomal system via the Golgi apparatus and perhaps also to the plasma membrane.

   The site of synthesis of VLDL and the intracellular pathway of VLDL transport in the liver cell *in vivo* were explored by pulse labelling of the TG moiety with [^{3}H-]glycerol, and by following the appearance of radioactivity in the VLDL (or precursors of VLDL) isolated from the various fractions, as is shown in Fig. 1. During

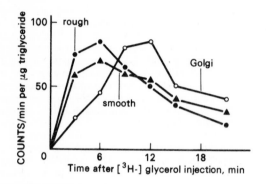

Fig. 1. [^{3}H-]Glycerol incorporation into triglycerides of VLDL precursors from rough and smooth microsomes and Golgi apparatus. The lipoprotein particles were isolated as described in MATERIALS & METHODS.

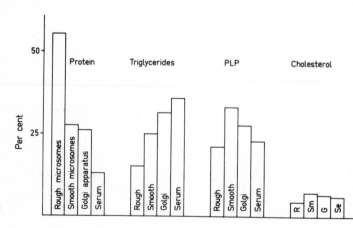

Fig. 2. Difference in chemical composition of VLDL precursors from microsomes and from the Golgi apparatus and of serum VLDL. The bars represent % of the total dry weight.

Table 2. Protein and lipid composition of VLDL-precursors from subcellular organelles and of serum VLDL. Values other than those for serum represent µg/g liver. Cholesterol values include cholesterol esters.

	Protein	Phospho-lipids	Choles-terol	TG
Total microsomes	90	58	15	32
Rough "	50	19	5.3	13
Smooth "	25	27	9.6	20
Golgi apparatus	17	18	5.7	29
Serum *(as µg/ml)*	83	109	56	236

the first interval of six min, the TG from rough and smooth microsomes displayed a much higher rate of labelling when compared with the Golgi fraction. Both rough and smooth microsomes displayed a similar rate of incorporation, with peak values at 6 min, followed by a fall-off, while the radioactivity of the Golgi fraction was still increasing, reaching maximal levels 10-20 min after the administration. The equally fast rate of incorporation of glycerol into rough and smooth microsomes, when compared to the Golgi fraction, indicates that TG are preferentially - if not exclusively - synthesized in the e.r. rather than the Golgi apparatus. The different localizations of the peak activities suggest a precursor-product relationship and also add biochemical support to the concept of a migration of at least a part of the TG pool from the e.r. to the Golgi apparatus before discharge into the circulation at the cell surface.

So as to obtain evidence on whether the lipoprotein particles are chemically altered during the intracellular migration or not, the VLDL precursors were isolated from the e.r. and the Golgi apparatus and were chemically characterized. The findings are summarized in Table 2 and Fig. 2. It is evident that the lipoprotein particles derived from the rough-surfaced e.r. are characterized by a very high protein content, whereas the lipid content is low in comparison with the completed VLDL in serum. As the particles make their way through the intracellular system, a concomitant change in composition occurs. TG and PLP are added in the rough- and smooth-surfaced e.r. The lipid composition of the Golgi VLDL particles was similar to that of VLDL in serum, indicating that the Golgi apparatus does not participate in the addition of lipid components to the VLDL. This is in agreement with the low rate of labelling of TG at short intervals, as described above.

Fig. 3 shows the *in situ* localization of the particles believed to represent VLDL. The particles were rare in the rough e.r., while they were comparatively frequent in the smooth e.r. The latter sites included glycogen areas as well as the transitional elements between rough e.r. and the Golgi apparatus, and the sub-plasmalemmal portions of the smooth e.r. connected with the 'distal' ends of rough e.r. In the Golgi apparatus, peripheral dilated portions of the cisternae ('concentrating vesicles') [10] and

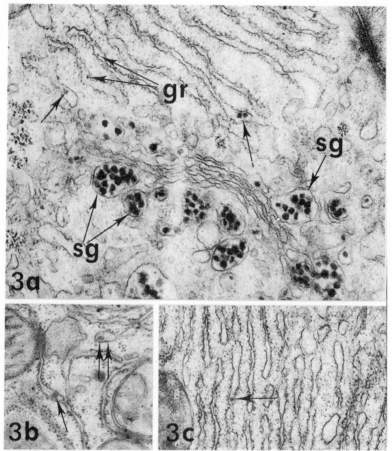

Fig. 3. Fine-structural distribution of VLDL particles in hepatocytes of immersion-fixed rat liver. Thin sections were stained with lead citrate and uranyl acetate.

(a) VLDL particles *(arrows)* in smooth-surfaced terminal ends of rough-surfaced e.r. near the convex side of the Golgi apparatus, and in 'secretory granules' (sg) [10]. *Note absence of similar granules within the flattened cisternae of rough e.r. which, however, contain smaller granules (forming particles ?) in two areas* (gr).   Maraglas. × 33,000.

(b) VLDL particle *(arrow)* in the lumen of rough e.r. and within an elongated vesicular e.r. element *(two arrows)* covered on one side with ribosomes. Epon. × 33,000.

(c) Cisternae of rough e.r., one of which contains a VLDL particle *(arrow)*. Maraglas. × 46,000.

---

vacuoles located at the 'emitting' face were filled with particles whilst, as a rule, the central, flattened regions of the cisternae were devoid of particles.

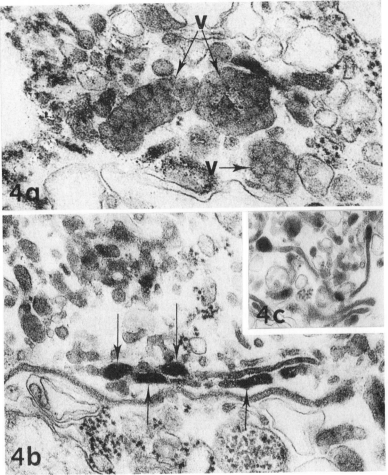

4a

4b

4c

Fig. 4.
Appearance of
representative
areas of a
glutaralde-
hyde-fixed
Golgi pellet
embedded in
Epon. Thin
sections
stained as
in Fig. 3.

(a) Vesicle (v)
containing
tightly
packed par-
ticles, 400
to 800 Å in
diameter.
× 84,000.
(b) Small
vesicular
and tubular
elements and
cisternal
structures
with areas
containing
dense material
(arrows)
(apparently
not VLDL).
× 67,000.
(c) Tubular
elements with
club-shaped
ends contai-
ning electron-
opaque mater-
ial as in (b).
× 33,000.

The appearance of the isolated Golgi complex has been described previously [3]. Additional features of special interest in the present context are illustrated in Fig. 4. The comparatively high purity of the Golgi fraction has been substantiated by biochemical analysis [3, 12]. With primary aldehyde fixation the VLDL particles were much less electron-opaque than following fixation in OsO_4, and usually appeared paler than the matrix substance (Fig. 4a). The particles were present in vesicular and vacuolar structures which as a rule were not in continuity with the cisternae. In general, there were no partic-

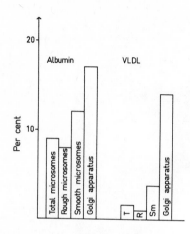

Fig. 5. Albumin and VLDL content as % of 'membranous' proteins *(cf. heading to Table 1)* in various subcellular organelles.

les present in the fenestrated plates or the cisternal tubules. On the other hand, some cisternae, or portions of cisternae, contained a highly electron-opaque material which was also observed in some of the vesicles and vacuoles surrounding the cisternal elements (Figs. 4b and c). These findings, together with the evidence obtained from the *in situ* observations, suggest that the VLDL particles are associated with spe-cial (mainly peripheral) portions of the Golgi system, while other macromole-cules are probably channelled through separate (often more centrally located) compartments of the same organelle.

At first sight, there appeared to be a discrepancy between the biochem-ical and the morphological findings since very few particles believed to represent VLDL were observed in thin sections of liver. However, this may be due to the relatively low lipid content in the lipoproteins isolated from the rough microsomes which may make the particles less prone to be 'stained' with osmium or, alternatively, with lead citrate or uranyl acetate, in comparison with the more lipid-rich VLDL in the Golgi apparatus.

The particles observed in the rough e.r. in thin sections of liver (Fig. 3) were 200-400 Å in diameter, and those in the Golgi apparatus 400-700 Å. Particles isolated from rough microsomes, embedded, and studied in thin sections were 200-400 Å in diameter; corresponding values for particles obtained from smooth microsomes and the Golgi apparatus were 400-700 Å and 500-800 Å, respectively. Thus, the changes in chemical composition are paral-leled by an increase in size of individual particles and augmented stainabi-lity in thin sections.

The data in Fig. 5 give further support to the morphological notion of an accumulation of the VLDL particles in the Golgi apparatus before discharge into the circulation. This concentration mechanism is much more conspicuous for the VLDL particles than for albumin, and agrees with our earlier propo-sal that some proportion of albumin is directly released to the sinusoid without first travelling through the Golgi apparatus [3].

Conclusions

The results presented are compatible with the view that TG and PLP are synthesized in both rough and smooth e.r. and that the formation of the VLDL takes place in a multi-site manner. In this way, the lipids are added to the apoprotein as the growing nascent macromolecule travels through the e.r. No lipids seem to be added in the Golgi apparatus. Instead the Golgi apparatus may be involved in the attachment of the carbohydrate moiety. Following conjugation, at least some of the VLDL is transported in particular form - via smooth-surfaced tubules and/or vesicles - to peripheral portions of Golgi cisternae where concentration and segregation occur.

Acknowledgement

This work was supported by grants from the Swedish Medical Research Council.

References

1. Glaumann, H. & Dallner, G., *J. Lipid Res.* 9 (1968) 720-729.

2. Morré, D.J., Hamilton, R.L., Mollenhauer, H.H., Mahley, R.W., Cunningham, W.P., Cheetham, R.D. & LeQuire, V.S., *J. Cell Biol.* 44 (1970) 484-490.

3. Glaumann, H. & Ericsson, J.L.E., *J. Cell Biol.* 47 (1970) 555-567.

4. Siekevitz, P., *Ann. Rev. Physiol.* 25 (1963) 15-40.

5. Chandra, S., *J. Micr.* 2 (1963) 297-312.

6. Stein, O. & Stein, Y., *J. Cell Biol.* 33 (1967) 319-339.

7. Hamilton, R.L., Regen, D.N., Gray, M.E. & LeQuire, V.S., *Lab. Invest.* 16 (1967) 305-319.

8. Jones, A.L., Ruderman, N.B. & Herrera, M.G., *J. Lipid Res.* 8 (1967) 429-446.

9. Parks, H.F., *Am. J. Anat.* 120 (1967) 253-280.

10. Claude, A., *J. Cell Biol.* 47 (1970) 745-766.

11. Mahley, R.W., Hamilton, R.L. & LeQuire, V.S., *J. Lipid Res.* 10 (1969) 433-439.

12. Morré, D.J., Cheetham, R.D., Nyquist, S.E. & Ovtracht, L., *Preparative Biochem.* 2 (1972) 61-69. (See also Article 21 in this book, by Morré et al.)

13. Dallner, G. & Nilsson, R., *J. Cell Biol.* 31 (1966) 181-193.

14. Glaumann, H., *Biochim. Biophys. Acta* 224 (1970) 206-218.

15. Arborgh, B., Glaumann, H. & Ericsson, J.L.E., *Lab. Invest.* (1974) in press.

16. Yunghans, W.N., Keenan, T.W. & Morré, D.J., *Exp. Molec. Path. 12* (1970) 36-45.

17. Thinès-Sempoux, D., Amar-Costesec, A., Beaufay, H. & Berthet, J., *J. Cell Biol. 43* (1969) 189-192.

18. Franke, W.W., Morré, D.J., Deumling, B., Cheetham, R.D., Kartenbeck, J., Jarasch, E.-D. & Zentgraf, H.-W., *Z. Naturforsch. 26B* (1971) 1031-1039.

23 LYSOSOMAL GLYCOSIDASES: CHEMISTRY OF β-D-GLUCURONIDASE AND A NEW α-D-MANNOSIDASE IN GOLGI MEMBRANES

Oscar Touster
Department of Molecular Biology
Vanderbilt University
Nashville, Tennessee, 37235, U.S.A.

Amongst the enzymological problems in the lysosome field which are discussed, with emphasis on glycosidases, are (1) multiple forms of hydrolases within the cell and within lysosomes, (2) analytical aspects in relation to the multiplicity of glycosidases and to the use of natural and synthetic substrates, (3) the chemical nature of lysosomal enzymes, and (4) possible biosynthetic pathways for these enzymes. The fact that few lysosomal enzymes have been isolated limits our understanding of (4). While it is likely that these enzymes are glycoprotein in nature, the suggestion that they contain neuraminic acid is based merely upon indirect evidence. Our recent work on the chemistry of β-glucuronidase is discussed. The β-glucuronidase of rat liver lysosomes has been isolated and yielded some new information, e.g. it is a tetrameric glycoprotein, but the limited amount of purified enzyme available made it desirable to investigate the β-glucuronidase of the preputial gland of the female rat, the richest known source. The enzymes from both sources are glycoproteins containing four subunits, each of ∿70,000 mol.wt., and they cross-react immunologically. Studies on the preputial gland enzyme indicate that the subunits are identical or very similar, that the enzyme has four catalytic sites, and that it has an unusual carbohydrate composition in that it contains mannose, glucosamine, and fucose, with little galactose and no sialic acid. The preputial enzyme is essentially the secretory form of β-glucuronidase, the lysosomes themselves containing very little in comparison to the total amount in the gland. Microheterogeneity was found, and several forms of the enzyme have been separated and are now being analyzed. Brief consideration is also given to our finding of a new α-D-mannosidase in rat liver Golgi membranes. This discovery stemmed from an examination of membrane sub-fractions for several glycosidases. The properties of the lysosomal, soluble cytoplasmic and Golgi mannosidases are quite different, for example, in stability, pH optima, and electrophoretic mobility. These studies highlight some problems in relation to the use of marker enzymes and to our understanding of the origin and role of some subcellular enzymes.

The aim of this contribution is to review problems in the lysosome area and to discuss some of our work on glycosidases that bear on studies of intracellular membranes as well as of lysosomes. Much of the fundamental biology of lysosomes remains to be uncovered. Little is known about their biosyn-

thesis and turnover, or about the chemistry and biosynthesis of their constituent enzymes. Nonetheless, it is very evident that these organelles are
important to the cell, for example, in intracellular digestive processes and
in a variety of developmental and immunological phenomena and in fertili-
zation mechanisms. The existence of the lysosomal storage diseases in man
is testimony to their importance. Recent interest in abnormalities in the
surface glycopolymers of tumour-transformed cells has focused interest on
glycosidases, which are of course abundant in lysosomes.

Lysosomes have clearly been demonstrated as having an extensive capacity to digest the whole range of cellular macromolecules, including proteins,
mucopolysaccharides, glycoproteins, glycolipids, and nucleic acids. Fig. 1
shows the complete degradation of sphingolipids as it is catalyzed by enzymes of mammalian lysosomes. Also indicated are diseases known to be associated with deficiencies of some of these enzymes. Since we shall emphasize
glycosidases in the present lecture, presented in Table 1 is a list of lysosomal diseases which are considered to result from particular glycosidase
deficiencies. This area is not only of obvious medical interest, but also
presents many fundamental genetic and biochemical questions. The two enzymes
on which we have worked for some time, β-glucuronidase and α-mannosidase,
have in recent years also been associated with storage diseases. Mannosidosis has been found in a human [1, 2] and more recently in cattle [3], and
in 1971 Sly and his associates [4] were able to demonstrate that a case of
atypical mucopolysaccharidosis in an infant could be associated with a lack
of β-glucuronidase. Lysosomal diseases offer the best opportunities for
enzyme therapy, since the enzymes should go directly to the site of the
defects. Published reports have not been encouraging, but it is said that
recent attempts have been more successful.

MULTIPLE FORMS OF THE ENZYMES

I emphasize at the outset that there are multiple forms of many lysosomal
enzymes. Before discussing this aspect more closely, let us consider possible bases for the presence of multiple forms in a tissue extract (Table 2).
One possibility is that there are isozymes, i.e. genetic variants in the
polypeptide itself. A second possibility is that, although the polypeptide
backbones are identical, there are minor differences, e.g. in the extent of
glycosylation. A common notion at the present time is that some enzymes
after synthesis on polysomes are glycosylated in a stepwise manner as they
pass through membrane systems, probably including the Golgi apparatus, and
are processed for secretion or for packaging into lysosomes. This concept
would dictate that there might be found enzymes at various stages of partial
glycosylation. The third possibility is that the multiple forms encountered
are artifacts. This has happened before, as may have been the experience
of some readers, due to the action of proteases on enzymes that were being
extracted, purified and analyzed in the laboratory. Finally, it is of course

Table 1. Inborn lysosomal diseases with assigned glycosidase deficiencies

Disease	Enzyme	Major substance accumulated
Pompe's disease (glycogen storage disease II)	α-Glucosidase	Glycogen
Hurler's syndrome, Scheie syndrome	α-L-Iduronidase	Dermatan and heparan sulphates
Sanfillipo syndrome Type B	α-N-acetyl hexosaminidase	Heparan sulphate
Fucosidosis	α-L-Fucosidase	Fucose-containing glycoproteins and glycolipids
Atypical mucopolysaccharidosis	β-Glucuronidase	Mucopolysaccharide
Mannosidosis	α-Mannosidase	Mannose-containing oligosaccharides & glycopeptides
Generalized gangliosidosis	β-Galactosidase	Ganglioside GM_1
Tay Sachs disease	β-N-acetylhexosaminidase A	Ganglioside GM_2
Sandhoff's disease	β-N-acetylhexosaminidase A and B	Globoside + ganglioside GM_1
Fabry's disease	α-Galactosidase A	Ceramide trihexoside
Lactosyl ceramidosis	β-Galactosidase	Ceramide dihexoside
Krabbe's leukodystrophy	β-Galactosidase	Ceramide galactoside
Gaucher's disease	β-Glucosidase	Ceramide glucoside

possible that the multiple forms do indeed represent truly different enzymes, in the sense that they are coded for by different genes and therefore have different properties and functions. They may simply have the ability to hydrolyze the same test substrate which is being used.

Other complexities exist. In Table 3 are listed some examples of enzymes which are known to occur both in lysosomes and in extralysosomal sites of rat liver. The lysosomal and microsomal forms of β-glucuronidase, while not identical, are very similar and are undoubtedly related to each other chemically, as is discussed more fully below. The aryl sulphatases in these two compartments are known to have different specificities. Little is known about the neuraminidases, although differences in pH optima, substrate

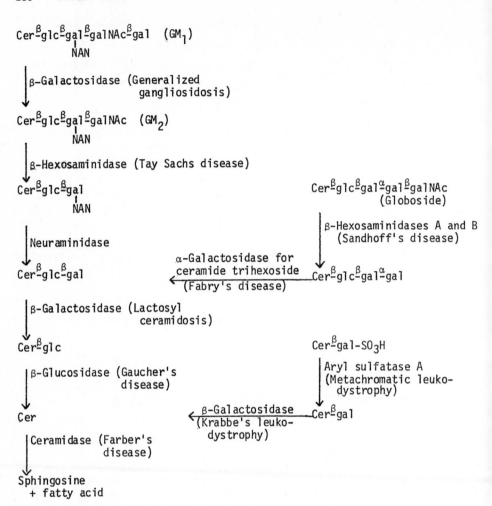

Fig. 1.   Catabolism of sphingolipids by lysosomal enzymes, and associated
genetically determined enzyme deficiency diseases.

specificity, and effects of metal ions have been reported for the lysosomal, soluble cytoplasmic, and plasma membrane activities. α-D-mannosidase occurs in three compartments, and the three forms have quite different properties, as we shall describe shortly. The galactosidases have also been found in two different subcellular fractions.

Table 2. Possible bases for multiple forms of lysosomal enzymes

---

1. Isozymes (i.e. genetic variants in polypeptide)

2. Biosynthesized 'multiple forms' derived from the same translation products (e.g. enzymes varying in extent of glycosylation)

3. Artifacts (e.g. products formed during extraction as a result of the degradative action of proteases or glycosidases)

4. Truly different enzymes which simply share a common test substrate

---

Table 3. Similar enzymatic activities in lysosomal and non-lysosomal sites of rat liver

---

β-Glucuronidase - 2/3 in lysosomes, 1/3 in endoplasmic reticulum

Aryl sulphatase A and B in lysosomes, C in endoplasmic reticulum

Neuraminidase - in lysosomes, soluble cytoplasm, and plasma membranes

α-D-Mannosidase - in lysosomes, soluble cytoplasm, and Golgi membranes

α- and β-D-Galactosidases - in lysosomes and in plasma membranes

---

Lysosomes themselves contain multiple forms of enzymes (Table 4). In regard to α-galactosidase, β-galactosidase, and β-glucosidase it has already been demonstrated that the different forms of these enzymes are associated with different genetic diseases in the human. We would expect, therefore, that the forms differ from each other in at least one subunit. There is now evidence that β-N-acetylhexosaminidases A and B contain a common subunit [5]. A few years ago we carried out a study on rat liver lysosomal phosphatases and chromatographically separated three of them, one of which was demonstrated to be a nucleotidase and not a general phosphatase [6]. These examples serve to illustrate the kinds of problems that arise in working with lysosomal enzymes.

These multiple forms lead to some analytical problems which are frequently neglected in carrying out biological studies. For example, in investigations designed to determine the effects of transformation by tumourviruses, routine assays of a number of glycosidases are frequently performed. Some of the enzymes being studied exist in more than one form and have different pH optima. Usually simple synthetic substrates are used at a single pH for convenience. It is obvious that important changes in the level of one particular form of the enzyme under analysis may not be detected. Examples

Table 4. Multiple forms of enzymes in liver lysosomes

Aryl sulphatase - A and B

α-Galactosidase - one with greater thermolability related to Fabry's disease

β-Galactosidase - A, B, C

β-Glucosidase - one with lower pH optimum related to Gaucher's disease

β-N-Acetylhexosaminidase - A and B

β-Glucuronidase - a major and a minor form

Phosphatase - at least 3, one a nucleotidase

DNase - a second very minor form also present?

---

may be quoted where it has been demonstrated that the action of a particular enzyme on simple synthetic substrates and on the natural substrates is quite different. Ultimately, it will be very important to be able to utilize the natural substrates at the optimal conditions for catalytic action on these substrates. Very recently the experiment shown in Fig.2 was reported [7]. Note that not only do the pH optima for the action of ceramide trihexosidase on the natural and synthetic substrates differ, but also that if a pH of 4.5 or 5 were being used in a survey, the natural and synthetic substrates would give very greatly different results. Fig. 3 shows an experiment from our own laboratory [8] illustrating how the conditions of the analysis drastically influence the results in the assay of lysosomal DNase. In regard to the lipid substrates, it is obvious that the need to disperse the substrate with detergents can lead to serious problems.

We have a very poor understanding about the chemistry of lysosomal enzymes, few such enzymes having been isolated in pure form for chemical studies. In view of the fact that there is often more than one intracellular form, it is frequently important that a purified lysosomal fraction be used as the raw material for the isolation of the enzyme. However, the preparation of such a fraction can be accomplished only in poor yield, partly because the lysosomal fraction of a cell is a relatively minor one. Consequently, inferences are frequently made about the composition of lysosomal enzymes on the basis of indirect evidence. In particular, they are considered by many to be glycoproteins on the basis of the staining of gels. It has been claimed that they are sialoproteins because neuraminidase alters their electrophoretic mobility in a manner that would be expected if neuramic acid were removed from the enzymes [9]. There is now increasing suspicion that such studies may be misleading. Some of the work cannot be repeated. In other cases alternate interpretations have been suggested.

β-GLUCURONIDASE

I now briefly review some of the work that we have done on β-glucuronidase, and firstly mention why we have studied this enzyme. It had long been considered that microsomal β-glucuronidase and lysosomal β-glucuronidase are identical, and two laboratories have reported that the microsomal form is the biosynthetic precursor of the lysosomal form [10, 11]. While this may in fact be the situation within the cell, the work that was reported was far from convincing. First, the forms are not identical and can be easily separated chromatographically. Second, the two laboratories did isotopic labelling experiments and counted the β-glucuronidase in impure form, a practice that has no validity when tracer experimentation is being employed. In Fig. 4 is shown a chromatographic analysis of a lysosomal-mitochondrial extract on DEAE-cellulose [12]. The enzyme activity plot shows that there is a minor lysosomal as well as a major lysosomal form. The microsomal forms actually elute in the area of the minor lysosomal form. Dr. Philip Stahl has purified the major liver lysosomal enzyme several thousandfold, as shown in Table 5, to the highest specific activity thus far obtained. The enzyme was pure by all criteria applied. As is summarized in Table 6, the molecular weight is ∿ 280,000, but the enzyme is composed of sub-units of ∿ 70,000, thus establishing that rat liver lysosomal β-glucuronidase is a tetramer. Hexosamine was also found to be a constituent of the enzyme, thus confirming previous reports on less pure bovine liver β-glucuronidase that the latter is a glycoprotein [13]. Although the isola-

[continued on p. 256

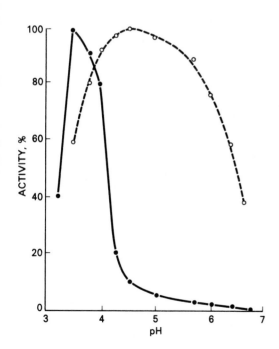

Fig. 2. The pH-activity curves of ceramide tri-hexoside α-galactosidase and methyl-umbelliferyl α-galactosidase, ●——● and o---o respectively. From M.W. Ho, ref. [7].

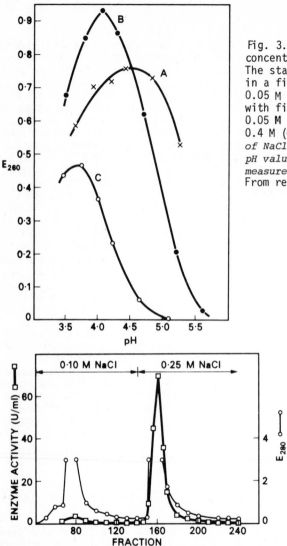

Fig. 3. Effect of pH and salt concentration on DNase activity. The standard assay was performed in a final concentration of 0.05 M acetate at various pH values, with final NaCl concentrations of 0.05 M (×——×), 0.25 M (●——●), or 0.4 M (o——o). *Use of KCl instead of NaCl gave the same results. The pH values shown are those actually measured in the assay mixtures.* From ref. [8].

Fig. 4 *(below).* DEAE-cellulose chromatography of β-glucuronidase ref.[12].

Elution pattern (o——o, A_{280}; □——□, enzyme activity) of β-glucuronidase (10,650 units, 2,695 mg) from a DEAE-cellulose column (3 x 100 cm). The column was washed sequentially with 0.10 M and 0.25 M NaCl in 0.005 M Tris-phosphate buffer, pH 7.8. The flow rate was 120 to 300 ml per h.

Fractions were 20ml. *The enzyme eluted with 0.25 M NaCl was pooled, dialyzed, and used in the next step of purification.*

Table 5. Purification of β-glucuronidase from rat liver lysosomes [ref.12].

Fraction	Protein, mg	Enzyme activity, units	Specific activity, units/mg	Yield, %	Purification ratio
Homogenate	130,000	46,500	0.36	100	1
Mitochondrial-lysosomal (ML) fraction	16,500	18,200	1.10	39	3
ML extract	9,660	16,000	1.65	34	5
Ammonium sulphate	2,695	10,650	3.95	23	11
Sigma DEAE-cellulose	708	10,550	14.9	23	41
Whatman DE-32	458	9,670	21.1	21	59
Whatman CM-cellulose	40.5	6,000	148	13	410
Sephadex G-200	14.3	4,800	336	10	933
5% Gel	1.63	3,940	2,400	8	6,660
7% Gel - polylysine	0.976	2,950	3,024	6	8,400
+ polylysine*	0.976	4,250	4,360		

assayed with 1 mg/ml of poly-D-lysine

Table 6. Molecular weight of rat liver lysosomal β-glucuronidase and its sub-unit. Data from ref. [12].

Technique for determination	Mol. wt.

Whole protein

1. Sucrose density gradient centrifugation
   - calculated on basis of yeast ADH and *E.coli* β-galactosidase — 340,000
   - calculated on basis of catalase — 265,000
2. Sepharose 4B filtration — 270,000
3. Gel electrophoresis, varying acrylamide concentrations — 280,000

Sub-unit

1. Sephadex G-150, 8 M urea, pH 5.0 — 68,000
2. SDS gel electrophoresis — 68,000 — 78,000*

* + urea → dimer of 155,000

Table 7. Purification of rat preputial gland β-glucuronidase. *Tissue from 50 rats was used.*

Fraction	Protein, mg	Total units	Specific activity	Recovery, %
Crude extract	556	120,000	215	100
70% $(NH_4)_2SO_4$	276	105,000	380	87
G-200	47	75,000	1600	62
DE-52	27	62,000	2300	52
5→20% Sucrose gradient	25	57,000	2300	48

tion procedure yielded enough enzyme for us to do an extensive study on the mechanism of catalysis and the active site of this enzyme [14, 15], details of which will not be given here, there was insufficient material for the kind of chemical study that is desirable. Moreover, at the present time the isolation of the pure microsomal form has not been reported in detail. However, we do know that pure form has been isolated by Dr. P.D. Stahl [16]. It has a specific activity considerably lower than that of the lysosomal enzyme, and the effect of pH differs to a small extent. The desired structural comparison of the microsomal and lysosomal forms has not yet been accomplished.

We therefore turned to the richest known source of β-glucuronidase, the preputial gland of the female rat. The enzyme can be isolated in high yield and purity by the method shown in Table 7, which is a slightly modified version of that originally described by Ohtsuka and Wakabayashi [17] Dr. Ken Keller found the β-glucuronidase from this gland to be very similar to the liver forms. Molecular weight and sub-unit molecular weight are identical to the lysosomal enzyme, and there is immunological cross-reactivity. From the number of peptides observed in the fingerprint analysis of tryptic hydrolysates, we can conclude that the four sub-units are identical or very similar. Binding studies with the potent competitive inhibitor 1,4-saccharolactone indicate that there are four binding sites per tetramer. I will not comment on the amino acid analysis of this enzyme but it is relevant to mention the unusual carbohydrate composition. In many analyses by gas-liquid chromatography the enzyme was found to contain mannose, glucosamine, and fucose in a ratio of about 10 to 5-6 to 1. There was only a trace of galactose. No sialic acid could be demonstrated by direct chemical analysis. It appeared that the carbohydrate occurred in an average of two clusters per sub-unit.

Recently Dr. Daulatram Tulsiani in our laboratory found that he could separate the highly purified preputial β-glucuronidase into several active forms by chromatography on hydroxyapatite. The main forms have been analyzed for sugar composition. They differ in total carbohydrate, and in fucose content, and they have more galactose than found in our earlier preparations.

Such microheterogeneity is common among glycoproteins but we are surprised that our rats are now producing β-glucuronidase richer in carbohydrate, particularly galactose. It should be pointed out, however, that most of the enzyme of the gland is in a soluble, easily extractable form. It must be acknowledged, therefore, that we are probably studying terminal secretory forms.

Different forms of glycopolymers complicate studies to elucidate biosynthetic processes. Relevant to therapeutic experiments, the work of Ashwell and his colleagues should be emphasized. Since they have demonstrated that the nature of the terminal sugars of glycoproteins is so decisive in determining binding to, and uptake by, the liver, the structure of the particular form of a lysosomal enzyme being used in therapeutic experiments may be crucial to the outcome of these experiments.

α-MANNOSIDASES

I now consider a study we have made on intracellular mannosidases [18]. In view of the current interest in glycosidases in connection with the occurrence of abnormal glycopolymers in tumour-transformed cells, we decided to examine intracellular membrane fractions for the possible presence of glycosidases. Initially we utilized the fractionation scheme which we published in 1970 for the preparation of plasma membranes [19]. Our method involved the homogen .ation of tissue in 0.25 M sucrose buffered to pH 8, and permitted the isolation of several different subcellular fractions, and the isolation from them, in a discontinuous sucrose density gradient, of the plasma membranes in the nuclear and microsomal fractions (Fig. 5). An appreciable portion of the plasma membranes separates with the nuclei from homogenates made in 0.25 M sucrose, and the partitioning of an enzyme between these two fractions constitutes strong evidence for its plasma membrane localization. When we presented this method we had found that all chemical and enzymatic comparisons of the plasma membranes isolated from the two subcellular fractions gave essentially the same results. Plasma membrane material from the nuclear fraction was designated N_2 and that from the microsomal fraction P_2. I refer again to these fractions below.

In our first experiment in the glycosidase survey we found that of eight glycosidases assayed, only three were present in appreciable quantities in P_2. One of them, α-glucosidase, had a pH optimum near 7 and was more abundant in P_4, the microsomal sub-fraction enriched in endoplasmic reticulum, than in P_2. This enzyme obviously was the neutral glycosidase originally reported in microsomal fractions by Lejeune *et al.* [20]. The second enzyme found in P_2 was N-acetyl-β-hexosamindase, a lysosomal enzyme well known to absorb readily into membranes. It could be removed from P_2 by salt washing. The third glycosidase found in P_2 was α-D-mannosidase, which was resistant to removal by salt.

With this interesting and unexpected result, Dr. Beatrice Dewald under-
took a systematic examination of intracellular fractions with the use of a
number of marker enzymes. The fractionation scheme of de Duve *et al.* [21]
was used when we wished to obtain intact lysosomes and mitochondria. I
should mention that in addition to the α-D-mannosidase with a pH 4.5 optimum
that is well known to occur in lysosomes, Gourlay and Marsh [22] reported in
1971 on the presence of a second α-D-mannosidase of pH optimum of about 6.5
in the soluble cytoplasmic fraction of rat liver. It is the lysosomal acid
mannosidase which is missing in the disease mannosidosis. In Fig. 6 is plotted the relative specific activity for several enzymes. As expected, the
α-mannosidase of pH 4.5 optimum is concentrated in the lysosomes with aryl
sulphatase. On the other hand, the pH 6.5 optimum α-mannosidase is in the
soluble fraction. β-Mannosidase, first shown to be in lysosomes by LaBadie
and Aronson [23], was also found in the lysosomal fraction. Since we had
observed in the original survey of glycosidase activities that the P$_2$ manno-

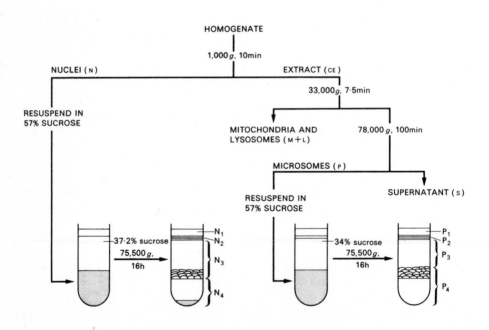

Fig. 5.   Isolation of plasma membrane fractions (N$_2$ and P$_2$) from nuclear
and microsomal fractions obtained from isotonic sucrose homogenate [ref. 19].

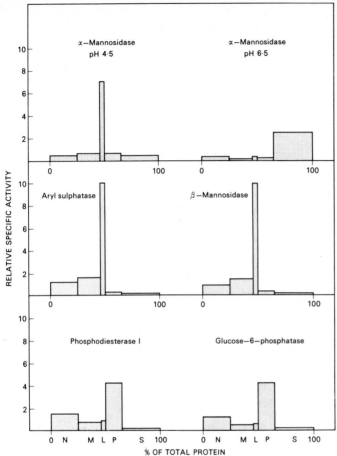

Fig. 6. Intracellular distribution of enzymes in rat liver (Dewald & Touster, ref. [18].) The results are presented by de Duve's method [21]. Relative specific activity = % of total activity /% of total protein. On the abscissa the fractions are represented by their relative protein content: N, nuclear; M, mitochondrial; L, lysosomal; P, microsomal; S, soluble cytoplasmic. α-Mannosidases (pH 4.5 and 6.5) were determined at the indicated pH values under standard conditions; β-mannosidase was assayed in 0.05 M acetate buffer, pH 4.5.

sidase had a pH optimum near 5.5, we examined a number of subcellular fractions for α-mannosidase activity at pH 4.5, 5.5 and 6.5. Fig. 7 (left side) shows that the pH 4.5 and 6.5 mannosidases are concentrated in the lysosomal and soluble fractions as expected, whereas the pH 5.5 form shows a different distribution pattern.

Further work on the microsomal subfractions, P_2, P_3, and P_4, was

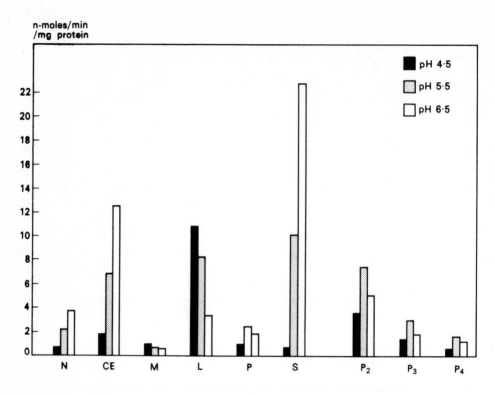

Fig. 7. α-Mannosidase activity measured at three pH values in different subcellular fractions of rat liver. The vertical bars represent the specific activity at pH 4.5, 5.5 and 6.5. CE, extract after removal of nuclei (cf. Fig. 6); P_2, P_3 and P_4, sub-fractions of microsomes (*see text*; cf. Fig. 5); other fractions are as designated in Fig. 6. (Dewald & Touster, ref. [18].)

carried out, these fractions being washed with salt to remove adsorbed manno-sidases and other enzymes. The mannosidase was again concentrated in fraction P, with the highest activity evident at pH 5.5. Table 8 shows the specific activity of a number of marker enzymes in P and in its sub-fractions P_2, P_3, and P_4. The high specific activity of glucose-6-phosphatase indicates that P_4 is enriched in endoplasmic reticulum, whereas the high 5'-nucleotidase and phosphodiesterase I indicates that P_2 is enriched in plasma membranes. Similarly, P_2 is enriched in α-mannosidase. On the other hand, N_2, the plasma membrane fraction isolated from the nuclear pellet, exhibited a much

Table 8. Specific activities of α-D-Mannosidase (pH 5.5) and of marker enzymes in rat liver microsomes and corresponding sub-fractions *(see Fig. 5)*. Values are expressed as units/mg protein.

Enzyme	P	P_2	P_3	P_4
Glucose-6-phosphatase (4)*	0.213 ±0.068	0.133 ±0.033	0.200 ±0.057	0.271 ±0.061
Esterase (6)	6.68 ±0.49	3.29 ±0.59	9.71 ±2.10	7.87 ±2.02
α-Glucosidase (3)	3.56 ±0.12	1.63 ±0.18	3.53 ±0.21	3 56 ±0.04
5'-Nucleotidase (3)	0.209 ±0.041	1.44 ±0.20	0.347 ±0.084	0.147 ±0.041
Phosphodiesterase I (7)	0.114 ±0.013	0.951 ±0.153	0.310 ±0.066	0.085 ±0.017
Galactosyl transferase (2)	0.562	2.84	0.907	0.219
α-Mannosidase (7)	2.04 ±0.23	†7.37 ±1.44	3.25 ±0.70	1.37 ±0.19

*No. of observations.
†N_2, the plasma membrane fraction isolated from the nuclear pellet, had a specific activity of only 0.92.

lower specific activity for the mannosidase. This was the first indication that P_2 and N_2 were not essentially identical. In seeking to determine the cause of the difference between P_2 and N_2 in mannosidase activity, one obvious explanation to test was that there might be Golgi membranes or other contaminants in the P_2 fraction. We therefore tested for the Golgi marker enzyme galactosyl transferase. Fraction P_2 was indeed enriched in this transferase, N_2 again having lower activity.

These results led us rather rapidly to a direct study of Golgi preparations, for which Dr. Dewald employed the method of Leelavathi *et al.* [24] (Fig. 8). In this procedure the post-nuclear extract is layered on 1.3 M sucrose and centrifuged at 105,000 g for 60 min, the crude membrane fraction being used for the isolation of Golgi membranes on a discontinuous sucrose gradient. In our work we added a third discontinuous sucrose centrifugation as well as a salt wash of the membranes to remove most of the adsorbed proteins. An electron micrograph of our preparation is shown in Fig. 9. Tubular profiles, secretory vesicles and cisternae are the dominating structures as originally observed by Leelavathi *et al.* [24]. Table 9 indicates that this Golgi preparation contains 39% of the galactosyl transferase present in the original post-nuclear extract. This yield is lower than that of Leelavathi *et al.* presumably because we added a third gradient centrifugation and

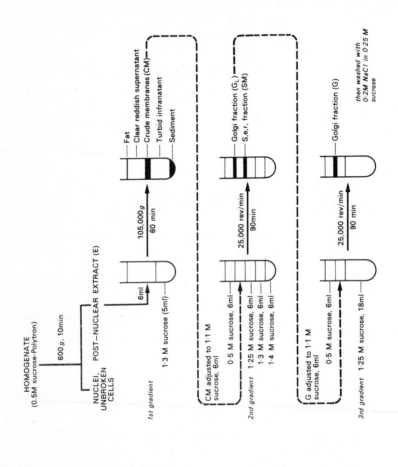

Fig. 8. Preparation of Golgi-rich membrane fraction, by method of Leelavathi & collaborators [24] with an additional centrifugation step. – *See text.*

Table 9. Enzyme activities in fractions obtained from the preparation
of Golgi membranes

*NOTE: The galactosyl transferase is enriched 48-fold in G as compared to E.
By comparing the α-mannosidase activities at pH 4.5, 5.5 and 6.5 in various
fractions and making reasonable assumptions about the distribution of the
pH 4.5 and 6.5 enzymes, an estimate of the enrichment of the pH 5.5 mannosidase in G over E yielded the value of 46-fold.*

	Post-nuclear extract (E) u/g liver	Crude membrane fraction (CM) %	Smooth membrane (SM) %	Final Golgi fraction (G) %
Protein (6)*	79.85 ± 8.60	12.6 ± 3.2	3.55 ± 1.0	0.7 ± 0.14
Glucose-6-phosphatase (3)	4.51 ± 0.48	13.2 ± 1.9	4.0 ± 0.4	0.7 ± 0.2
5'-Nucleotidase (3)	2.88 ± 0.28	33.0 ± 4.8	14.5 ± 2.7	4.7 ± 1.3
Phosphodiesterase I (6)	2.00 ± 0.07	30.7 ± 3.9	14.3 ± 4.5	6.1 ± 1.3
Galactosyl transferase (6)	15.92 ± 2.92	71.6 ±18.8	4.6 ± 1.6	39.5 ±11.5
α-Mannosidase (6)	469.4 ± 70.4	11.8 ± 3.4	1.45 ± 0.4	2.4 ± 0.7
	u/mg protein	u/mg protein	u/mg protein	u/mg protein
Glucose-6-phosphatase (3)	0.053 ± 0.004	0.062 ± 0.019	0.071 ± 0.021	0.049 ± 0.004
5'-Nucleotidase (3)	0.034 ± 0.002	0.099 ± 0.029	0.156 ± 0.029	0.245 ± 0.016
Phosphodiesterase I (6)	0.026 ± 0.003	0.063 ± 0.011	0.103 ± 0.018	0.218 ± 0.024
Galactosyl transferase (6)	0.220 ± 0.046	1.14 ± 0.203	0.254 ± 0.038	10.6 ± 1.27
α-Mannosidase (6)	5.95 ± 1.22	5.58 ± 1.50	2.48 ± 0.57	19.1 ± 1.27

* *No. of observations*

salt-washed the final product. The yield of mannosidase is misleading because the original extract contains all three mannosidases. The overall enrichment of the transferase was 48-fold. By comparing the α-mannosidase
activity at pH 4.5, 5.5 and 6.5 in various fractions, it was possible to obtain a rough estimate of 7% for the contribution of the Golgi membrane mannosidase to the total activity in E. By making other assumptions regarding the
distribution and amount of the soluble cytoplasmic enzyme and the lysosomal
enzyme, we estimated that a 46-fold increase in the specific activity of the

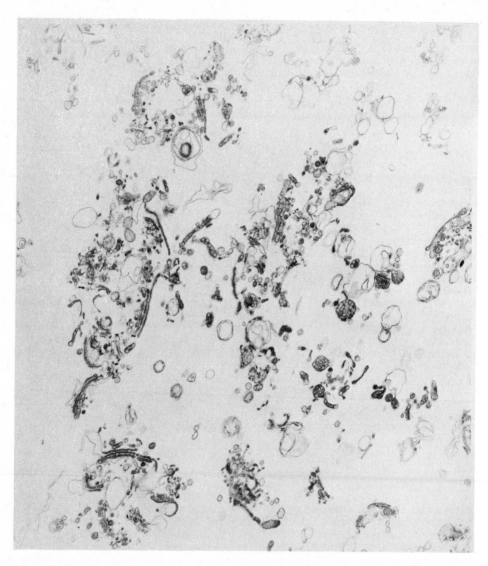

Fig. 9. Electron micrograph of a Golgi membrane fraction of rat liver prepared by modification of the method of Leelavathi *et al.* [24]. To an aliquot

[Legend continued opposite

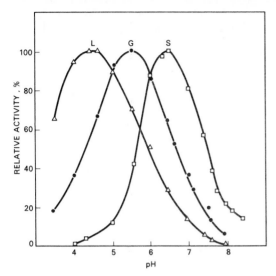

Fig. 10. Effect of pH on α-mannosidase activity of the lysosomal (L, △), Golgi (G, ●) and cytoplasmic (S, □) fractions. The buffer was varied as follows: 0.05 M sodium acetate, pH 3.5 to 5.5; 0.05 M sodium cacodylate, pH 5.0 to 7.5; 0.05 M sodium phosphate, pH 6.0 to 8.0; and 0.05 M Tris-HCl, pH 7.5 to 8.5. Optimal activity of each fraction is expressed as 100%. *At a given pH, different buffers gave very similar values.* (Dewald and Touster, ref.[18].)

Golgi enzyme was effected by preparing fraction B from E, which is very similar to the 48-fold enrichment obtained for galactosyl transferase.

Both fraction P_2 and the Golgi-rich preparation were used in solubilization experiments. Neither salt nor Triton X-100 released the enzyme from the membrane. However, deoxycholate at a concentration of 0.5% released 80% to 95% of the α-mannosidase activity from either of the membrane preparations; but this extracted activity did not appear to be in pure solution. The activity remained in the supernatant after centrifugation at 105,000 *g* for 60 min, but 60% of it was found in the pellet after centrifugation at 165,000 *g* for

Legend to Fig. 9 (opposite) cont'd.

of the Golgi-rich fraction (approximately 0.4 mg of protein) was added an equal volume of 5% glutaraldehyde in 0.2 M sodium cacodylate buffer, pH 7.4. Before pelleting the membranes, this mixture was kept at 4° overnight. The pellet was post-fixed in 1% osmium tetroxide in 0.1 M veronal-acetate buffer, pH 7.4, containing 2.4 mM $CaCl_2$ and 0.06 M NaCl for 2 h at 4°. All samples were block-stained with 0.5% uranyl acetate in veronal-acetate buffer, pH 7.4, containing 2.4 mM $CaCl_2$ and 0.06 M NaCl for 2 h at 4°, and then with 0.5% uranyl acetate in veronal-acetate buffer, pH 6.0, for 2 h at room temperature before dehydration in a series of increasing ethanol concentrations. The samples were embedded in Araldite. Thin sections were cut on an LKB Ultratome (LKB Instruments, Inc. Rockville, Md.) with diamond knives (Du Pont de Nemours). The mounted sections were doubly stained with 1% uranyl acetate followed by lead citrate and were examined in a Hitachi Hu-11B electron microscope. Magnification x20,000. (Dewald and Touster, ref. [18].).

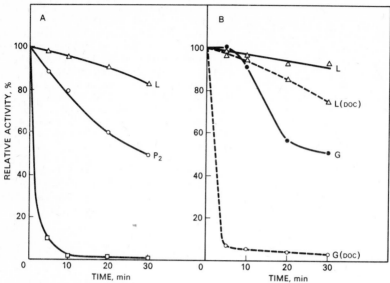

Fig.11. Heat inactivation at 55° of α-mannosidase activity of the lysosomal (L), Golgi (P₂ or G) and cytoplasmic (S) fractions.
A.- L, P₂ and S were incubated at 55° for 5, 10, 20 and 30 min in 0.25 M sucrose containing 5 mM Tris-HCl, pH 8.0, and 0.58 mg of protein per ml. (L contained in addition 0.1% Triton X-100.) Appropriate aliquots were then assayed under standard conditions.
B.- Incubation and assays were done as in A. Protein concentration in L was 0.82 mg per ml, in G 0.70 mg per ml. L (DOC) and G (DOC) contained in addition 0.5% DOC. (Dewald and Touster, ref. [18].)

3 h. Even more of the activity was sedimented if much of the detergent was removed by dialysis prior to the higher speed centrifugation. It is evident, therefore, that the detergent is required to maintain the enzyme in an unsedimentable form.

The pH-activity curves of the lysosomal, Golgi, and soluble cytoplasmic α-mannosidases are shown in Fig. 10. In addition to the clear differences in pH optima and the tightly bound nature of the Golgi membrane enzyme, other differences among the mannosidases are encountered. K_m values toward two substrates are shown in Table 10. The high K_m values of the membrane enzyme are not particularly disturbing in connection with possible biological function. We are of course using synthetic substrates. K_m values have been reported of similar magnitude, for mammalian α-mannosidases, and there are glycosidases for which natural substrates have greater affinity than do artificial substrates. It will therefore be of considerable interest to determine the substrate activity of the mannosidases on naturally occurring metabolites, including glycoproteins and glycopeptides. The differing stabilities of the three enzymes are shown in Fig. 11. The lysosomal form is rather stable at 55°; the soluble form is quite unstable. The deoxycholate-released membrane enzyme is also quite unstable.

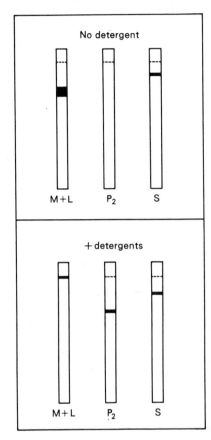

Fig. 12. Electrophoresis of lysosomal (M + L), Golgi (P$_2$) and cytoplasmic (S) α-mannosidases in polyacrylamide gel. The preparation of the samples, the electrophoresis conditions, and the staining of the gels have been described (Dewald and Touster, ref. [18].)
The dashed line marks the junction between stacking gel and separating gel.
A. No detergent was present in either the gels or the electrophoresis buffer.
B. The separating gel, stacking gel and electrophoresis buffer contained 0.1% Triton X-100 and 0.1% DOC. In all samples the concentration of both detergents was adjusted to 0.5%.

Electrophoretic comparisons of the three mannosidases are shown in Fig. 12. We had a great deal of trouble with the Golgi enzyme in that we were at first unable to get it to migrate in polyacrylamide gel electrophoresis. As shown in the upper panel, both lysosomal and soluble forms migrate in the absence of detergent. However, it was necessary to add two detergents to the system in order to get the Golgi enzyme to move. Both Triton X-100 and deoxycholate were present in the acrylamide gels and in the electrophoresis buffer in the experiment shown in the lower panel. Various other differences amongst the three enzymes were found. For example, methyl-α-D-mannoside was unable to inhibit the hydrolysis of *p*-nitrophenyl-α-D-mannoside by the Golgi enzyme, in distinction to its ability to inhibit this hydrolysis by the other two α-mannosidases.

Much work needs to be done with these rat liver mannosidases. They

a)   syn. in r.e.r. ——➤ s.e.r. ——➤ Golgi ——➤ lysosomes

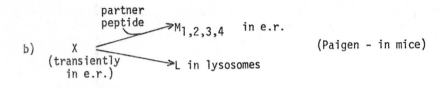

c)   By uptake from extracellular fluid
     (Hickman and Neufeld fibroblast study of lysosomal
     enzymes of I-cell disease)

Fig. 13.   Speculations on the origin of lysosomal enzymes.
Rough and smooth endoplasmic reticulum are denoted r.e.r. and s.e.r.

have not been purified, they have not been extensively studied in relationship to substrate specificity, and there is no sound basis on which to speculate about the function of the Golgi or the soluble cytoplasmic enzyme. It
is possible that the Golgi enzyme is the precursor of the lysosomal form, but
a considerable change in the properties of the enzyme would be required if
this biosynthetic relationship existed. Also, in view of the evidence of the
involvement of the Golgi in glycoprotein metabolism and of the abundance of
mannose in glycoproteins, it is possible that the membrane mannosidase has
some role in the biosynthesis or turnover of glycoproteins. Again, this is
speculation.

CONCLUDING REMARKS

A few remarks on the origin of lysosomal enzymes are appropriate (Fig. 13).
The most general notion involves passage of lysosomal enzymes from the rough
endoplasmic reticulum through various membrane systems, the enzymes becoming
increasingly glycosylated during this transport process. A second scheme ori-
ginates from the work of Paigen on the β-glucuronidase in mouse kidney. He
has found a mutant which has little microsomal β-glucuronidase but the normal
amount of lysosomal enzyme. He suggests that the precursor X cannot be converted to the normal microsomal forms because of a genetic deficiency, that
might lead to a deficiency of a partner peptide required for the conversion
of X into the normal microsomal forms. Indeed, Swank and Paigen [25] have
found that if microsomal β-glucuronidase is treated with urea or trypsin it
is converted from a variety of higher molecular weight into a single form of
molecular weight ∿260,000. The native microsomal forms, M_1, M_2, M_3 and M_4,

Table 10. Apparent K_m values of the lysosomal (L), Golgi (G or P_2), and soluble cytoplasmic (S) α-mannosidases

Substrate	K_m			
	L	G	P_2	S
	mM	*mM*	*mM*	*mM*
p-Nitrophenyl α-D-mannopyranoside	8.5	∿28	∿25	0.16
4-Methylumbelliferyl α-D mannopyranoside	3.2	7.4	7.7	0.11

contain, according to their experimental data, from one to four subunits of the 50,000 mol. wt. partner peptide. According to this scheme (Fig. 13) the microsomal forms which are normally found in these membranes are not the precursors of the lysosomal forms, although of course a common precursor X would exist transiently in the rough endoplasmic reticulum where it was being biosynthesized. Lastly, mention should be made of the recent work in Neufeld's laboratory on the nature of the defect in the lysosomal storage disease known as I-cell disease. The low activity of many lysosomal enzymes in these cells had been speculated as perhaps being due to the existence of leaky lysosomes. Hickman and Neufeld [26] have presented evidence indicating that I-cell lysosomes are not leaky. Moreover, they have found that lysosomal enzymes prepared from normal tissue are taken up readily by normal fibroblasts, whereas enzymes isolated from I-cell tissue are not so readily incorporated by these cells. They therefore speculate that the I-cell enzymes themselves are abnormal and that, at least in the case of certain tissues, perhaps a deficiency of lysosomal enzymes may be due to inadequate uptake from extracellular compartments. Further consideration of the possible extracellular origin of lysosomal hydrolases will obviously require supporting evidence of a more physiological nature.

In conclusion I mention that, although there has been a great advance in our understanding of the role of lysosomes in biological and pathological processes, and although we are gradually learning more about their constituent enzymes, there are really major problems still ahead. Aside from the question of the mechanism of formation of these organelles and of their constituent enzymes, there is little known about intracellular vacuolar fusion, which almost certainly involves interesting membrane phenomena as well as chemical determinants of the rate and specificity of membrane fusion events.

Acknowledgements

Previously unpublished work mentioned in this paper was supported in part by Grant CA-07489 of the National Cancer Institute, U.S. Public Health Service, and Grant No. GB33176X of the National Science Foundation.

References

1. Öckerman, P.A., *Lancet ii* (1967) 239-241.

2. Kjellman, B., Gamstorp, I., Brun, A., Öckerman, P.A. & Palmgren, B., *J. Pediat. 75* (1969) 366-373.

3. Hocking, J.D., Jolly, R.D. & Batt, R.D., *Biochem. J. 128* (1972) 69-78.

4. Quinton, B.A., Sly, W.S., McAlister, W.H., Rimoin, D.L., Hall, C.W. & Neufeld, E.F., *Abstracts, 81st Annual Meeting American Pediatric Society, Inc.,* and *41st Annual Meeting of the Society for Pediatric Research,* 1971 (April 28 - May 1).

5. Srivastava, S.K. & Beutler, E., *Nature (Lond.) 241* (1973) 463.

6. Arsenis, C. & Touster, O., *J. Biol. Chem. 243* (1968) 5702-5708.

7. Ho, M.W., *Biochem. J. 133* (1973) 1-10.

8. Dulaney, J.T. & Touster, O., *J. Biol. Chem. 247* (1972) 1424-1932.

9. Goldstone, A., Konecny, P. and Koenig, H., *FEBS Lett. 13* (1971) 68-72.

10. van Lanker, J.L. & Lentz, P.L., *J. Histochem. Cytochem. 18* (1970) 529-541.

11. Kato, K., Hirohata, I., Fishman, W.H. & Tsukamoto, H., *Biochem. J. 127* (1972) 425-435.

12. Stahl, P.D. & Touster, O., *J. Biol. Chem. 246* (1971) 5398-5406.

13. Plapp, B.V. & Cole, R.D., *Biochemistry 6* (1967) 3676-3681.

14. Wang, C.-C. & Touster, O., *J. Biol. Chem. 247* (1972) 2644-2649.

15. Wang, C.-C. & Touster, O., *J. Biol. Chem. 247* (1972) 2650-2656.

16. Stahl, P., Owens, J.W., Boothby, M. & Gammon, K., *this volume,* Art. 24.

17. Ohtsuka, K. & Wakabayashi, M., *Enzymologia 39* (1970) 109-124.

18. Dewald, B. & Touster, O., *J. Biol. Chem., in press.*

19. Touster, O., Aronson, N.N., Jnr., Dulaney, J.T. & Hendrickson, H., *J. Cell Biol. 47* (1970) 604-618.

20. Lejeune, N., Thinès-Sempoux, D.and Hers, H.G., *Biochem. J. 60* (1963) 16-21.

21. de Duve, C., Pressman, B.C., Gianetto, R., Wattiaux, R. & Appelmans, E., *Biochem. J. 60* (1955) 604-617.

22. Gourlay, G.C. & Marsh, C.A., *Biochim. Biophys. Acta 235* (1971) 142-148.

23. LaBadie, J. & Aronson, N.N., *Fed. Proc. 31* (1972) 281 *Abs.*

24. Leelavathi, D.E., Estes, L.W., Feingold, D.S. & Lombardi, B., *Biochim. Biophys. Acta 211* (1970) 124-138.

25. Swank, R.T. & Paigen, K., *J. Mol. Biol. 77* (1973) 371-389.

26. Hickman, S. & Neufeld, E.F., *Biochem. Biophys. Res. Commun. 49* (1972) 992-999.

24 STUDIES WITH β-GLUCURONIDASE FROM RAT LIVER MICROSOMES

P. Stahl, J.W. Owens, M. Boothby & K. Gammon
Department of Physiology and Biophysics
Washington University School of Medicine
St. Louis, Missouri, 63110, U.S.A.

Rat liver β-glucuronidase is characterized by a dual subcellular localization, a large portion of the enzyme being associated with the microsomal fraction. Purified microsomal membranes were prepared by a modification of a published method. The β-glucuronidase associated with the repeatedly washed membranes was non-latent and could be solubilized with Triton X-100 or deoxycholate. Polyacrylamide gel electrophoresis of the detergent-solubilized enzyme showed a single band when stained for enzyme activity. The solubilized enzyme was purified ∿100-fold using an anti-β-glucuronidase antibody-Sepharose column. Pure rat preputial gland β-glucuronidase served as antigen. When enzyme in Tris buffer pH 7.5 containing 0.2% Triton X-100 was passed over the antibody column, β-glucuronidase was adsorbed while the bulk of the protein passed through unretarded. Following elution with 6 M urea, dialysis, subsequent chromatography on Sephadex G-200 and preparative gel electrophoresis, the enzyme was purified to apparent homogeneity. The purified enzyme has a pI of 6.7 in 6 M urea, acts optimally at pH 5 and displays kinetics very similar if not identical to lysosomal β-glucuronidase.

Rat liver β-glucuronidase is unique because in addition to its typical lysosomal localization, a significant portion of the enzyme is associated with microsomes [1]. In 1960, Moore and Lee [2] detected multiple forms of β-glucuronidase in rat liver and later, Sadahiro et al. [3] suggested the presence of three forms of the enzyme which were associated with different cellular organelles.

Lysosomal β-glucuronidase from rat liver is a glycoprotein of molecular weight 280,000 and apparently composed of four subunits [4]. Whilst the precise relationship of microsomal to lysosomal β-glucuronidase has not delineated, the experiments of Kato and his colleagues [5] in mice, using ^{14}C-amino acid incorporation, have suggested microsomal β-glucuronidase to be the direct precursor of lysosomal β-glucuronidase. More recently, however, Swank and Paigen [6] have proposed that the bulk of mouse kidney microsomal β-glucuronidase is not transported to lysosomes.

The purpose of the present investigation was three-fold; (i) to prepare the best possible microsomal membranes (with respect to lysosomal contamination) for the preparation of microsomal β-glucuronidase, (ii) to establish the multiplicity of forms of microsomal β-glucuronidase and (iii) to effect its purification.

METHOD, PARTICULARLY FOR STARTING MATERIAL

Preputial gland β-glucuronidase was purified to apparent homogeny by a modification of the method of Ohtsuka and Wakabayashi [7]. Antisera to preputial gland β-glucuronidase were raised by administering mg quantities of the pure enzyme in complete Freund's adjuvant to goats and rabbits. The immunoglobulin fraction was isolated from serum by ammonium sulphate fractionation and DEAE-cellulose chromatography as outlined by Weir [8]. Immunoglobulin Sepharose columns were prepared by the cyanogen bromide method as described by Cuatrecasas [9] with coupling at pH 6.5 in 0.2 M sodium citrate. β-Glucuronidase, β-galactosidase, glucose-6-phosphatase and protein were assayed by standard techniques. Electrofocusing in urea was carried out by methods described by LKB.

Washed microsomes were prepared by two methods. At first, a modification of the method of Weihing *et al.* [10] was employed. The original procedure includes several washings, using high-speed centrifugation, with 0.15 M and 1.0 M NaCl followed by washing with an alkaline bicarbonate solution. We modified the procedure by excluding the alkaline bicarbonate wash, which caused severe losses of β-glucuronidase activity from the membranes, and including two washes with 0.005 M Tris Cl pH 7.5. The latter was included to lyse lysosomes contaminating the preparations. A second procedure, adapted from the method of Kamath and Rubin [11], was more extensively used in these studies because of its convenience; the results are qualitatively the same. It involves precipitation of microsomes (post-mitochondrial supernatant) by the addition of 8 mM $CaCl_2$. The presence of 8 mM Ca^{2+} permits the collection and washing of microsomes by low-speed centrifugation. Again, the employment of washing with low ionic strength Tris-Cl was made to rid the preparation of contaminating lysosomes.

RESULTS AND DISCUSSION

Subcellular fractionation

Rat liver from female Wistar rats was homogenized in 0.25 M sucrose. All subsequent procedures were carried out at 4°. The homogenate was subjected to differential centrifugation by a modification of the methods of de Duve [1] and Kamath and Rubin [11]. The scheme followed appears in Fig. 1. Briefly, the crude extract was prepared by homogenizing minced liver in 3 vol of cold sucrose using a Potter-Elvehjem homogenizer. The low-speed sediment following centrifugation was twice re-suspended and re-centrifuged. All the supernatants were pooled and 0.25 M sucrose was added to make a final volume of 9 × the original weight of liver (9 ml/g). The resulting crude extract was centrifuged at 1.5×10^5 g min in an IEC refrigerated centrifuge. The supernatant was poured off and the pellet, re-suspended in 2 vol of sucrose, was

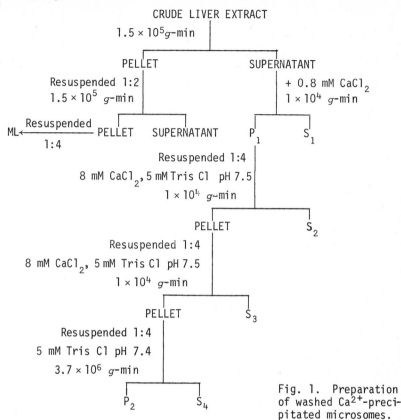

CRUDE LIVER EXTRACT

1.5×10^5 g-min

PELLET — SUPERNATANT

Resuspended 1:2
1.5×10^5 g-min

+ 0.8 mM $CaCl_2$
1×10^4 g-min

ML ← Resuspended 1:4 — PELLET — SUPERNATANT — P_1 S_1

Resuspended 1:4
8 mM $CaCl_2$, 5 mM Tris Cl pH 7.5
1×10^4 g-min

PELLET S_2

Resuspended 1:4
8 mM $CaCl_2$, 5 mM Tris Cl pH 7.5
1×10^4 g-min

PELLET S_3

Resuspended 1:4
5 mM Tris Cl pH 7.4
3.7×10^6 g-min

P_2 S_4

Fig. 1. Preparation of washed Ca^{2+}-precipitated microsomes.

centrifuged a second time at 1.5×10^5 g-min. The supernatants from both spins were pooled (post-mitochondrial supernatant) and enough 1 M $CaCl_2$ was added to produce a final concentration of 8 mM. After 5-10 min the solution was distinctly turbid and was centrifuged in an IEC 870 rotor as shown, giving a clear supernatant (S_1). The pellet (P_1) was re-suspended, using a loose fitting Dounce homogenizer and repeatedly washed in pH 7.5 Tris buffer as shown. The membranes, suspended in Ca^{2+}-free buffer, were spun at high speed in a Spinco 50.1 rotor yielding a final pellet (P_2) which was re-suspended 1:4 in 5 mM Tris Cl for further processing.

The distribution of marker enzymes glucose-6-phosphatase, β-glucuronidase and β-galactosidase is summarized in Table 1. All supernatants generated after S_1 were pooled and have been designated WASH. The data in Table 1 have

Table 1.   Summary of fractionation experiments

Enzyme	Absolute values* (CE + N)	Proportion following fractionation							Recovery, %	
		% of CE + N: A					% of P_1: B		A	B
		CE + N	N	ML	P_1	S	P_2	WASH		
β-Glucuroni-dase	73.1	100	36.1	30.8	28.8	4.9	76.4	19.6	100.6	96.0
β-Galacto-sidase	379.6	100	33.1	41.3	17.4	10.1	52.7	64.9	101.9	117.6
Glucose-6-phosphatase	724.7	100	30.0	4.4	40.6	9.9	91.9	5.6	85.0	97.6
PROTEIN	203.0	100	27.9	14.3	14.3	29.1	81.8	28.1	85.6	109.9

*   μmoles/h/g or mg/g

been presented as absolute values for each marker enzyme in units/g/tissue. The enzyme in each fraction is expressed as a per cent of the total activity in the original homogenate (CE + N). The fate of marker enzymes associated with P_1, following the washing procedure, can be found in the final column, which gives the per cent of enzyme that remains firmly bound to membranes (P_2).

The results indicate that β-glucuronidase, unlike β-galactosidase, displays dual subcellular distribution. Further, a much larger percentage of the β-galactosidase was solubilized by the washing procedure. This finding is not unexpected since acid β-galactosidase is largely lysosomal in distribution. The microsomal marker, glucose-6-phosphatase, on the other hand, remains entirely associated with the membrane fraction (P_2).

In other experiments, P_1, P_2 and ML were assayed in a 0.25 sucrose medium. To some of the assays, 0.2% Triton X-100 was added. Since the incubation mixture was isotonic, any increase in enzymatic activity produced by the detergent would reflect latency or the increased access of substrate to enzyme. The results from these experiments indicated that β-glucuronidase from ML was activated 83% and P_1 18%. P_2 was not activated by detergent. The finding that P_2 was non-latent suggested the virtual absence of intact lysosomes. Electron micrographs corroborated this finding.

Extraction of P_2

β-Glucuronidase can be solubilized from P_2 (washed microsomes) suspended in 0.005 M Tris Cl pH 7.5 by treatment with detergents, high pH or acetone. All of these procedures, followed by high-speed centrifugation, solubilized over 85% of the membrane-bound β-glucuronidase. We have routinely used Triton X-100 at 0.2-0.3% (v/v); however, 0.1% (w/v) Na deoxycholate is equally effective.

Microsomal β-glucuronidase

Electrophoresis of Triton X-100 solubilized enzyme on gels of polyacrylamide followed by staining for enzyme activity yields a single band, and electrofocusing of acetone solubilized enzyme from P_2 membranes reveals a single enzyme peak with a pI of 6.7.

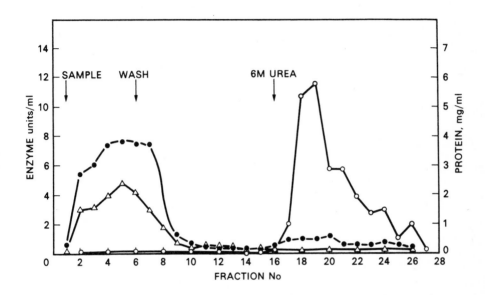

Fig. 2. Antibody-Sepharose chromatography of detergent-solubilized microsomal β-glucuronidase. —o—, β-Glucuronidase; —△—, N-acetyl-β-D-hexosaminidase.

Table 2. Purification of rat-liver microsomal β-glucuronidase

Procedure	β-Glucuronidase units (μmoles/h)	Protein, mg	Specific activity, units/mg protein	Yield %
Solubilized microsomes	969	1384	0.7	100
Antibody-Sepharose chromatography	809	7.1	114	84
Sephadex G-200	575	1.5	391	60
Preparative electrophoresis	266	0.21	1240	27

Purification of microsomal β-glucuronidase

Purification of detergent solubilized microsomal β-glucuronidase was achieved using an anti-rat preputial gland
β-glucuronidase antibody-Sepharose column. Fig. 2 shows a
typical experiment where β-glucuronidase solubilized by
Triton X-100 was passed over a rabbit antibody-Sepharose
column. Protein and *N*-acetyl-β-D-hexosaminidase are essentially unretarded by the column while β-glucuronidase was
completely adsorbed. Following extensive washing with
buffer (0.005 M Tris Cl pH 7.5) or saline, the enzyme can
be eluted with 6 M urea. The column, having been washed
free of urea, is ready for re-use and the β-glucuronidase
activity, following dialysis, is fully recovered. The
yield in this step has varied from 60 to 90%. Polyacrylamide gel electrophoresis of the product at this stage
yielded a single band when stained for protein. However,
much protein remained at the top of the gel, unable to penetrate, presumably because of its large size. Subsequent
Sephadex G-200 chromatography (0.01 M Tris Cl, ph 8.0 + 0.01
M Nac1) and preparative electrophoresis removed this material. The final product has shown a single protein band
when electrophoresed at pH 9.5 (Fig. 3) or 4.3 and has a
specific activity of 1240 units/mg (1 unit = μmole phenol-
phthalein glucuronide hydrolyzed per h at pH 5.0). A
summary of the purification appears in Table 2.

Characterization of the purified enzyme

Purified enzyme has a pH optimum of 5.0 when assayed in
0.05 M sodium acetate. The enzyme shows a K_m of
5.3×10^{-5} M, a figure similar to if not identical with that
found for lysosomal β-glucuronidase [4]. The finding of
identical kinetics presents some difficulty in that the
purified microsomal β-glucuronidase has a specific activity
$\sim\frac{1}{2}$ that of the purified lysosomal β-glucuronidase [4].
Studies are in hand to resolve this question, *viz,* whether
there are fewer substrate binding sites on the microsomal
enzyme or whether the purification of an antigenically reactive but enzymatically inactive species has been
achieved.

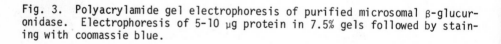

Fig. 3. Polyacrylamide gel electrophoresis of purified microsomal β-glucuronidase. Electrophoresis of 5-10 μg protein in 7.5% gels followed by staining with coomassie blue.

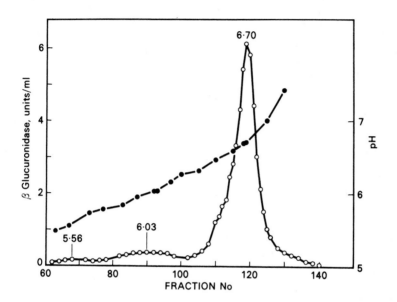

Fig. 4. Electrofocusing of purified microsomal β-glucuronidase in 6 M urea. pH 5-7. 50 units enzyme; 40 hours focusing time.

Multiplicity of microsomal β-glucuronidase

Electrofocusing in 6 M urea provides a notably valuable tool to resolve multiple forms of β-glucuronidase, because the inactivation of the enzyme with urea is completely reversible. Following electrofocusing in 6 M urea, the pH should be corrected by subtracting 0.4 units from the observed reading [12]. β-Glucuronidase from lysosomes can be resolved into 4 or 5 forms by this technique [3]. Microsomal β-glucuronidase, purified by antibody-affinity chromatography, clearly electrofocuses as a single form with a pI of 6.7 (Fig. 4).

An observation made several years ago [4] suggested that microsomal β-glucuronidase may be protease-sensitive. Electrofocusing in 6 M urea has

permitted a re-evaluation of this question. Hydrolysis of purified β-glucuronidase with trypsin modifies the enzyme and yields a variety of forms
which electrofocus with pI value ranging from 6.2 to 6.5. This modification
does not alter enzymatic activity; whilst its full significance is not yet
certain, it does underline the potential importance of cellular proteases in
modifying proteins in the process of their isolation. Moreover, the trypsin
sensitivity of the microsomal enzyme raises some interesting questions in
relation to the membrane localization of this hydrolase. In recent years a
number of membrane-bound proteins and intracellular precursors (i.e. pro-
parathyroid hormone) have been demonstrated which display protease-sensitive
hydrophobic tails. The latter are apparently involved in protein-membrane
interactions. Of particular relevance is the work with cytochrome b_5 whose
trypsin-sensitive hydrophobic tail is involved in the membrane binding of
this molecule [14]. Whether the trypsin-sensitive region of microsomal
β-glucuronidase plays any role in its membrane-bound localization is now
under investigation.

References

1. de Duve, C., Pressman, B.C., Gianetto, R., Wattiaux R. & Appelmans, F.,
    Biochem J. 60 (1955) 604-617.

2. Moore, B.W. & Lee, R.H., *J. Biol. Chem. 235* (1960) 1359-1364.

3. Sadahiro, R., Takanishi, S. & Kawada, M., *J. Biochem. (Tokyo) 58* (1965)
    104-106.

4. Stahl, P.D. & Touster, O., *J. Biol. Chem. 246* (1971) 5398-5406.

5. Kato, K., Ide, H., Shirahama, T. & Fishman, W.H., *Biochem. J. 117* (1970)
    161-167.

6. Swank, R.T. & Paigen, K., *J. Mol. Biol. 77* (1973) 371-389.

7. Ohtsuka, K. & Wakabayashi, M., *Enzymologia 39* (1970) 109-124.

8. Weir, D.M., *Handbook of Experimental Immunology,* Davis, F.A.,
    Philadelphia (1967).

9. Cuatrecasas, P., *J. Biol. Chem. 245* (1970) 3059-3065.

10. Weihing, R.R., Manganiello, V.C., Chiu, R. & Phillips, A.H., *Biochemistry
    (U.S.A.) 11* (1972) 3128-3135.

11. Kamath, S.A. & Rubin, E., *Biochem. Biophys. Res. Comm. 49* (1972) 52-59.

12. Ui, N., *Biochim. Biophys. Acta 229* (1971) 567-581.

13. Gammon, K. & Stahl, P., *Unpublished observations* (1973).

14. Strittmatter, P., Rogers, M.J. & Spatz, L., *J. Biol. Chem. 247* (1972)
    7188-7194.

25 FRACTIONATION OF RABBIT SKELETAL MUSCLE INTRACELLULAR MEMBRANE SYSTEMS

Denis R. Headon, Henry Keating, Edward J. Barrett and [†]Patrick F. Duggan
Departments of Biochemistry and of [†]Medicine
University College
Cork, Ireland

Fractionation of rabbit skeletal muscle subcellular components has been carried out by density-gradient techniques. The isolation of the various subcellular components was dramatically influenced by the choice of homogenization and gradient media. A method for the preparation of concentrated microsomal suspensions from large volumes of solution using the B-XIV zonal rotor has been developed. Some 40 to 50% of muscle microsomal material was lost in low-speed sediments which were discarded in the sample preparation procedure. Preliminary results obtained by density-gradient fractionation of a low-speed sediment indicate the presence of large microsomal components.

The skeletal muscle cell is composed of three main structural components. These are firstly the myofibrils or contractile elements, secondly the various subcellular components such as nuclei, mitochondria and lysosomes [1, 2] and finally the membrane systems which consist of the external membrane or sarcolemma [3] and the intracellular membrane systems [4]. The intracellular membranes consist of the transverse tubule system and the sarcoplasmic reticulum. The transverse tubule system runs transversely through the muscle cell and arises by invagination of the sarcolemma [5]. To develop effective mechanical force all the myofibrils of the muscle cell should develop active tension simultaneously. The transverse tubules are ideally arranged to conduct inwards the impulse of contraction which arrives at the sarcolemma [6, 7]. The sarcoplasmic reticulum runs longitudinally through the muscle cell. This reticulum consists of three distinct regions all of which are interconnected - these regions are the terminal cisterenae, the longitudinal elements and the fenestrated collar [8]. The sarcoplasmic reticulum regulates the intracellular calcium concentration [9].

A pre-requisite to the undertaking of a study on the composition and function of both the sarcoplasmic reticulum and the transverse tubule system and a study of their interdependent role in the excitation-contraction-relaxation cycle of skeletal muscle is the isolation of these components in a purified state. In this laboratory, as in others [10-17], skeletal muscle subcellular components have been isolated by density gradient techniques.

Markers

Subcellular fractionation of skeletal muscle involves the use of some marker constituents which while fulfilling the criteria of suitable markers are not ubiquitous. Table 1 lists suitable marker constituents for muscle subcellular fractionation studies together with references to their use and methods of assay. To the list might be added the characteristic electrophoretic patterns obtained with solubilized subcellular components.

Table 1.   Suitable markers in muscle subcellular fractionation

Muscle cell component	Marker
Plasma membrane	Na^+- and K^+-ATPase inhibited by ouabain; 5'-nucleotidase [15].
Internal membrane systems	ATP-dependent calcium uptake[17; *see also* 18]; K^+-stimulated ATPase [19]; calcium-independent ATPase [20, 21]; acetylcholinesterase [22]; AMP deaminase [23]; phospholipid synthesis [24]; incorporation of ^{3}H-leucine into protein [15; *see also* 25].
Mitochondria	Succinate dehydrogenase [26]; azide-sensitive ATPase, oligomycin-sensitive ATPase [27]; cytochrome oxidase, NADH-cytochrome c reductase, spectrophotometric estimation of cytochrome content [28].
Lysosomes	Acid phosphatase, β-glucuronidase [2].
Soluble	Lactate dehydrogenase.

Media

The choice of homogenization and gradient media in muscle subcellular fractionation critically influences the separations attainable. This influence is detectable at two levels. Firstly, some muscle proteins, especially myosin and to an extent actomyosin, are readily extracted by ionic solutions. The undesirable results obtained, due to protein extraction, using ionic solutions in the fractionation of skeletal muscle have been considered by Uchida *et al*. [29]. Muscle subcellular fractionation procedures are also complicated because of the tendency of the subcellular components to clump and agglutinate [18, 30-33]. This clumping is extensive in sucrose solutions containing ionic constituents and in unbuffered sucrose solutions. Buffering of sucrose solutions and the addition of dispersing agents such as heparin [13, 18] reverse the clumping process as shown in Fig. 1.

From earlier results [18, 34] it appears that a certain degree of

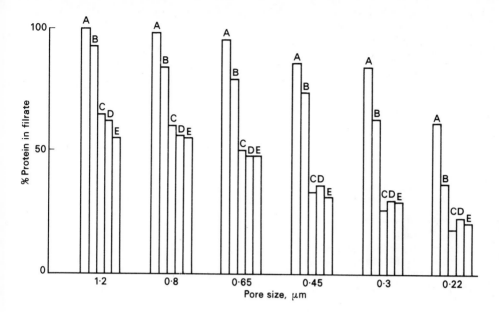

Fig. 1. Effects of various media on the aggregation of muscle microsomal vesicles. Ultrafiltration studies, using Millipore filters, were carried out on a concentrated microsomal suspension prepared as described in text. Volumes filtered were such as to ensure that clogging of the filters did not occur. The media A to E contained 0.5 M sucrose and 100 μg/ml protein, with the following variants: A, 1 mM Imidazole + 50 I.U. heparin/ml; B, 1 mM Imidazole; C, Unbuffered sucrose; D, 1 mM Imidazole + 100 mM KCl; E, 1 mM Imidazole + 100 mM NaCl.

clumping occurs between different subcellular particles thereby preventing their separation in density gradients. However, as particle movement in a centrifugal field is governed by particle size and to a lesser extent by particle density it is possible that aggregates which form from the clumping of similar type particles acquire sedimentation characteristics similar to different subcellular organelles, thus preventing their separation by rate centrifugation. The poor resolution obtained in such systems would appear to result from varying extents of both types of clumping and agglutination outlined above.

Some of the media which have been used in muscle fractionation procedures are listed in Table 2.

Table 2. Media in muscle subcellular fractionation

A. *Homogenization media*

(I) Sucrose media

Sucrose alone - 0.25 M, 0.3 M, 1.0 M
Sucrose buffered with either histidine, imidazole, HEPES*, triethanolamine or phosphate to pH 7.0-7.4
Sucrose containing either KCl, NaCl, $K_2(COO)_2$, CH_3COOK, or mixtures of these

(II) Ionic media - used in the preparation of intracellular membrane systems by differential centrifugation

0.1 M KCl
0.1 M KCl buffered with histidine to pH 7.4
0.12 M NaCl buffered with imidazole to pH 7.4

B. *Density gradient media*

Sucrose alone
Buffered sucrose (buffered with histidine, imidazole or triethanolamine)
Sucrose and dispersing agents such as heparin, dextran sulphate, polyethylene sulphonate, potassium oleate [13]
Sucrose and ionic media such as LiBr, CsCl, $MgCl_2$, KCl, $K_2(COO)_2$, CH_3COOK,

---

*N-2-hydroxyethylpiperazine-N'-2-ethanesulphonic acid

MATERIALS AND METHODS

Young adult rabbits weighing approximately 2.5 kg were killed by an injection of 250 mg Nembutal into an ear vein. The white skeletal muscle was removed immediately after death and trimmed of connective tissue and fat. The trimmed muscle was minced and the mince homogenized in either 0.25 M or 1.0 M sucrose pH 7.4 forming a 25% w/v homogenate.

Density gradient centrifugations were carried oun in a B-XIV zonal rotor which was used in a Super Speed 40 centrifuge or in an HS zonal rotor operated in a High Speed 18 centrifuge (equipment from M.S.E. Ltd., Crawley, U.K.). Gradient profiles which were used in these rotors were calculated using a simple computer programme and were produced using a modified exponential gradient former *(see below)*. The separatory conditions were optimized by simulating the separation in the rotors using a computer programme. All the calculations were carried out on an IBM 1130 computer.

Assays for the various enzymic activities and for protein were carried out as previously described [18]. Refractive indices were obtained using an Abbé 60 refractometer (Bellingham & Stanley Ltd., London). Sucrose concentrations were determined from the refractive indices by consulting standard tables.

SDS[†]-polyacrylamide gel electrophoresis of the membrane fragments was carried out as follows: samples for electrophoresis were prepared by overnight treatment of the microsomal suspensions at a concentration of 2 mg protein/ml with 0.4% w/v SDS at 30°. The protein samples were concentrated either by sedimentation and resuspension or by using Lyphogel (Gelman Instrument Co. Ltd.). The composition of the polyacrylamide gels, the conditions of separation and methods of band detection are shown in Table 3.

Table 3.   Polyacrylamide gel electrophoresis

Polyacrylamide gels:

Size 0.5 x 6.5 cm

   containing 5.0% w/v acrylamide,
           0.2% w/v N, N^1 - methylene - bisacrylamide,
           0.2% v/v N, N, N^1, N^1 - tetramethylethylenediamine
           0.075% w/v ammonium persulphate,

   in 0.1 M sodium phosphate buffer pH 6.0, 0.1% SDS

Electrode buffers:

   0.1 M sodium phosphate buffer pH 6.0 containing 0.1% SDS

Sample application:

   0.05 ml containing 7% w/v sucrose and bromophenol blue (10 µl of 0.5% bromophenol blue per ml sample) as tracker dye

Current:

   Initially 0.5 mA/ gel, then 5 mA/gel until dye reaches end of gel (about 2½ h)

Band detection:

   (a) Removal of SDS by soaking in methanol:water:acetic acid (50:43:7 v/v).
   (b) Staining in 0.1% amido black in methanol:water:acetic acid (50:43:7 v/v)
   (c) Destaining in methanol:water:acetic acid (50:43:7 v/v)
   (d) Storage in 7% v/v acetic acid.

Concentrated microsomal suspensions

The preparation of concentrated muscle microsomal suspensions became necessary because in our earlier work [8] we were hampered by the relatively small amount of microsomal protein present in the post-1,000 g-30 min supernatant which served as the starting material in these fractionation studies. The method developed furnishes concentrated microsomal suspensions from large volumes of material and avoids a sedimentation-resuspension step, this being undesirable if the material is to be fractionated by subsequent density gradient centrifugation [13]. The method of preparing concentrated microsomal

† *Sodium dodecyl sulphate*

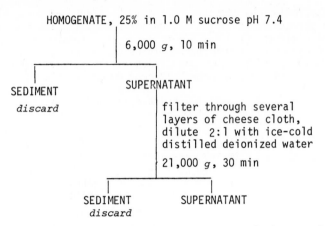

HOMOGENATE, 25% in 1.0 M sucrose pH 7.4

6,000 *g*, 10 min

SEDIMENT   SUPERNATANT

discard   filter through several
layers of cheese cloth,
dilute 2:1 with ice-cold
distilled deionized water

21,000 *g*, 30 min

SEDIMENT   SUPERNATANT
discard

Fig. 2.   Sample preparation procedure for obtaining concentrated microsomal
suspension. *For amplification see text.*

suspensions used in these studies is essentially that reported earlier
[34, 35]; as outlined in Fig. 2; 1.0 M sucrose was used as homogenization
medium to reduce the degree of mitochondrial fragmentation which might occur
during homogenization.   From the post-21,000 *g*-30 min supernatant 450 ml was
pumped to the edge wall of a B-XIV rotor rotating at 3,000 rev/min.   This
was followed by 100 ml 1.0 M sucrose, and 2.0 M sucrose to fill the rotor.
After centrifugation at 30,000 rev/min for 1 h, the rotor was unloaded at
3,000 rev/min by displacing the contents through the centre line.   After
collection of 450 ml the remaining volume was collected in 25 ml fractions.
The 100 ml of 1.0 M sucrose which was pumped into the rotor was collected as
a gradient in four 25 ml fractions.   The results of analysis of these fractions for physical, enzymic and chemical parameters are shown in Fig. 3.   The
$S_{20,w}$ values reported in this figure are used only as a guide, since most of
the components in the fractions would have reached their isopycnic position.
Thus the $S_{20,w}$ values indicate only the most slowly sedimenting species which
would be present in these concentrated microsomal suspensions.

Some 30-40% of the total microsomal protein of skeletal muscle is
sedimented in a 600 *g*-10 min centrifugation step.   Thus the material used to
prepare concentrated microsomal suspensions which is centrifuged at gravi-
tational forces up to 21,000 *g* is not a random sample of the various microsomal components.   As the 600 *g*-10 min pellet contains a large amount of
microsomal protein, density gradient fractionation of this low-speed sediment
has been undertaken.

The 600 *g*-10 min pellet contains almost all of the myofibrillar

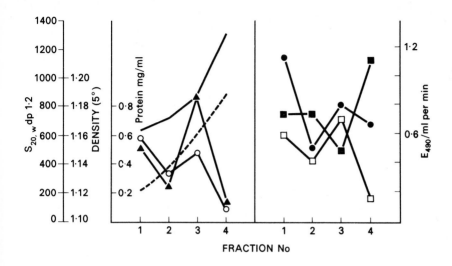

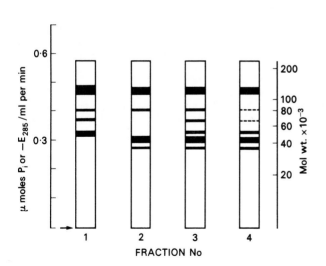

Fig. 3. Distribution of protein, K^+-stimulated ATPase (■——■), basal ATPase (□——□), succinate-INT reductase (▲——▲), AMP deaminase (●——●), sucrose density and $S_{20,w}$ values in the four concentrated microsomal fractions prepared as described in text. o——o denotes protein. Also shown are SDS-polyacrylamide patterns for these four fractions. The *arrow* indicates the position of the tracker dye at the end of run.

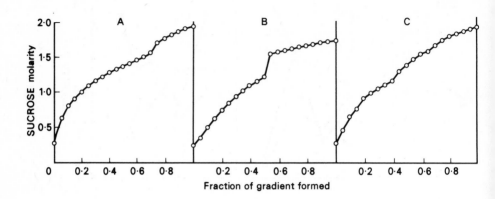

Fig. 4.   Diagram of modified exponential gradient former. The lines marked a and b connect with the reservoirs (1, 2) as shown. Line c serves as an air inlet (if solution in mixing vessel is reduced in volume or removed) or as an air outlet (if solution in mixing vessel is increased in volume or replaced by another solution). Line g serves as a bubble trap and, in addition, with line f clamped and line g open solution can be removed from the mixing vessel by pumping through line g. The volume of solution removed can be measured by collecting the effluent in a graduated cylinder.

material and, in addition, most of the other subcellular components with the exception of soluble material. The fractionation of this 600 *g* pellet by density-gradient centrifugation requires the construction of rather complex gradient profiles. As has been pointed out by Anderson [36] and by Birnie [37], exponential gradients because of their high load-carrying capacity in the region of the sample zone are the most desirable gradients for density-gradient centrifugation in zonal rotors. The simple exponential gradient maker [38], whilst producing a range of exponential gradient profiles which are determined by the concentrations of the solutions in the mixing vessel and reservoir and by the ratio of the volumes of these solutions, is not very versatile.

The equation shown below relates molarity and volume in an exponential gradient generating system.

$$M_{(v)} = M_{(m)} + (M_{(r)} - M_{(c)})(1 - e^{-v/v(m)})$$

where

$M_{(v)}$ molarity at volume v

$M_{(m)}$ initial molarity of solution in mixing vessel

$M_{(r)}$ molarity of solution in reservoir

v   volume of gradient formed

v(m)   volume of solution in mixing vessel.

In a conventional exponential gradient maker neither v(m) nor M(m) are variable without dismantling the apparatus. This inability to vary these parameters limits the range of gradient profiles that can be produced. We have modified the simple exponential gradient maker thereby enabling the production of a wider range of gradient profiles. A diagram of the modified gradient former is shown in Fig. 4.

Such a gradient former enables one to readily change the molarity and/ or the volume of the solution in the mixing vessel, thereby resulting in a wider range of gradient profiles. Changes in the molarity of the reservoir solutions are made more convenient by using two reservoirs.

Some of the gradient profiles produced, according to the data in Table 4, with this gradient former are in Fig. 5.

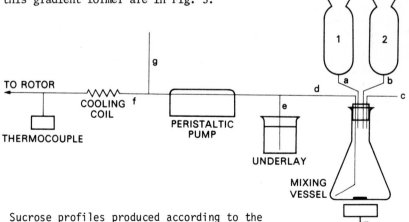

Fig. 5. Sucrose profiles produced according to the data in Table 4 by the gradient former described in text.

Table 4. Conditions for obtaining particular gradient profiles using the apparatus shown in Fig. 4 (with switching between reservoirs where the molarity has to be changed). All values are in ml.

	Profile 'A' Volume	Profile 'A' Molarity	Profile 'B' Volume	Profile 'B' Molarity	Profile 'C' Volume	Profile 'C' Molarity
Initial content of mixing vessel (M.V.):	*150*	0.25	*100*	0.25	*250*	0.25
Additions from reservoir: (i)	50	2.0	(i)50	2.00	(i) 200	2.0
(ii)	90	1.50				
(iii)	90	2.00				
M.V. solution reduced by:	*50*					
M.V. solution replaced by:			*200*	1.00		
M.V. solution replaced by:					*400*	1.50
Additions from reservoir: (iv)	170	2.00	(ii)200	2.00	(ii) 200	2.00
M.V. solution reduced by:			*200*			
Addition from reservoir:			(iii)150	2.00		
Volume of gradient:	400		400		400	

A 600 *g*-10 min pellet was resuspended by gentle homogenization in 0.25 M sucrose pH 7.4, forming a 25% w/v suspension. This suspension was filtered through several layers of cheese cloth and applied to a gradient in an HS zonal rotor. After centrifugation at 9,500 rev/min for 100 min the rotor contents were collected in 20 ml fractions. The results obtained from the analysis of these fractions are shown in Fig. 6. Calcium uptake activity is localized in a narrow region at a density of 1.10. The specific activity of calcium uptake is low, indicating the presence of other microsomal components which are not of relaxing factor origin.

CONCLUSIONS

The preliminary results obtained using the methods outlined above, together with information obtained by distinguishing, through repeated washings, between positional and normal sedimentation in the 600 *g*-10 min pellet, indicate the presence of large microsomal components. From the structural appearance of the intracellular membrane systems as seen in the electron microscope [4, 8, 39, 42], it is not surprising that differential rupturing of the various regions of the intracellular membranes might occur during homogenization. This differential rupturing would be expected to give rise to vesicles of

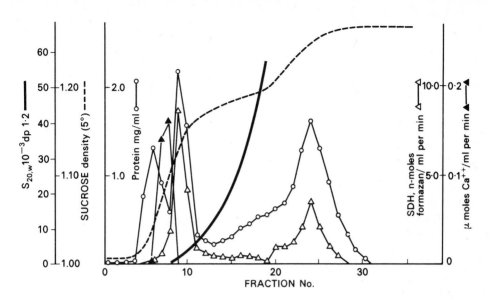

Fig. 6. Distribution of protein, calcium uptake, succinate-INT reductase, sucrose density and $S_{20,w}$ values after centrifugation of a 40 ml sample on a 400 ml sucrose gradient in the HS zonal rotor. The gradient ranged in density from 1.10 to 1.25. Overlay, 0.05 M sucrose (100 ml); 2.0 M sucrose to fill the rotor. Centrifugation was at 5° and 9,500 rev/min for 100 min. Fractions of 20 ml volume were collected.

differing size and consequently different sedimentation characteristics.

The use of buffered sucrose solutions would appear to be the medium of choice in the fractionation of skeletal muscle by density gradient techniques. However in certain cases we have accumulated evidence to indicate a preferential clumping of some muscle subcellular components in unbuffered sucrose solutions. In these circumstances this clumping would appear exploitable to advantage in an initial separation followed by a disaggregation through buffering the medium and then by separation of the components present.

Acknowledgments

E.J.B. is a holder of a Research Scholarship of the Department of Education (Dublin). Part of this work was supported by a grant from the Muscular Dystrophy Associations of America to PFD. The M.S.E. centrifuge was kindly donated by the Wellcome Trust.

References

1. Canonico, P.G. & Bird, J.W.C., *J. Cell Biol.45* (1970) 321-334.

2. Stagni, N. & De Bernard, B., *Biochim. Biophys. Acta 179* (1968) 129-139.

3. Mauro, A. & Adams, W.R., *J. Biophys. Biochem. Cytol. 170* (1961) 177-185.

4. Porter, K.R. & Palade, G.E., *J. Biophys. Biochem. Cytol. 3* (1957) 269-300.

5. Huxley, H.E., *Nature (Lond.) 202* (1964) 1067-1071.

6. Eisenberg, B. & Eisenberg, R.S., *J. Cell Biol. 39* (1968) 451-467.

7. Huxley, A.F. & Taylor, R.E., *J. Physiol. 144* (1958) 426-441.

8. Peachey, L.D., *J. Cell Biol. 25* (1965) 209-231.

9. Ebashi, S. & Endo, M., *Prog. Biophys. Mol. Biol. 18* (1968) 123-183.

10. Schuel, H., Lorando, L., Schuel, R. & Anderson, N.G., *J. Gen. Physiol. 48* (1965) 737-752.

11. Heuson-Stiennon, J.A., Wanson, J.C. & Drochmans, P., *J. Cell Biol. 55* (1972) 471-488.

12. Meissner, G., Conner, G.E. & Fleishcher, S., *Biochim. Biophys. Acta 298* (1973) 246-269.

13. Wheeldon, L.W. & Gan, K., *Biochim. Biophys. Acta 233* (1971) 37-48.

14. Hasselbach, W. & Makinose, M., *Biochem. Z. 339* (1963) 94-111.

15. Kidwai, A.M., Radcliffe, M.A., Lee, E.Y. & Daniel, E.E., *Biochim. Biophys. Acta 298* (1973) 593-607.

16. Kinoshita, S., Andoh, B. & Hoffmann-Berling, H., *Biochim. Biophys. Acta 79* (1964) 88-99.

17. Seraydarian, K. & Mommaerts, W.F.H.M., *J. Cell Biol. 26* (1965) 641-656.

18. Headon, D.R. & Duggan, P.F. in *Separations with Zonal Rotors* (Reid, E., ed.), Wolfson Bioanalytical Centre, University of Surrey, Guildford (1971) pp. V-2.1—V-2.12.

19. Ohtsuki, I., *J. Biochem. 66* (1969) 645-650.

20. Duggan, P.F., *Life Sci. 7* (1968) 1265-1271.

21. Weber, A., Herz, R. & Reiss, I., *Biochem. Z. 345* (1966) 329-343.

22. Karnofsky, M.J., *J. Cell Biol., 23* (1964) 217-233.

23. de Duve, C., Wattiaux, R. & Baudhuin, P., in *Advances in Enzymology* (Interscience) *24* (1962) 291-358.

24. Pennington, R.J. & Worsfold, M., *Biochim. Biophys. Acta 176* (1969) 774-782.

25. Martonosi, A. & Halpin, R.A., *Arch. Biochem. Biophys. 152* (1972) 440-450.

26. Pennington, R.J., *Biochem. J. 80* (1961) 649-654.

27. Fanburg, B. & Gergely, J., *J. Biol. Chem. 244* (1965) 613-624.

28. Meissner, G. & Fleischer, S., *Biochim. Biophys. Acta 104* (1965) 287-289.

29. Uchida, K., Mommaerts, W.F.H.M. & Meretsky, D., *Biochim. Biophys. Acta 104,* (1965) 287-289.

30. Allfrey, V., in *The Cell* (Brachet, J. & Mirsky, A.E., eds), *Vol. 1,* Academic Press, New York (1960) pp. 193-256.

31. Dallner, G. & Ernster, L., *J. Histochem Cytochem. 16* (1968) 611-632.

32. Dallner, G. & Nilsson, R., *J. Cell Biol. 31* (1966) 181-193.

33. Katz, A.M., Repke, D.I., Upshaw, J.E. & Polascik, M.A., *Biophys. Acta 205* (1970) 473-493.

34. Headon, D.R. & Duggan, P.F. in *The European Symposium of Zonal Centrifugation in Density Gradient* (Raye, J.C., ed.), Editions Cité Nouvelle, Paris, *in press.*

35. Headon, D.R. & Duggan, P.F., *Trans. Biochem. Soc. 1* (1973) 294-297.

36. Anderson, N.G., in *Methods of Biochemical Analysis* (Glick, D., ed.). *Vol. 15,* Interscience, New York (1967) pp. 271-310.

37. Birnie, G.D., in *Methodological Developments in Biochemistry* (Reid, E., ed.), *Vol. 3,* Longman, London (1973) pp. 17-28.

38. Birnie, G.D. & Harvey, D.R., *Anal. Biochem. 22* (1968) 171-174.

39. Anderson-Cedergren, E., *J. Ultrastruct. Res., Suppl. 1* (1959) 5-191.

40. Porter, K.R., *J. Biophys. Biochem. Cytol., 10 Suppl.* (1961) 219-226.

41. Fawcett, D.W. & Revel, J.P., *J. Biochem. Biophys. Cytol., 10 Suppl.* (1961) 89-109.

42. Revel, J.P., *J. Cell Biol. 12* (1962) 571-587.

26 ISOLATION OF PLASMA MEMBRANE AND OTHER SUBCELLULAR FRACTIONS FROM DIFFERENT TYPES OF MUSCLE

A.M. Kidwai *
Department of Pharmacology
University of Alberta
Edmonton
Alberta, Canada

Three major subcellular fractions were separated from smooth, skeletal and cardiac muscle by centrifugation on a continuous sucrose density gradient, namely plasma membrane (p.m.), endoplasmic reticulum (e.r.) and mitochondria [1-3]. The fractions were characterized by electron microscopy and by chemical and enzymic markers.

Isolation methods for liver plasma membrane (p.m.) have been studied extensively, and various versions are in use [4]. However, not many attempts have been made to isolate and characterize muscle p.m. Methods for isolation of skeletal muscle p.m. require extraction by salts or at 37° [5, 6]. These treatments might render the membrane unsuitable for enzymic and structural studies. We were able to separate p.m., endoplasmic reticulum (e.r.) and mitochondria using a single sucrose density-gradient step. The method was found suitable for skeletal muscle, smooth muscles and cardiac muscle [7].

METHODS AND RESULTS

The sucrose gradient centrifugation is intended to eliminate contractile proteins and recover all the subcellular components from the same preparation.

In general the muscle is homogenized in 0.25 M sucrose using a 'Polytron PT 20' (Brinkmann Inc.) for a short time, followed by filtration of the contractile proteins as described elsewhere [7]. The filtrate is centrifuged at 100,000 g for 30 min to collect the particulate fraction, which is then subjected to density-gradient centrifugation.

Isolation of skeletal muscle p.m., e.r. and mitochondria

Muscles from the hind legs of rats were dissected free of nerves and connective tissue, and homogenized in 5 g batches (15,000 rev/min, 10 sec). The sediment from the centrifuged filtrate was suspended in 0.25 M sucrose and layered on top of the gradient prepared by using 2 M and 0.25 M sucrose solutions in an ISCO density gradient former [7-9]. Centrifugation for $1\frac{1}{2}$-2 h at 111,000 g in a swinging-bucket rotor gave three bands (Figs. 1 & 2), namely p.m., equilibrated at the sample zone/gradient boundary; e.r. as a diffused

---

* *Present address: Faculty of Medicine, University of Libya, Benghazi.*

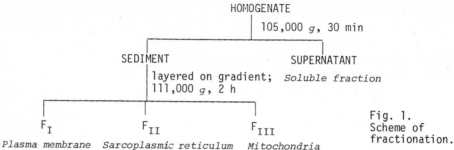

HOMOGENATE

105,000 *g*, 30 min

SEDIMENT   SUPERNATANT

layered on gradient; *Soluble fraction*
111,000 *g*, 2 h

F$_I$ F$_{II}$ F$_{III}$

Plasma membrane   Sarcoplasmic reticulum   Mitochondria

Fig. 1.
Scheme of
fractionation.

turbid area, and mitochondria in the middle of the tube. Nuclei and unbroken cells along with some contractile proteins were settled at the bottom of the tube.

Characterization

The subcellular fractions were characterized by chemical, enzymatic and electron-microscopic means. The p.m. fraction showed enrichment in Na$^+$/K$^+$-ATPase and 5'-nucleotidase, whilst [^{3}H]leucine incorporation was maximal in fraction F$_{II}$ designated as e.r. The mitochondrial fraction was enriched in cytochrome *c* oxidase. The phospholipid:cholesterol ratio was highest in the p.m. fraction.

By electron microscopy both p.m. and e.r. were vesicular in nature, but the thickness of the p.m. was significantly more than that of the e.r. measured at high magnifications [3].

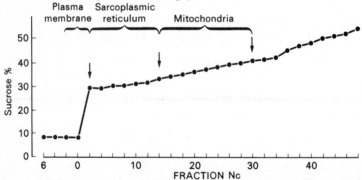

Fig. 2. Representative values of sucrose concentration and approximate position of each band after density-gradient equilibration. The sucrose concentration was determined refractometrically; up to Fraction '0' it was 8% (loading medium). 'Sacroplasmic reticulum' = e.r. (From Kidwai *et al.* [3]).

Cardiac muscle

The homogenate (10,000 rev/min for 4 sec; 10% w/v) from rat cardiac muscle was centrifuged as for skeletal muscle, but omitting the filtration [2]. The density-gradient centrifugation time can be reduced to 90 min without significant effect on separation of the subcellular components.

Smooth muscle

Oestrogenized rat myometrium or rabbit aorta was used, the former being easily homogenized in 4 sec at 15,000 rev/min whereas aorta required at least 20 sec for complete homogenization. The homogenate was filtered and centrifuged as described; for density-gradient centrifugation a suitable time was 2 h. Occasional mitochondria were seen in the p.m. fraction by electron microscopy; therefore it was centrifuged at 13,000 rev/min for 15 min, thereby reducing the mitochondrial contamination. The p.m. fraction was vesicular in nature and possessed Na^+/K^+-ATPase, 5'-nucleotidase and other marker enzymes. The e.r. fraction was heterogeneous by electron microscopy and possessed [^{3}H]-leucine incorporating activity [1]. The mitochondrial fraction possessed maximum cytochrome c oxidase activity [1].

Mesenteric arteries

P.m. was isolated from mesenteric arteries by a modification of the above method [8]. Rat mesenteric arteries were removed and the fat cells separated by differential homogenization, being selectively homogenized by use of a loose-fitting Teflon homogenizer for 2 min with 0.25 M sucrose. The arteries were separated by filtration through a gauze cloth and homogenized (5%) in the Polytron at 15,000 rev/min for 2 min. The particulate fraction obtained by centrifuging at 100,000 g was re-suspended in 0.25 M sucrose and layered on top of a discontinuous gradient prepared by using 29, 33 and 40% sucrose. After centrifugation for 2 h at 112,000 g the p.m. was found to be equilibrated between the loading medium and 29% sucrose. Other fractions can also be recovered at the interfaces. The p.m. was vesicular in nature under the electron microscope and possessed Na^+/K^+-ATPase and 5'-nucleotidase activities.

CONCLUDING REMARKS

The present method for isolating p.m. and other subcellular components is fast and gentle, and yields fairly pure fractions which can be used for biochemical and pharmacological studies.

References

1. Kidwai, A.M., Radcliffe, M.A., Lee, E.Y. & Daniel, E.E., *Biochim. Biophys. Acta 233* (1971) 538-549.

2. Kidwai, A.M., Radcliffe, M.A., Duchon, G. & Daniel, E.E., *Biochim. Biophys. Res. Commun. 45* (1971) 901-910.

3. Kidwai, A.M., Radcliffe, M.A., Lee, E.Y. & Daniel, E.E., *Biochim. Biophys. Acta 298* (1973) 593-607.

4. Steck, T.L. & Wallach, D.F.R., in *Methods in Cancer Research, Vol. 5* (Busch, H., ed.), Academic Press, New York (1970) pp. 93-153.

5. Kono, T., Kakuma, F., Homma, M. & Fukuda, S., *Biochim. Biophys. Acta 88* (1964) 155-176.

6. McCollester, D.L., *Biochim. Biophys. Acta 57* (1962) 427-437.

7. Kidwai, A.M., in *Methods in Enzymology, Vol. 31* (Colowick, S.P. & Kaplan, N.O., eds.), Academic Press, New York, *in press.*

8. Wei, J.W., Kidwai, A.M. & Daniel, E.E., *Proc. Canad. Fed. Biol. Soc. 16* (1973) 75.

27 THE PREPARATION OF RAT KIDNEY PLASMA MEMBRANES BY ISOPYCNIC ZONAL CENTRIFUGATION IN A B-XIV ROTOR

S.A. Kempson, R.G. Price and J.L. Stirling
Department of Biochemistry
Queen Elizabeth College
Campden Hill, London,
W8 7AH, U.K.

A previously described procedure used for the preparation of plasma membranes (p.m.) from rat kidney cortex in the A-XII zonal rotor has been successfully adapted to the B-XIV titanium zonal rotor using a discontinuous gradient. The preparation was characterized by electron microscopy and chemical and enzyme analysis. Alkaline phosphatase, 5'-nucleotidase, L-leucine p-nitro-anilidase and maltase were found to be selectively associated with the p.m.

The kidney nephron selectively retains materials from the glomeular filtrate and controls the excretion of waste products. The understanding of the role of plasma membranes (p.m.) from different regions of the nephron [1] will depend on their chemical and biochemical characterization. Such studies will require rhe preparation of large quantities of membranes. In an earlier study [2] the large-scale preparation and the characterization of rat kidney cortical p.m. in an A-XII zonal rotor was described. This procedure involved an 18 h centrifugation step to achieve isopycnic separation. We have now adapted the procedure to the B-XIV titanium rotor (M.S.E. Ltd.).

METHODS AND RESULTS

The enzyme assays and chemical analysis have previously been described [2]. The membrane pellets for electron microscopy were fixed in 1% osmium tetroxide for 1 h at room temperature, washed 3 times in water, dehydrated with ethanol, and finally embedded in Spurr's resin [3]. Sections were cut with an L.K.B. Ultratome, stained with lead citrate and observed with an A.E.I. EM 6B electron microscope.

Initially the previous procedure [2] for the separation of p.m. in an 'A' rotor was used unaltered. It was found, however, that a change in the homogenization medium from 20 mM to 60 mM $NaHCO_3$ resulted in a better yield of alkaline phosphatase, 5'-nucleotidase, L-leucine p-nitroanilidase and maltase, the p.m. markers. Their yield was further enhanced by increasing the initial centrifugation from 1,000 g for 10 min to 1,500 g for 10 min.

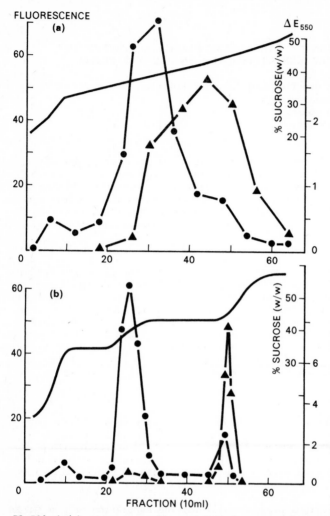

Fig. 1. Distribution of alkaline phosphatase (-●-) and cytochrome oxidase (-▲-) activities across the gradient after centrifugation in the B-XIV rotor for

(a) 1 h at 30,000 rev/min in a linear 30-50% (w/w) sucrose gradient;

(b) 3 h at 46,000 rev/min in a discontinuous gradient.

The recoveries of alkaline phosphatase and L-leucine p-nitroanilidase rose from 24.5% to 42.0% (mean of 4 determinations) of the total activity of these enzymes in the homogenate. The extent of the homogenization and the diameter clearance of the hand Potter-Elvehjem homogeniser were found to be critical. A clearance of 0.26 mm and five downward passes gave maximum recoveries in the 15,000 g-min pellet.

In the previous procedure [2] with the A-XII rotor a linear 30-50% (w/w) sucrose gradient (pH 7.4) was used with a 55% sucrose cushion and a 20% sucrose overlay. After 18 h centrifugation at 5,000 rev/min (2,500 g_{av}) rat kidney cortical p.m. had reached their isopycnic density. Initially a similar gradient and total g-min were used in an attempt to separate cortical p.m. in the B-XIV titanium rotor with an M.S.E. Superspeed 65 centrifuge. Centrifugation of the 15,000 g-min pellet was carried

out for 1 h at 30,000 rev/min (44,000 g_{av}). When the distributions of the
marker enzymes for lysosomes (acid phosphatase) peroxisomes (catalase), endoplasmic reticulum (glucose 6-phosphatase) and mitochondria (cytochrome oxidase) were compared with those of the p.m. marker enzymes, considerable overlap of the mitochondrial and p.m. marker enzymes was found (Fig. 1a). L-leucine p-nitroanilidase and alkaline phosphatase activities were confined to a
single broad peak occurring between 34% and 42% sucrose with a peak at 39%
sucrose (d 1.172 g/ml). Cytochrome oxidase activity extended from 38% to
48.5% sucrose with a peak at 42% sucrose (d 1.188 g/ml). In the 'A' rotor
the mitochondria had sedimented at a density of 1.21 g/ml after centrifugation for the same g-min. Extension of the centrifugation time to 5 h,
at 30,000 rev/min, failed to separate the mitochondria from the p.m. and 12%
of the total cytochrome oxidase activity of the cortical homogenate was still
associated with the p.m. fraction.

Several attempts were made to reduce the mitochondrial contamination
of the p.m. In one, the same gradient was used but the run time was reduced
to 4 h and the rotor speed increased to 46,000 rev/min (108,000 g_{av}). The
alkaline phosphatase and L-leucine p-nitroanilidase activities still occurred
as a single peak (39.3% sucrose, d 1.173 g/ml) but the cytochrome oxidase
activity separated into two distinct peaks. A major sharp peak of activity
was present at 46% sucrose (d 1.20 g/ml) corresponding to the cytochrome
oxidase peak separated in the 'A' rotor. A second broad peak of activity,
overlapping the p.m. fractions, was found at 41% sucrose (d 1.182 g/ml).
The major peak represented 18% and the smaller peak 11% of the total cytochrome oxidase activity of the cortical homogenate.

The incorporation of 1 mM $NaHCO_3$ pH 7.6 or 10 mM $MgCl_2$ and 10 mM tris-
HCl pH 7.6 into sucrose gradients failed to improve the separation of the less
dense mitochondrial population from the p.m. fractions. The inclusion of
1 mM ATP and 10 mM succinate in the gradient [4] decreased the cytochrome
oxidase activity associated with the plasma membrane fractions from 8% to 6%
of the activity of the cortical homogenate. This procedure was rejected
because the small improvement in resolution did not justify the complications
introduced. For example, phosphate ions liberated by hydrolysis of ATP inter-
fered with a number of enzyme assays, particularly 5'-nucleotidase, glucose-6-
phosphatase and (Na^+, K^+-)ATPase.

Various gradient profiles were tried and the best separation of p.m.
from mitochondria was attained in a discontinuous gradient (Fig. 1b). This
was prepared as follows: the rotor was loaded at 3,000 rev/min with 210 ml
of 34.0% (w/w) sucrose pH 7.4 followed by 300 ml of 41.0% sucrose pH 7.4.
The remaining rotor capacity was filled with 61.0% sucrose pH 7.4. The
15,000 g-min fraction was loaded and followed by an overlay of 40 ml of 10.0%
sucrose pH 7.4. Other conditions are given in Fig. 2. After the 3 h centrifugation the activities of the p.m. marker enzymes were confined to a narrow
peak at 39% sucrose (d 1.172 g/ml). Most of the cytochrome oxidase activity

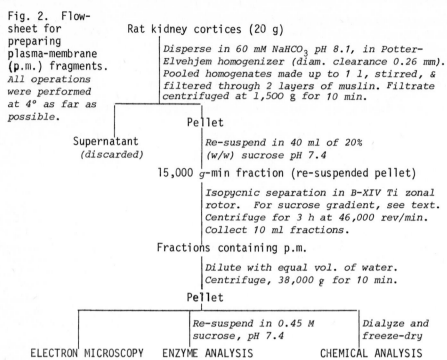

Fig. 2. Flow-sheet for preparing plasma-membrane (p.m.) fragments. *All operations were performed at 4° as far as possible.*

Rat kidney cortices (20 g)

Disperse in 60 mM NaHCO₃ pH 8.1, in Potter-Elvehjem homogenizer (diam. clearance 0.26 mm). Pooled homogenates made up to 1 l, stirred, & filtered through 2 layers of muslin. Filtrate centrifuged at 1,500 g for 10 min.

Pellet

Supernatant (discarded)

Re-suspend in 40 ml of 20% (w/w) sucrose pH 7.4

15,000 g-min fraction (re-suspended pellet)

Isopycnic separation in B-XIV Ti zonal rotor. For sucrose gradient, see text. Centrifuge for 3 h at 46,000 rev/min. Collect 10 ml fractions.

Fractions containing p.m.

Dilute with equal vol. of water. Centrifuge, 38,000 g for 10 min.

Pellet

Re-suspend in 0.45 M sucrose, pH 7.4   *Dialyze and freeze-dry*

ELECTRON MICROSCOPY   ENZYME ANALYSIS   CHEMICAL ANALYSIS

occurred in a narrow peak at 46% sucrose (d 1.208 g/ml) but a small residual amount of activity (2% of total) was found to be associated with the p.m. fractions. Assays across the gradient for marker enzymes indicated that this procedure separated the p.m. from lysosomes, endoplasmic reticulum and peroxisomes. The inclusion of 0.5 mM $CaCl_2$ in the homogenization medium [5] decreased the residual cytochrome oxidase activity to 1% but did not increase the recovery of the p.m. marker enzymes.

Electron microscopy on the membrane pellet showed well-defined brush-borders with many radiating microvilli. Mitochondria were rarely seen and no other organelles could be identified. The specific activities of the p.m. marker enzymes relative to the cortical homogenate values were: alkaline phosphatase 7.7, 5'-nucleotidase 4.0, L-leucine p-nitroanilidase 8.9 and maltase 5.7. The recovered activities of these enzymes represented 15-33% of their activities in the cortical homogenate. Low recoveries (1-2%) of other marker enzymes indicated only slight contamination by lysosomes, pero-xisomes, endoplasmic reticulum and mitochondria. From 20 g of cortex 50-60 mg of freeze-dried p.m. was obtained. Chemical analysis showed that the membrane contained 55.8% protein, 26.7% phosphatide, 8.4% cholesterol and 8.8% carbohydrate.

DISCUSSION

Methods developed for the isolation of kidney plasma membranes in the A-XII rotor evidently cannot be directly transferred to the B-XIV rotor: a completely different gradient profile was necessary to obtain a comparable separation. The greater *g* forces possible with the 'B' rotor enable much shorter run times to be used, and p.m. can be isolated in a single working day. The procedure described here involves fewer centrifugation steps than methods already available for the preparation of kidney [6, 7] and liver [8] p.m. using a B-XIV rotor.

The main obstacle in adapting the method to the 'B' rotor was the presence of two mitochondrial populations of different density. This heterogeneity may be caused by disruption of a homogeneous population in concentrated sucrose solutions [9] or by damage caused by the high hydrostatic pressure generated during isopycnic zonal centrifugation [10]. However, two mitochondrial populations have been isolated from rabbit kidney cortex [11] under conditions that are thought to minimize disruption [12] and may indeed exist in the tissue. Whatever the origin of these populations, a considerable reduction in contamination by the less dense mitochondria was achieved by the introduction of a discontinuous gradient.

This method is being applied to the separation of p.m. from the kidney medulla with the object of preparing and characterizing p.m. from normal and diseased kidneys.

Acknowledgment

The authors acknowledge the financial assistance of the Nuffield Foundation.

References

1. Rouiller, C. & Muller, A.F. (eds.), *The Kidney, Vol.1,* Academic Press, New York (1969).
2. Price, R.G., Taylor, D.G. & Robinson, D., *Biochem. J. 129* (1972) 919-928.
3. Spurr, A.R., *J. Ultrastruct. Res. 26* (1969) 31-43.
4. Wong, D.J., Van Frank, R.M. & Horng, J.S., *Life Sci. 9* (1970) 1013-1020.
5. Ray, T.K., *Biochim. Biophys. Acta 196* (1970) 1-9.
6. George, S.G. & Kenny, A.J., *Biochem. J. 134* (1973) 43-57.
7. Quirk, S.J. & Robinson, G.B., *Biochem. J. 128* (1972) 1319-1328.

8. Evans, W.H., *Biochem. J. 116* (1970) 833-842.

9. Beaufay, H., Jacques, P., Baudhuin, P., Sellinger, O.Z., Berthet, J.
    & de Duve, C., *Biochem. J. 92* (1964) 184-205.

10. Wattiaux, R., Wattiaux-De Coninck, S. & Ronveaux-Dupal, M.F., *Eur. J.
    Biochem. 22* (1971) 31-39.

11. Bondi, E.E., Devlin, T.M. & Ch'ih, J.J., *Biochem. Biophys. Res. Commun.
    47* (1972) 574-580.

12. Pollak, J.K. & Morton, M., *Biochem. Biophys. Res. Commun. 52* (1973)
    620-626.

28 DISTRIBUTION OF ^{67}Ga IN CELLS: I. LOCALIZATION IN ZONALLY FRACTIONATED SUBCELLULAR PARTICULATES OF L1210 CELLS

Richard B. Ryel, George B. Cline, Jerry D. Glickson & Richard A. Gams
Department of Medicine
Division of Hematology-Oncology and Department of Biology
University of Alabama in Birmingham
Birmingham, Alabama 35294, U.S.A.

Gallium-67 has been widely employed in the detection of a broad range of solid tumours and malignant lymphomas. To determine its subcellular distribution in L1210 leukaemic cells containing ^{67}Ga, fractionation was performed in a Ti-14 zonal rotor using a discontinuous sucrose gradient. Seven distinct banding zones were isolated from cell homogenates. Gel filtration of the soluble phase with Sephadex G-75 indicated association of the radioisotope with macromolecules. Correlation between the ^{67}Ga activity and protein content of each banding zone suggests a probable protein binder.

Considerable interest has been shown in the use of ^{67}Ga citrate as a scanning agent for a wide variety of solid tumcurs and malignant lymphomas [1, 2]. Diverse techniques have been employed to investigate the subcellular binding sites of ^{67}Ga in a number of tumour systems. Investigators at Oak Ridge [3, 4, 5], using autoradiography and differential and zonal centrifugation of cell homogenates, reported "lysosomal-like" granules as the primary site of ^{67}Ga localization in various non-osseus rodent tumours. Higasi [6] however, reported that 40% of the ^{67}Ga was bound to the nuclear fraction of Ehrlich ascites tissue homogenates. Similarly, Ito *et al.* [7] found that 33% of the ^{67}Ga activity was in the nuclear fraction from rabbit carcinoma homogenates. Orii [8], on the other hand, observed no localization of ^{67}Ga in a specific subcellular organelle of Yoshida sarcoma cells. Most of the ^{67}Ga activity was in the cell sap.

In the present study, L1210 leukaemic cells grown in the peritoneal cavity of BDF1 mice were zonally fractionated in order to determine the distribution of ^{67}Ga among subcellular components. The experimental conditions employed for optimal ^{67}Ga binding were based on an extensive investigation recently conducted in this laboratory which defined the effects of various experimental parameters on ^{67}Ga binding to L1210 tumour cells in both *in vivo* [9] and *in vitro* [10] systems.

METHODS AND MATERIALS

Female BDF1 mice (6-10 weeks old) were sacrificed by cervical fracture six days after intraperitoneal injection of 10^5 L1210 cells. Harvesting of the cells was accomplished by injecting 6 ml of 0.9% NaCl into the peritoneal cavity, perforating the abdominal cavity, and dripping the intraperitoneal exudate into iced 0.9% NaCl. The cells were pelleted, exposed to brief hypotonic saline haemolysis, washed, and resuspended in 20 ml 0.9% NaCl. Stained smears of the cell suspension demonstrated 99%-100% L1210 leukaemic lymphoblasts. The 20 ml cell suspension was then incubated with 5.20×10^{-7} μmoles of ^{67}Ga citrate for one hour at 37°. After incubation, the cells were washed three times in 0.9% NaCl to remove unbound isotope, resuspended in 20 ml 0.9% NaCl (2.8×10^7 cells/ml) and homogenized by 40 double strokes with a Dounce glass homogenizer.

The 20 ml homogenate represented the starting sample for zonal fractionation. The techniques employed (all at 37°) are described in detail by Cline and Ryel [11]. A discontinuous sucrose gradient consisting of 9%, 25%, 35%, 43% 47% and 50% sucrose with a 55% sucrose cushion was loaded into a spinning Ti-XIV Spinco zonal rotor (2,000 rev/min) in order of increasing density. The starting sample was layered over the gradient and centrifuged 45,000 rev/min for 11 min. Then the gradient containing the fractionated components was displaced from the spinning rotor (2,000 rev/min) with 57% sucrose. The effluent was continuously monitored with a Beckman DB-GT spectrophotometer at 280 nm, and collected in 20 ml fractions. The sucrose concentration of the individual fractions was determined with the use of a Bausch and Lomb Abbe 3-L refractometer.

Each fraction was assayed for ^{67}Ga activity, total protein nitrogen, cytochrome oxidase and AMPase. ^{67}Ga activity was determined with an automatic well-scintillation counter (Nuclear Chicago Model 1185). Total protein nitrogen was determined by the Elrod modification of the Lowry technique [12]. AMPase was assayed by the method of Yoda and Hokin [13] and cytochrome oxidase by that of Smith [14].

RESULTS

The results of the zonal fractionation study on leukaemic homogenates are shown in Fig. 1. Seven distinct banding zones are seen. The data for each are summarized in Table 1. ^{67}Ga activity, which approximately follows the profile for protein nitrogen, is present in each banding zone, with most of the activity located in zones 1 and 7 (Fig. 2).

Fig. 3 shows the distribution of AMPase and cytochrome oxidase. Banding zone 3 contains 37.5% of the AMPase activity and zone 4 contains 77.2% of the cytochrome oxidase activity. Light microscopy indicated that banding zones 5 and 6 were primarily composed of partially homogenized cells. Cells of normal appearance were present in zone 7.

The profile of the Sephadex G-75 gel filtration performed on fraction 4

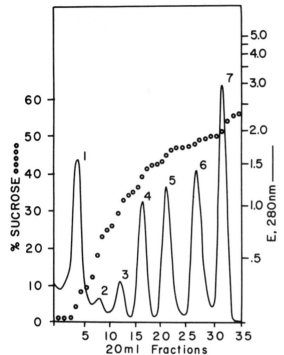

Fig. 1. Zonal absorbance profile of fractionated homogenates of L1210 cells. *Banding zones are shown numbered 1 - 7 in order of increasing density. The discontinuous sucrose gradient is indicated as unloaded from the Ti-XIV zonal rotor.*

(banding zone 1) is shown in Fig. 4. Approximately 20.5% of the recovered ^{67}Ga was associated with the void volume which contained 82.3% of the recovered protein nitrogen. Fractions 16-20 contained 66.0% of the recovered ^{67}Ga and 3.6% of the recovered protein nitrogen. This, however, is the elution volume of free ^{67}Ga citrate (insert on Fig. 4).

DISCUSSION

The fractionation studies on L1210 leukaemic cells indicated a general distribution throughout the subcellular components. There does appear to be a relationship between the amounts of protein nitrogen and ^{67}Ga activity (Fig. 2). That ^{67}Ga citrate penetrates the cell membrane is clearly demonstrated by its presence in banding zones 1 (the soluble phase) and 4. Cytochrome oxidase activity establishes the presence of mitochondria in banding zone 4, where 5.3% of the ^{67}Ga activity is localized (Table 1). The large localization of ^{67}Ga in zone 1 is partially due to free ^{67}Ga citrate as indicated by Sephadex G-75 gel filtration. The association of 20.5% of the ^{67}Ga activity with the void volume indicates binding to macromolecular species of at least 70,000 daltons.

AMPase activity suggests ^{67}Ga binding to plasma membranes. Excluding whole and partially disrupted cells (in zones 5-7), which contain plasma membranes, AMPase (37.5%) is primarily localized in banding zone 3 with some activity also seen in zone 2. It is quite possible that zones 2 and 3 represent light and heavy plasma membrane fractions similar to those reported for human lymphocytes [15].

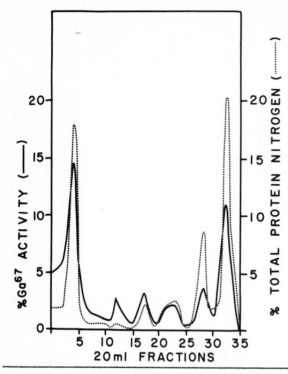

Fig. 2. ^{67}Ga activity (%) an and total protein nitrogen (%) in relation to the banding zones indicated in Fig. 1.

Binding of ^{67}Ga by L1210 leukaemic cells cannot be attributed to a particular subcellular organelle. Rather, the radioisotope is present in a number of subcellular fractions, at least some of which contain macro-molecular ^{67}Ga binders. These results demonstrate the usefulness of zonal centrifugation on discontinuous sucrose gradients for the study of subcellular localization of various agents in experimental tumour systems.

BANDING ZONE No.	Centre fractions	Sucrose %	Density g/ml	Proportion, % of total		Integrated AMPase activity, μmoles/min	Integrated cytochrome oxidase activity [*]
				^{67}Ga activity	Protein nitrogen		
1	4, 5	4.2	1.015	27.49	25.13	0.036	0.00
2	8, 9	18.2	1.073	2.89	1.28	0.43	0.00
3	12, 13	30.7	1.130	4.21	0.87	1.45	0.073
4	16, 17	39.2	1.172	5.28	4.05	.07	1.29
5	21, 22	45.0	1.203	6.25	7.06	.65	0.186
6	26, 27	48.0	1.219	8.09	12.35	0.50	0.057
7	31, 32	51.3	1.237	20.67	32.69	0.36	0.066

[*] *The unit of activity* [14] *was the first order rate constant for oxidation of cytochome c divided by the wt. of protein N (in μg). Integrated values of this quantity for each banding zone are presented here.*

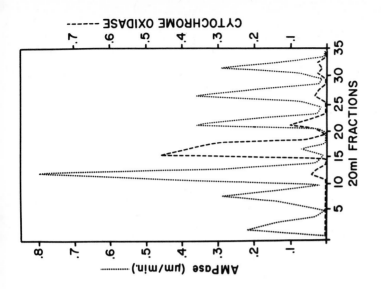

Fig. 3. AMPase and cytochrome oxidase activity determined on the fractions. *The pattern corresponds with the zonal absorbance profile seen in Fig. 1.*

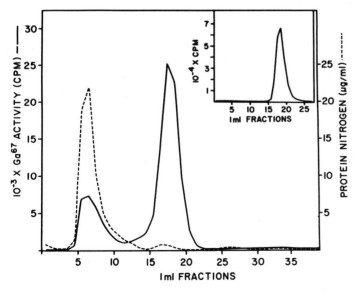

Fig. 4. Elution pattern from Sephadex G-75 gel filtration of 1 ml aliquot taken from the soluble phase (banding zone 1, fraction 4) following zonal fractionation of the homogenates. *Insert: ^{67}Ga citrate elution pattern obtained from the same column.*

References

1. Edwards, D.L. & Hayes, R.L., *J. Amer. Med. Assoc., 212* (1970) 1182-1190.

2. Langhammer, H., Glaubitt, G., Grebe, S.F., Hampe, J.F., Haubold, V., Hor, G., Kaul, A., Koeppe, P., Koppenhagen, J., Roedler, H.D. & van der Schoot, J.B., *J. Nucl. Med., 13* (1971) 25-30.

3. Hayes, R.L., Nelson, B., Schwartzendruber, D.C., Brown, D.H., Carlton, J.E. & Byrd, B.L., *J. Nucl. Med. 12* (1971) 364.

4. Schwartzendruber, D.C., Nelson, B. & Hayes, R.L., *J. Nat. Cancer Inst. 46* (1971) 941-952.

5. Brown, D.H., Carlton, E., Byrd, B., Harrell, B. & Hayes, R.L., *Arch. Biochem. Biophys. 155* (1973) 9-18.

6. Higasi, T., Skemoto, S., Nakayama, Y. & Hisada, T., *J. Nucl. Med. 6* (1969) 217-226.

7. Ito, Y., Okuyama, S., Sato, K., Takahashi, T. & Kanno, I., *Radiology 100* (1971) 357-362.

8. Orii, H., *Strahlentherapie, 144 (2)* (1972) 192-200.

9. Glickson, J.D., Ryel, R.B., Bordenca, M.M., Kim, K.H. & Gams, R.A., *Cancer Res. 33* (1973) 2706-2713.

10. Gams, R.A., Ryel, R.B., Glickson, J.D., unpublished data.

11. Cline, G.B. & Ryel, R.B., *Methods in Enzymology* (Colowick, S.P. & Kaplan, N.O. eds; Academic Press) *22* (1971) 168-204.

12. Elrod, H., in *USAEC Report ORNL-4171* (Oak Ridge National Laboratory) (1967) 123.

13. Yoda, A. & Hokin, L.E., *Biochem. Biophys. Res. Commun., 40* (1970) 880-886.

14. Smith, L., *Methods in Biochemical Analysis* (Glick, D., ed., Interscience, Wiley) *2* (1955) 427-434.

15. Demus, H., *Biochim. Biophys. Acta 291* (1973) 93-106.

29 PREPARATION OF GASTRIC CELL MEMBRANES BY DENSITY GRADIENT CENTRIFUGATION WITH A ZONAL ROTOR

Jerry G. Spenney, A. Strych, A.H. Price, H.F. Helander and G. Sachs
*University of Alabama in Birmingham
and Veterans Administration Hospital
Birmingham, Alabama 35294, U.S.A.*

*The subcellular components in gastric mucosal homogenates have been fractiona-
ted with sucrose density gradients in a zonal rotor. The method is designed
to give cell membranes free from mitochondrial contamination. It allows of
large-scale preparation of smooth-walled vesicular membranes containing HCO_3^--
ATPase activity free from mitochondrial contamination as assessed by electron
microscopic morphology, with no detectable succinic dehydrogenase or monoamine oxidase activity.*

Investigation of the biochemical basis of acid secretion has led to various
postulated mechanisms to explain secretion of H^+ against 10^6-fold concentration gradient. The most popular mechanisms are an ATPase mechanism and a
redox-linked mechanism. Previous studies of gastric mucosal homogenate have
shown that both a HCO_3^--stimulated, SCN^--inhibited ATPase and various redox
enzymes are present in the membrane-rich fraction obtained by differential
centrifugation. Clearly mitochondrial contamination was present and precluded
any conclusion regarding the origin of the ATPase or redox enzymes from cell
membranes or mitochondrial membranes.

Density gradient centrifiguation resolved a band of cell membranes
free of redox enzymes and another band of membranes rich in mitochondrial
enzymes. The preparation of cell membranes in a high state of purity
and in sufficient quantity to allow other biochemical studies necessitated
the development of techniques of zonal density gradient fractionation of homogenates from gastric mucosa.

METHODS

The stomach is removed from a fasted dog and immediately opened along the
greater curvature, washed with H_2O, and cooled in ice. After 3-5 min any
remaining oesophagus and antrum is cut away. The stomach is laid flat on a
cooled pyrex dish and with a microscope slide the mucosa is scraped as a sheet
from the underlying muscular layer. The mucosa suspended in 700 ml of buffer
is chopped into small fragments in a Waring blender with a dulled blade at
medium speed. In a Parr 'bomb' under 950 psi (67 kg/cm^2) nitrogen it is allowed to equilibrate with magnetic stirring for 20 min. The mucosal fragments
are disrupted on decompression. After 10 min standing at 0° the floating

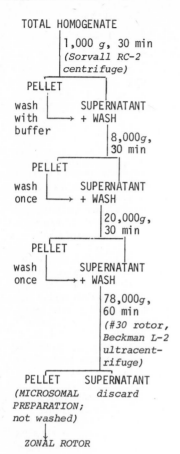

TOTAL HOMOGENATE

1,000 *g*, 30 min
(Sorvall RC-2 centrifuge)

PELLET

wash with buffer — SUPERNATANT + WASH

8,000*g*, 30 min

PELLET

wash once — SUPERNATANT + WASH

20,000*g*, 30 min

PELLET

wash once — SUPERNATANT + WASH

78,000*g*, 60 min
(#30 rotor, Beckman L-2 ultracentrifuge)

PELLET   SUPERNATANT
(MICROSOMAL   discard
PREPARATION;
not washed)

ZONAL ROTOR

Fig. 1. Preparative steps in purification of material prior to zonal density gradient centrifugation.
For the 78,000 rev/min spin the g value at $R_{max} = 105,000$.

mucus layer is aspirated and discarded. The remaining homogenate is filtered through 4 layers of cheesecloth and subjected to differential and zonal density-gradient centrifugation.

The buffer used in each instance is 40 mM Tris-HCl, pH 7.4, in 0.25 M sucrose. The addition of 3 mM $MgCl_2$ or EDTA or of 1 mM dithiothreitol seemed to have no effect either on the levels of activity obtained or on the separations achieved, hence there was no additive in later work.

Differential centrifugation was performed as shown in Fig. 1. Then the requisite gradient was formed in a Beckman Ti-14 rotor spinning at 3,000 rev/min in a Beckman L-2 ultracentrifuge. *Step gradients* are formed using 100 ml 20% sucrose, 200 ml 24%, 150 ml 34% and 150 ml 47% sucrose (w/w). The solutions are sequentially pumped into the zonal rotor beginning with the 20% sucrose. The sample is added last, usually at the centre of the rotor. The *continuous gradient* is formed from 0.25 M sucrose containing 7.5% Ficoll (w/w) and 37% sucrose (w/w) or alternatively from 20% sucrose (w/w) and 50% sucrose (w/w) using a Beckman gradient marker calibrated to deliver a 600 ml gradient which is linear with volume delivered from the pump. All solutions contain 20 mM Tris-Cl, pH 7.4.

The sample as obtained in pellet form during differential centrifugation may be prepared for injection at the periphery of the rotor by suspending the pellet in the most dense sucrose solution. When the sample is injected into the centre of the rotor it is suspended in 0.25 M sucrose. We most commonly use central injection. We examined the efficacy of injection into the centre of the *gradient*. In this instance 200-350 ml of the gradient is allowed to form. The pump is stopped and 10-20 ml of gradient is pumped from the side-arm of the delivery tube. The pellet is resuspended in this solution and re-injected into the rotor via the side-arm of the delivery tube. The remainer of the gradient is then formed.

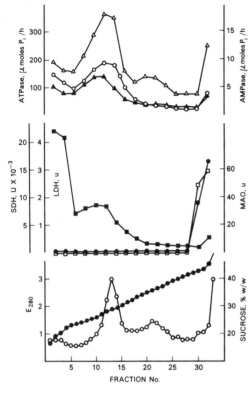

Fig. 2. Fractionation of the microsomal fraction on a 20-45% sucrose gradient.
Top: $\triangle$——$\triangle$, HCO_3^--ATPase; $\blacktriangle$——$\blacktriangle$, Mg^{2+}-ATPase; o——o, AMPase.
Centre: ●——●, SDH; □——□, MAO; ■——■, LDH.
Bottom: ●——●, sucrose; o——o, protein.

For zonal fractionation, the loaded rotor is closed and brought to 47,000 rev/min. Isopycnic centrifugation, our usual practice, is allowed to continue for 6-8 h. Step gradients are allowed to continue at 47,000 rev/min for 2 h. The gradient is unloaded at 3,000 rev/min by central displacment by 60% sucrose pumped into the periphery of the rotor, and 33 fractions, each of 20 ml, are collected for assays and further biochemical characterization.

Sucrose concentration was measured with an Abbe refractometer. (In Ficoll-sucrose gradients the density is likewise expressed as % sucrose.) Mg^{2+}- and HCO_3^- ATPase and 5'-nucleotidase were measured as previously described [1]. Succinate dehydrogenase activity was measured by the method of King [2], and monoamine oxidase according to the method of Tabor *et al.* [3]. *p*-Nitrophenyl phosphatase is assayed in a 1 ml volume with these final concentrations: 50 mM Tris-Cl, pH 7.5, 6 mM $MgCl_2$, 6 mM *p*-nitrophenyl phosphate as either the Na^+ or the Tris salt. The enzyme activity is assessed in the presence and absence of 25 mM KCl. The reaction is terminated with 1 ml of 1 M NaOH and after centrifugation the absorbance at 405 nM is measured on a Gilford 2400 spectrophotometer. Results are expressed as μmoles *p*-nitrophenol liberated per h per 20 ml fraction.

For electron microscopy, a 1-3 ml portion of one of the fractions is diluted to 8.6% sucrose with 40 mM Tris-HCl, pH 7.4, and placed in centrifuge tubes in which a flat agar cushion has been formed. The material is centrifuged at 39,000 rev/min in a Beckman SW-39 rotor for 60 min. The supernatant is decanted and buffered 1% osmium tetroxide is layered on the agar

Table 1. Enzyme enrichment of the two main protein peaks *(cf. Fig. 2)*. C is the peak total activity in the region 28-32% or 40-44% (w/w) sucrose. C_0 is the hypothetical total activity per sample if activity were evenly distributed on the gradient.

Activity	C/C_0, 28-32%	C/C_0, 40-44%
Mg^{2+}-ATPase	2.2	1.1
HCO_3^--ATPase	2.1	1.4
AMPase	2.2	0.9
SDH	-	8.5
MAO	--	11.0

Table 2. Activity of bands from the zonal rotor *(cf. Fig. 6)*.

Band	HCO_3^--ATPase specific activity moles $P_i.mg^{-1}$
Total homogenate	3.20
1	0.25
2	3.73
3	4.50
4	26.55
5	11.91

base for 12 h. The material is subsequently dehydrated and embedded in Vestopal W. Thin sections contrasted with lead hydroxide and uranyl acetate were studied in a Phillips EM 200 or EM 300 electron microscope at 80 kv.

RESULTS

Fractionation of the microsomal preparation, the 78,000 *g* pellet from the 20,000 *g* supernatant, is shown in Fig. 2. The sample was injected into the centre of the rotor, but in other runs the sample was injected at about the 30% point in the gradient. In this instance a 20-45% sucrose gradient was used. This preparative technique furnishes a protein peak (280 nm absorbance) at 28-31% sucrose about equal to a peak at the periphery of the rotor. In addition a small peak usually appears at 36-39% sucrose. HCO_3^- & Mg^{2+}-ATPase distribution is trimodal, with major peaks at 28-31% sucrose and at the periphery of the rotor. A small tertiary peak occurs at 36-39% sucrose. AMPase activity is bimodal coinciding with the two major ATPase peaks. Succinate dehydrogenase (SDH) and monoamine oxidase (MAO) are undetectable below 41.5% and peak at the periphery of the rotor. Lactic dehydrogenase (LDH) is most active at the site of sample application, but there is a peak coincident with the lighter density HCO_3^--ATPase, besides a lesser peak at the periphery of the rotor. Table 1 shows the enrichment of the two major protein peaks with respect to these enzymes.

Electron microscopic morphology of the material recovered at 28-31% sucrose is shown (Fig. 3). In five experiments the fractions at 28-31% sucrose displayed smooth-surfaced vesicular membrane structures of very high purity. The peak at 40-44% sucrose shows predominantly mitochondria with some vesicular membrane structures.

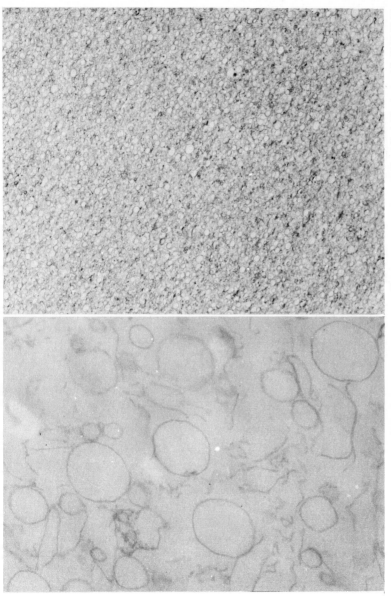

Fig. 3. Electron microscopic morphology of the membrane material isolated from the microsomal preparation at 28-32% sucrose on linear gradients.
3A *(top):* × 6,500
3B *(bottom):* × 65,000

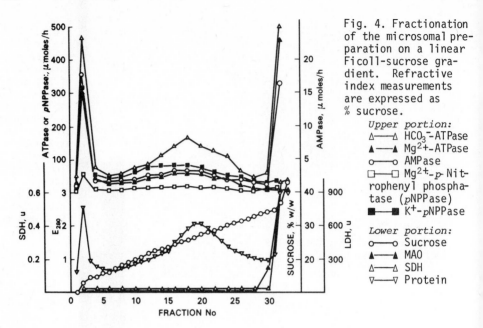

Fig. 4. Fractionation of the microsomal preparation on a linear Ficoll-sucrose gradient. Refractive index measurements are expressed as % sucrose.

Upper portion:
△——△ HCO_3^--ATPase
▲——▲ Mg^{2+}-ATPase
○——○ AMPase
□——□ Mg^{2+}-*p*-Nitrophenyl phosphatase (*p*NPPase)
■——■ K^+-*p*NPPase

Lower portion:
○——○ Sucrose
▲——▲ MAO
△——△ SDH
▽——▽ Protein

        On many linear sucrose gradients of the microsomal fraction there
appeared to be a separation of one or two tubes between the peak HCO_3^--ATPase
and AMPase activities. Gradients consisting of 7.5% Ficoll-0.25 M sucrose to
0% Ficoll-37.5% sucrose were utilized to separate the ATPase and AMPase activities. Fig. 4 shows one such gradient. Mg^{2+}-& HCO_3^--ATPase and AMPase show
trimodal distributions. The least dense fraction at 13% sucrose is greatly
enriched in AMPase and ATPase activities. Morphologically this fraction
(Fig. 5) consists of virtually homogeneous membranes interconnected in a
syncytium. During centrifugation lasting as long as 18 h in a Beckman SW-25
rotor, this membrane fraction fails to leave the top of the gradient. A
sharp peak of protein coincides with the lowest density ATPase and AMPase
peak. A broad protein peak from 23.5 to 29% sucrose is also associated with
a peak of Mg^{2+}-& HCO_3^--ATPase and AMPase activities. At the edge of the rotor
a third protein peak is located. Not only ATPase and AMPase activities, but
also succinate dehydrogenase and monoamine oxidase activities are associated
with this protein peak. Mg^{2+}-*p*-nitrophenyl phosphatase (*p*NPPase) activity
is restricted almost solely to the least dense protein peak. K^+-stimulated
p-nitrophenyl phosphatase activity is seen not only in this least dense peak
but also in the protein peak from 23.5-29% sucrose. It is significant that
the peak of mitochondrial enzyme activity while associated with a prominent

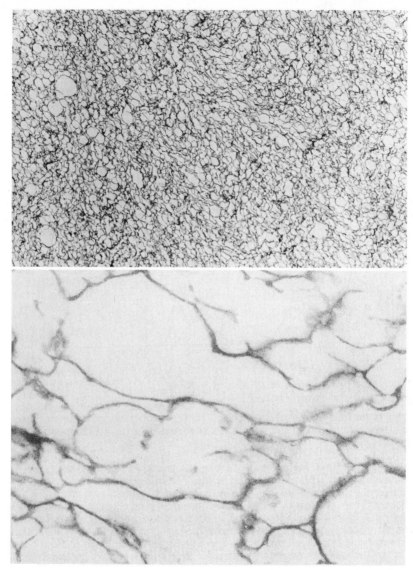

Fig. 5. Electron microscopic morphology of the membrane material isolated at '13%' in the Ficoll-sucrose gradient. A *(top)*, ×6,500; B *(bottom)*, ×65,000.

Table 3.   Enzyme enrichment of peaks obtained in
Ficoll-sucrose gradient *(expressed as in Table 1).*

Activity	C/C_0, 13%	C/C_0, 26-27%	C/C_0, 35%
Mg^{2+}-ATPase	3.8	0.73	5.8
HCO_3^--ATPase	3.3	1.1	3.4
AMPase	4.7	0.72	4.2

peak of Mg^{2+} and HCO_3^- activity, is accompanied by only a small amount of K^+-stimulated *p*-nitrophenyl phosphatase activity.

Step gradient fractionation of the gastric mucosa has been investigated. Fig. 6 shows the protein fractionation and Table 2 the ATPase activities obtained in these protein peaks. Protein is accumulated at each step in the gradient, and HCO_3^--ATPase preferentially in the fourth protein peak at about 28% sucrose. Succinate dehydrogenase and monoamine oxidase activity were not measured.

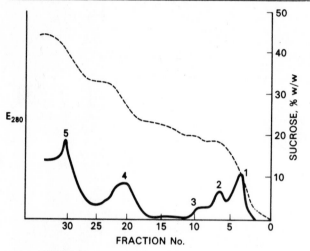

Fig. 6.   Fractionation of the microsomal pellet
on a sucrose step gradient.

DISCUSSION

Development of zonal rotors has made preparative scale density-gradient centrifugation feasible. We have utilized this technique to develop methods of fractionation of gastric mucosal homogenates that furnish gastric cell membranes in high purity and in large quantity.

Fractionation of the microsomal fraction by rate-zonal or step-gradient centrifugation has resulted in notably rapid purification of the vesicular membrane fraction free from soluble protein contamination. Our greatest interest has, however, been in the greater purity of the smooth vesicular membrane fraction obtained by isopycnic density-gradient centrifugation of the microsomal preparation. Membranes prepared in this manner are free from mitochondrial contamination as assessed by electron microscopic morphology and by lack of detectable succinate dehydrogenase or monoamine oxidase activity. We do, however, routinely assay the former to monitor the freedom of these membranes from mitochondrial contamination since mitochondrial ATPase also shows HCO_3^- stimulation. It is this preparative technique that we routinely use to prepare gastric membranes for solubilization with Triton X-100 and for further characterization of transport properties.

The observation that AMPase activity frequently peaks one to two tubes after HCO_3^--ATPase led to efforts to resolve these two activities. Rate-zonal centrifugation, however, failed to resolve these into two peaks of activity. On Ficoll-sucrose gradients a third peak of ATPase activity was resolved. Whilst AMPase activity accompanies each peak, there is enrichment in the peak at 13% sucrose (Table 3). The very light isopycnic density of these membranes is certainly not expected for plasma membranes. In long runs high frictional resistance should be overcome, but these membranes are found at the same density after 6 or 10 h zonal runs or after 18 h of centrifugation in an SW-25 rotor. We feel that osmotic intactness could best explain the failure of these membranes to sediment. It is consistent that this very light peak is non-existent or quite small on sucrose gradients and is greatly increased when Ficoll is used in the gradient.

With hypotonic lysis of isolated oxyntic cells from Necturus, we are able to prepare large vesicular 'ghost-like' structures which float on 5% Ficoll. On sucrose gradients these structures are reduced to tiny granules when viewed under the phase microscope. Thus the possibility remains that the upper peak of HCO_3^--ATPase activity resides in osmotically intact membrane structures.

On Ficoll-sucrose density gradients the K^+-stimulated *p*-nitrophenyl phosphatase activity closely parallels that of HCO_3^--ATPase with the exception of the most dense peak which is associated with the mitochondrial marker enzymes. This is consistent with the known stimulation of mitochondrial ATPase by HCO_3^- and with the electron microscopic morphology which shows predominantly mitochondria with relatively few membrane structures. The K^+-*p*-nitrophenyl phosphatase may reflect cell membrane contamination in this area of the gradient better than HCO_3^--ATPase activity.

Acknowledgements

This work was supported by NIH Grants AM08541, AM15878 & CA13158 and NSF Grant GB31075. A.S. was a trainee on NIH Grant TIAM05286. Credit is given to Project No. 8059-01, VA Hospital in Birmingham, Alabama.

References

1. Sachs, G., Shah, G., Strych, A., Cline, G. & Hirschowitz, B.I., *Biochim. Biophys. Acta 266* (1972) 625-638.

2. King, T.E. & Howard, R.L., in *Methods in Enzymology, 10* (1967) 275-294 [Academic Press, New York].

3. Tabor, C.W., Tabor, H. & Rosenthal, S.M., *J. Biol. Chem. 208* (1954) 645-661.

30 THE SUBCELLULAR FRACTIONATION OF PIG BLOOD PLATELETS BY ZONAL CENTRIFUGATION

D.G. Taylor and N. Crawford
Department of Biochemistry
University of Birmingham
P.O. Box 363
Birmingham B15 2TT, U.K.

The subcellular structures of blood platelets can be separated by simple frac-
tionation in tube sucrose gradients into three major fractions; the soluble
phase, the membranes and the granular elements [1, 2]. The fractions obtain-
ed are, however, rather heterogeneous; the membrane fraction, although essen-
tially free from granules, contains both intracellular and surface-membrane
components, whilst the granule fraction consists of mitochondria, amine stor-
age bodies and lysosome-like granules. Efforts to improve the separation and
yield of platelet organelles, using linear sucrose gradients of various pro-
files with the B XIV titanium zonal rotor, have led to conditions that seem
to allow not only a cleaner separation of the membrane vesicles from the gran-
ules, but also a simultaneous separation of the granular components into frac-
tions enriched in lysosome-like organelles, mitochondria, and amine-storage
bodies. The criteria have been assays for various marker enzyme activities,
the determination of 5-hydroxy-tryptamine (5-HT) and electron microscopic ob-
servations.

Using a different gradient procedure we have been able to separate two
distinct phosphodiesterase (PDE) activities, both associated with membrane
vesicles. One of these enzymes can be assayed with bis-p-nitrophenyl phos-
phate at pH 5.5, and the other with 5'-thymidine-p-nitrophenyl phosphate at
pH 7.9. This has been achieved by an extended centrifugation time in the B-
XIV rotor (18 h) with very shallow linear sucrose gradients. The subfraction-
ation of the membrane vesicles into two distinct populations on the basis of
PDE activity, together with a procedure for labelling the surface membrane by
exposing whole platelets to an [125]iodide-labelled sheep antibody raised to pig
platelet membrane fractions, has indicated that the bis-p-nitrophenyl phos-
phate PDE activity is associated with surface-membrane components and that the
second PDE activity is probably a feature of membranes of intracellular origin.

The blood platelet, although not nucleated, contains within its cytoplasm a
variety of organelles and membrane structures, many of which are known to be
important in the cell's secretory and haemostatic activities (for reviews see
[3, 4]). In the circulation platelets are discoid in shape and measure ∿2-3μm
in diameter, but in response to certain stimuli, e.g. ADP, thrombin and coll-

agen, they form spiny pseudopodia which assist the cells to adhere to each other and to foreign or natural surfaces. Up to a certain point in this shape change the process is reversible, but to fulfil their haemostatic role in the repair of injured blood vessels, the platelets then undergo a highly specific release of intracellular granule-bound components, followed by membrane fusion and loss of single-cell identity, finally resulting in the formation of a consolidated platelet plug which seals off the site of injury.

The importance of the platelet and its surface interactions in the normal haemostatic process, and in the pathological events leading to the formation of an intravascular thrombus, is now well recognised, yet the precise chemical and enzymic nature of the various intracellular structures seen in electron micrographs and their role in these activities is still largely unknown.

Platelets possess several membrane systems and in addition to the plasma membrane, there is a surface-connected intracellular membrane complex - the open canalicular system, and a dense tubular membrane system, believed to be derived from megakaryocyte endoplasmic reticulum [5]. Also seen in electron micrographs are a peripheral ring of microtubules, and several granular organelles including mitochondria, lysosome-like granules thought to contain acid hydrolases [6], and small dense osmiophilic bodies, known to be concerned with 5-hydroxy-tryptamine (5-HT) storage and release [1]. In addition, various fibrillar or filamentous structures are seen in the cytoplasm of the cell and these are now considered to represent some of the components of the contractile protein system [7].

By subcellular fractionation of platelets using tube sucrose density gradients it has been possible to separate platelet homogenates into a soluble-phase fraction, and two major particulate zones [1,2,8 & 9]. However, the fractions so obtained are heterogeneous mixtures of organelles, since the low density particulate fraction that contains membrane vesicles, although essentially free of granular bodies, is almost certainly derived from both surface and intracellular membranes. The heavier granule fraction includes the mitochondria, the amine-storage granules, many lysosome-like organelles and a number of other less well identified features.

We now report some results obtained in attempts to improve the resolution and yield of the platelet subcellular organelles with use of a B-XIV titanium zonal rotor with various gradient profiles.

METHODS AND RESULTS

The procedures we have used routinely for the isolation of platelets from freshly collected pig blood, and for the preparation of homogenates for subcellular fractionation are summarized in Fig. 1. Homogenization of the final whole platelet pellet (suspended in ∿3 vol of 0.3 M sucrose buffered with

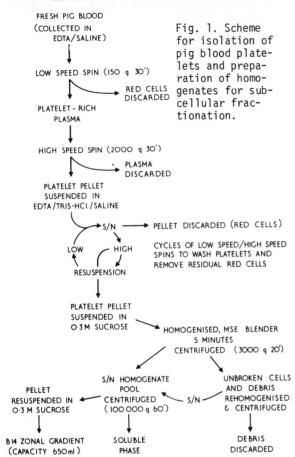

FRESH PIG BLOOD
(COLLECTED IN EDTA/SALINE)

LOW SPEED SPIN (150 g 30')

RED CELLS DISCARDED

PLATELET - RICH PLASMA

HIGH SPEED SPIN (2000 g 30')

PLASMA DISCARDED

PLATELET PELLET SUSPENDED IN EDTA/TRIS-HCl/SALINE

S/N ——→ PELLET DISCARDED (RED CELLS)

LOW   HIGH

RESUSPENSION

CYCLES OF LOW SPEED/HIGH SPEED SPINS TO WASH PLATELETS AND REMOVE RESIDUAL RED CELLS

PLATELET PELLET SUSPENDED IN 0·3 M SUCROSE

HOMOGENISED, MSE BLENDER 5 MINUTES CENTRIFUGED (3000 g 20')

PELLET RESUSPENDED IN 0·3 M SUCROSE

S/N HOMOGENATE POOL CENTRIFUGED (100 000 g 60')

S/N

UNBROKEN CELLS AND DEBRIS REHOMOGENISED & CENTRIFUGED

B14 ZONAL GRADIENT (CAPACITY 650 ml)

SOLUBLE PHASE

DEBRIS DISCARDED

Fig. 1. Scheme for isolation of pig blood platelets and preparation of homogenates for subcellular fractionation.

0.005 M Tris/HCl, pH 7.4) was performed with a top-drive homogenizer (M.S.E. Ltd.) running at full speed for a total of 5 min at 4°. Each homogenization period of 1 min was followed by an interval of 1 min for re-cooling. The homogenate was centrifuged at 3,000 g for 20 min at 4°, the supernatant was removed and retained, and the pellet of unbroken cells and debris re-suspended in 2 vol of buffered 0.3 M sucrose, for a second homogenization under similar conditions. The final pellet was discarded and the pooled supernatants constituted the starting homogenate for subsequent fractionation. In most experiments, the soluble phase was removed by centrifuging the homogenate at 100,000 g for 60 min at 4°. The resultant particulate pellet was re-suspended in buffered 0.3 M sucrose, pH 7.4, and used for the zonal fractionation.

We first sought to increase the resolution of the major subcellular components in order to obtain simultaneously, in good yield and reasonable purity, as many of the different organelles as possible. The best system devised is now described, and the results are depicted in Fig. 2. A sample of 50 ml of platelet particulate phase in 0.3 M sucrose was applied to a 500 ml linear gradient of 30-55% w/w sucrose (pH 7.4) in a titanium B-XIV rotor (M.S.E. Ltd.), with a cushion of 60% w/w sucrose and an overlay of 40 ml 0.2 M sucrose. After running at 47,000 rev/min for 6 h at 4°, 20 ml fractions were collected and assayed for protein and for marker enzyme activities.

The protein ·distribution across the gradient (Fig. 2) shows three peaks. Peak I occurs at a density range of 1.09-1.12 and has associated with it the major peak of PDE activity towards bis-p-nitrophenyl phosphate measured at the

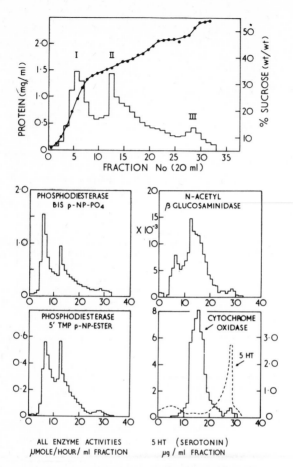

Fig. 2. Zonal fractionation of platelet particulate phase, after centrifugation at 47,000 rev/min for 6h at 4° in a linear gradient of 30-55% w/w sucrose. Enzyme activities are expressed as μmole/h/ml fraction and 5-hydroxytryptamine (5-HT) as μg/ml fraction.

pH optimum of 5.5. This enzyme activity was included in our studies since plasma membrane marker enzymes characteristic for other cells, e.g. 5'-nucleotidase and alkaline phosphatase, show very low activities with platelet membrane fractions. PDE activity has been previously reported to be present in human platelet membranes [10], and we have observed high activities in our pig membrane fractions. A second PDE activity was also measured, using 5'-thymidine-p-nitrophenyl phosphate as substrate, at the optimum pH of 7.9. This enzyme also showed high activity associated with protein peak I, and to a lesser extent with peak II.

The distribution of several acid hydrolases, including acid phosphatase (using ρ-nitrophenyl phosphate and 4-methyl-umbelliferyl phosphate) and several glycosidases, viz. β-galactosidase, β-glucuronidase, and N-acetyl-β-glucosaminidase (all assayed using 4-methylumbelliferyl derivatives) showed, in all cases, a predominant peak in the region of the protein peak II, occurring in the density range 1.16-1.17. Only the N-acetyl-β-glucosaminidase activity has been presented in Fig. 2 since this enzyme profile is typical for these glycosidases and acid phosphatase.

Also found in this region of the gradient is a sharp peak of cyto-chrome oxidase activity, partially resolved from the acid hydrolases peak, with a density range of 1.18-1.19.

When the zonal fractions were assayed for their 5-HT content by an automated extraction and fluorimetric assay procedure [11], two peaks in the zonal gradient were detected, the major one at density of 1.22-1.23 corres-ponding to the protein peak III, and a minor component at the lighter end of the gradient which is non-particulate and can be accounted for by residual soluble-phase material applied to the gradient.

Electron microscopy showed that the fractions coinciding with the peaks of PDE, acid hydrolase and cytochrome oxidase activities, and of 5-HT were enriched respectively with membrane vesicles, lysosome-like granules, mitochondria, and small osmiophilic dense bodies. We therefore concluded that a partial separation had been achieved of the major subcellular compo-nents of platelets. However, from further experiments with minor variations in this gradient profile, it became apparent that in order to improve the purity of the fractions further, one must consider separately either the low-density or the higher-density organelles.

Firstly we directed our attention to investigating the mixed membrane fractions, with the object of isolating and purifying plasma membrane frag-ments. In several zonal experiments, the two PDE activities showed slightly different distributions within the membrane-containing zones. We therefore attempted to separate these two activities for further study. At the same time, since there have been no reports of a satisfactory platelet plasma-membrane marker enzyme, we investigated various fluorescent and radioactive labelling techniques, to assist with the identification of surface-membrane fragments. The technique we now employ routinely, involves the use of an [125]iodide-labelled γ-globulin fraction prepared from a sheep antiserum raised to a pig platelet membrane fraction. Whole platelets were incubated with this radioactively labelled antibody for 5 min at 20°, then mixed with un-labelled platelets (in a ratio of 1:4), homogenized as described earlier and fractionated. A sample of 50 ml of the labelled platelet particulate phase in 0.3 M sucrose (pH 7.4) was applied to a 500 ml linear gradient of 20-45% w/w sucrose, with a cushion of 60% sucrose and an overlay of 40 ml 0.2 M sucrose. The rotor was centrifuged at 47,000 rev/min for 18 h at 4°, and 20 ml fract-ions collected for assay.

It can be seen (Fig. 3) that the two PDE activities were now well sepa-rated and that the distribution of the [125]iodide label followed closely the distribution of the bis-*p*-nitrophenyl phosphate PDE activity. This result indicated a possible subfractionation of the membranes, and electron micro-scopy confirmed that the fractions possessing the highest activities of the two PDE's both contained membrane vesicles.

However, the PDE activity towards 5'-thymidine-*p*-nitrophenyl phosphate

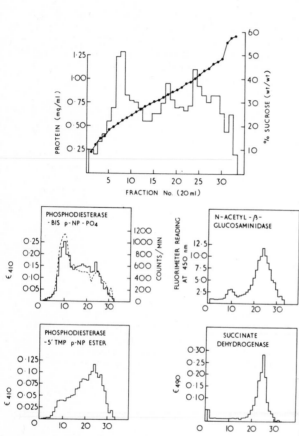

now overlapped the acid hydrolase and mitochondrial enzyme activities (Fig. 3). In the next experiments the above conditions were repeated exactly except that the sample applied to the rotor was a crude platelet membrane fraction separated from granular organelles by the tube density-gradient procedure of Harris and Crawford [2]. This membrane fraction is low in acid hydrolase and mitochondrial enzyme activities. With this modification it has now been possible to separate the two PDE activities reasonably discretely from each other [Fig. 4]. Both enzymes showed specific activity increases 6-8 fold with respect to the original homogenate, confirming a high degree of purification.

DISCUSSION

The separation of platelet homogenates by zonal centrifugation into the major subcellular components, viz. membrane vesicles, lysosome-like granules, mitochondria and 5-HT storage bodies, simultaneously, represents a significant improvement over existing fractionation procedures. The advantages are not only increased resolution and reproducibility but also ease of operation and procedural simplicity. We consider that by a careful choice of gradient profiles

Fig. 3. Zonal fractionation of platelet particulate phase prepared from platelets pre-labelled with ^{125}iodide-γ-globulin. Centrifugation was at 47,000 rev/min for 18 h at 4° in a linear gradient of 20-45% w/w sucrose. Enzyme activities are expressed as μmole/h/ml fraction, and ^{125}iodide label as counts/min/mg protein (---).

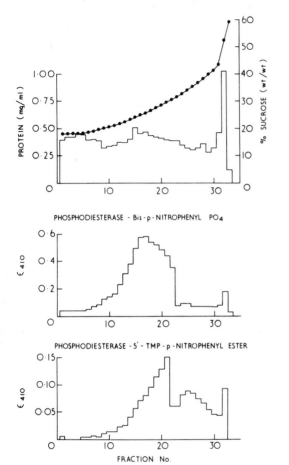

PHOSPHODIESTERASE - Bis - p - NITROPHENYL PO₄

PHOSPHODIESTERASE - 5' - TMP - p - NITROPHENYL ESTER

FRACTION No.

Fig. 4. Zonal fractionation of platelet membrane fraction prepared by tube density gradient centrifugation. Conditions of zonal centrifugation were as for Fig. 3.

and centrifugation conditions, it should now be possible to use the B-XIV rotor for the large-scale preparation of very pure fractions of platelet sub-cellular organelles for more definitive study.

Initially we focussed our attention on the vesiculated membrane fraction, at the light end of our zonal gradients, with the intention of purifying plasma-membrane fragments. Several fractionation techniques have been described in which the separation of fractions enriched in platelet plasma membranes can be achieved, although the precise nature of the various isolated membrane fractions has not been definitively examined due to the lack of a range of good marker enzymes. The possible association with the platelet plasma-membrane vesicles of a PDE activity acting upon bis-*p*-nitrophenyl phosphate has been suggested by this and other studies [10], and our preliminary experiments with platelets pre-labelled with ^{125}iodide-γ-globulin appear to confirm this observation.

Our finding that a second PDE activity towards 5'-thymidine-*p*-nitrophenyl phosphate can be well separated from the bis-ρ-nitrophenyl phosphate activity, by increasing the time of centrifugation, illustrates the heterogeneous nature of the membrane components of the platelet and suggests that further fractionation into different membrane populations may now be possible. This zonal separation can also be achieved using as the starting material a crude platelet membrane fraction prepared by conventional tube centrifugation in a sucrose density gradient.

Our current investigations are concerned with the use of other surface-membrane labelling techniques to confirm that the plasma membranes can be separated in a relatively pure state from intracellular membranes, and with identifying the nature and origin of the membrane associated with the 5'-thymidine-*p*-nitrophenyl phosphate PDE. The natural substrates for both these PDE activities are, however, as yet unknown.

References

1. Minter, B.F. & Crawford, N., *Biochem. Pharmacol. 20* (1971) 783-802.

2. Harris, G.L.A. & Crawford, N., *Biochim. Biophys. Acta 291* (1973) 701-709.

3. Johnson, S.A., (ed.), *The Circulating Platelet*, Academic Press, New York (1971).

4. Brinkhous, K.M., Shermer, R.W. & Mostofi, F.K.(eds.), *The Platelet*, Williams & Wilkins, Baltimore (1971).

5. White, J.G., *Amer. J. Pathol. 66* (1972) 295-312.

6. Holmsen, H. & Day, H.J., *Nature (Lond.) 219* (1968) 760-761.

7. Bettex-Galland, M. & Lüscher, E.F., *Biochim. Biophys. Acta 49*(1961) 536-547.

8. Day, H.J., Holmsen, H. & Hovig, T., *Scand. J. Haematol. Suppl. 7*(1969) 1-35.

9. Marcus, A.J., Zucker-Franklin, D., Safier, L.B. & Ullman, H.L., *J. Clin. Invest. 45* (1966) 14-28.

10. Barber, A.J. & Jamieson, G.A., *J. Biol. Chem. 245* (1970) 6357-6365.

11. Taylor, D.G. & Crawford, N., *Analyt. Biochem., in press.*

31 EFFECTS OF ANTI-LYSOSOME SERUM ON CULTURED FIBROBLASTS

P. Tulkens
Laboratoire de Chimie Physiologique
Université Catholique de Louvain
Dekenstraat 6, B - 3000 Louvain, Belgium

Cultured fibroblasts have been treated with antisera raised against a soluble fraction from lysosomes. The antibodies inhibit several lysosomal hydrolases; they were shown to reach the lysosomal space, to impair intracellular digestion, and to provoke morphological alterations reflecting dysfunction.

Lysosomes exhibit the remarkable property of being accessible from both the exoplasmic and the endoplasmic spaces. Heterophagy carries to lysosomes aliquots of the external milieu; autophagy will bring into them cytoplasmic constituents. Both materials will be handled by lysosomal hydrolases. It therefore seemed of interest to impair their action in order to evaluate the real role of lysosomes in cellular physiopathology. Inhibitors of lysosomal enzymes can be easily directed towards lysosomes through the heterophagic pathway. Amongst many possible molecules, antibodies were chosen, as they are powerful inhibitors, target-selective and easily endocytozed [1, 2]. Furthermore, they are rather resistant to lysosomal proteolysis [3, 4].

Success with the use of antisera directed against either purified lysosomal enzymes or a soluble lysosome fraction has already been reported [5, 6]. This paper describes the action of the second type of antiserum on cultured fibroblasts at both toxic and non-toxic levels.

ANTISERUM PRODUCTION

Lysosomes were isolated from rats previously injected with Triton WR-1339, a non-haemolytic detergent [7-9]. As extensively discussed by Beaufay [10], pending further developments only modified lysosomes, namely lysosomes resulting from treatment with Triton WR-1339, can be produced on a reasonable scale. Modifications to lysosomes through this technique seem to be minimal. Isolated lysosomes were thereafter ruptured by hypotonicity, and a soluble fraction was separated from the membrane material by centrifugation. This fraction (henceforth referred to as LS-fraction) is injected into rabbits. Frequent injections and stimulation with Freund's adjuvant are necessary to obtain satisfactory responses.

From the immune-sera were prepared either globulin fractions by ammonium sulphate precipitation or pure IgG fractions by ion-exchange chromatography. For labelling purposes, both preparations were reacted with fluorescein isothiocyanate or with [^{3}H-] or [^{14}C-]acetic anhydride.

FRACTIONATION OF RAT EMBRYO FIBROBLASTS (TARGET CELLS)

Rat embryo fibroblasts were selected as target cells. Their subcellular constituents were characterized by means of differential and isopycnic centrifugation. Optimum conditions for assay of enzymes (pH, activators, calibration curves in respect of time and concentration) were carefully established.

We first demonstrated a structural latency for three acid hydrolases. This latency is abolished and the enzymes solubilized by treatment with digitonin (Table 1). In comparison, non-acid-hydrolases behave differently in this respect, suggesting already that the two types of enzyme are associated with different structures [11].

Cytoplasmic extracts (homogenate *minus* nuclear fraction) were thereafter analyzed by density-gradient centrifugation. We used the special zonal rotor of Beaufay [9] which allows complete equilibration of most subcellular constituents in a single 3 h run thanks to the low radial distance of its chamber. Fig. 1 shows the result of a typical experiment, plotted according to Beaufay *et al.* [12].

Four patterns of enzyme can be observed. The three acid hydrolases show an assymetrical distribution around a similar modal density of 1.21 g/ml and, despite some heterogeneity, behave essentially in a similar manner. We may reckon that they serve to characterize the lysosomes of these cells.

Catalase equilibrates in essentially the same region of the gradient. The shape of its distribution is, however, largely different from that of the hydrolases. It is probably associated with peroxisome-like structures.

The third type of pattern is displayed by the marker enzyme for mitochondria, cytochrome oxidase. This population of granules is remarkably homogeneous with respect to density. A fourth pattern is given by 5'-nucleotidase, nucleoside diphosphatase and cholesterol. 5'-Nucleotidase is a typical marker for plasma membrane in liver [13]. Its association with light elements, its lack of latency and its insolubility suggest a similar localization in fibroblasts. Moreover, it shows the main characteristic physicochemical features (activation by Mg^{2+}, inhibition by F^-, insensitivity to tartrate) of the hepatic 5'-nucleotidase. Cholesterol, known to be largely associated with liver plasma membrane [14], is essentially distributed like 5'-nucleotidase. This suggests an identical subcellular localization in fibroblasts. Fibroblasts do not possess the typical microsomal inosine diphosphatase, which is a latent enzyme, activated and solubilized by detergents [15]. The nucleoside diphosphatase activity of the fibroblasts seems to belong to the plasma membrane, as does a non-latent nucleoside diphosphatase in liver [16].

Table 1. Influence of digitonin on free and soluble activities of enzymes. The homogenate (0.25 M sucrose)was treated with digitonin to 0.04% (cholesterol/digitonin molar ratio = 3.97).
FREE ACTIVITY: activity measured during 10 min under isotonic conditions, as % of the activity measured in the pre snce of 0.1% Triton X-100.
SOLUBLE ACTIVITY: activity non-sedimentable at 3×10^6 *g* × min (Beckman/Spinco 40 rotor, 40,000 rev/min for 30 min), as % of the activity of the original homogenate. *Recoveries ranged from 85% to 105%.*

	CONTROL HOMOGENATE		TREATED HOMOGENATE	
	FREE ACTIVITY	*SOLUBLE ACTIVITY*	*FREE ACTIVITY*	*SOLUBLE ACTIVITY*
Acid phosphatase	18.6	9.1	95.2	73.8
Acid deoxyribonuclease	17.5	10.3	92.0	79.9
Cathepsin D*	15.8	7.5	89.7	72.6
Catalase	25.7	3.1	24.6	7.7
5'-Nucleotidase	102.5	1.7	105.7	2.6
Inosine diphosphatase	92.7	21.2	98.6	23.7

* *Free activity determined at pH 5*

The distribution of the proteins of the cell sap is indicated by the neutral pyrophosphatase (Mg^{2+}-dependent), a readily soluble and non-adsorbable enzyme.

Isopycnic centrifugation therefore allows unambiguous resolution of the fibroblast homogenate to give lysosomes, mitochondria, peroxisome-like structures, plasma membrane, and cell sap. One should, however, note that the separation is essentially analytical, as each type of pattern considerably overlaps the others. Microsomes (smooth and rough) are more difficult to characterize. On one hand, the distribution of NADH : cytochrome *c* reductase reflects its association with both mitochondria and endoplasmic reticulum; on the other hand fibroblasts contain numerous free ribosomes and polysomes; the distribution of the rough endoplasmic reticulum can therefore not be clearly established. Other microsomal enzymes or constituents could not be assayed satisfactorily, owing to the limited amount of material available.

The technique of isopycnic centrifugation was thereafter used in order to study the localization of two proteins susceptible to be taken up by endocytosis: fluorescein-labelled rabbit IgG and horseradish peroxidase. In two different experiments these proteins were incorporated in the culture fluid for 30 h. The results of the isopycnic centrifugation of the cytoplasmic extract (which contained more than 85% of the total amount taken up by the cells) are shown in Figs. 2 and 3. In both cases, the bulk of the protein is distributed essentially like lysosomal hydrolases, and it is highly probable that the proteins become effectively localized in lysosomes. Furthermore,

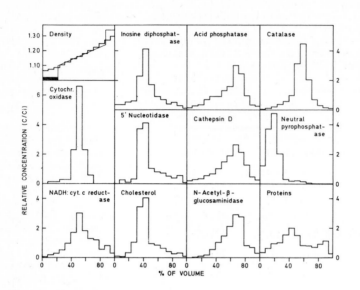

Fig. 1. Isopyc-
nic equilibration
of a post-nuclear
supernatant (ex-
tract) of rat
embryo fibro-
blasts. Centri-
fugation was per-
formed in Beau-
fay's rotor [8]
for 3 h at 35,000
rev/min, and the
results plotted
according to
Beaufay *et al.*
[12]. The *upper
left* diagram
shows the shape
of the injected
gradient (thin
line: *sample,*
10 ml of cyto-
plasmic extract
in sucrose of
density 1.034
g/ml; *sucrose
gradient* 1.10-1.24 g/ml, 32 ml; *sucrose cushion* 1.34 g/ml, 6 ml) and the
actually recovered gradient (thick histogram).
*Volume recovery was 98.2%. Recoveries of constituents from the original ex-
tract varied from 85.3 to 106.2%.*

fractions rich in fluorescent material (Fig. 3) were shown, by double immunodiffusion with anti-rabbit goat antiserum, to indeed contain rabbit immunoglobulins.

Thus, access of exogeneous proteins to the lysosomal space is clearly
demonstrated to occur efficiently in fibroblasts.

IN VITRO CHARACTERIZATION OF ANTI-LYSOSOME SERUM

Anti-lysosome serum (ALS) reacts with proteins of fibroblast lysosomes.
This can be readily demonstrated by immunofluorescence: intensively stained
discrete dots appear in fibroblasts exposed to labelled anti-lysosome serum
[9]. Furthermore, several enzymes, including cathepsin D are strongly inhibited [6]. The degree of inhibition varies, however, from one serum to
another, as illustrated for cathepsin D in Fig. 4. Absence of inhibition
is accompanied by lack of precipitation. This suggests that there are at

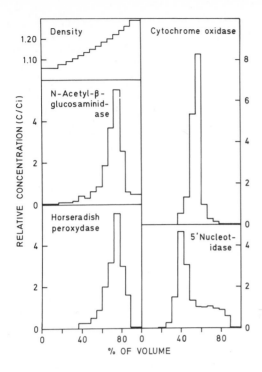

Fig. 2. Isopycnic equilibration
of a post-nuclear supernatant of
rat embryo fibroblasts cultivated in the presence of horseradish peroxidase (10 μg/ml) for
24 h. The experimented conditions and mode of representation
are as in Fig. 1.
*Recoveries of constituents va-
ried between 76.3% and 105.2%.*

least two different antigenic
forms of the enzyme, one of which
did not elicit an immunological
response in serum TS_B.

Finally the reaction of
the anti-lysosome serum is specific for the lysosomal components
of fibroblasts. This was shown
by performing a double-diffusion
test, against anti-lysosome serum, on fractions obtained by
isopycnic equilibration (Fig. 5).
Reaction occurred only in fractions containing significant
amounts of acid hydrolases. Conversely, anti-microsome serum
or anti-cell sap serum failed to
react significantly with lysosome-rich fractions.

This clearly shows that an antigenic community exists between lysosomes of rat cultured fibroblasts and rat hepatocytes. The same holds true for
microsomal soluble constituents and for the cell sap. Antibodies raised
against liver organelles can therefore largely be used with fibroblasts.

EFFECTS OF ANTIBODIES ON LYSOSOME FUNCTION (AT TOXIC LEVELS)

When cells are incubated for more than 24 h with a sufficient amount of anti-
lysosome serum, they show evident signs of cell damage and eventually detach
from glass and die. This toxicity is complement-independent and is probably
related to lysosomal dysfunction, as will be evidenced below. At sublethal
doses, cells survive for a few days and their pathology may be studied. Fig.6
summarizes the main effects observed when fibroglasts, given the occasion to
accumulate and to digest fluorescein-labelled globulins, are incubated with
anti-lysosome IgG.

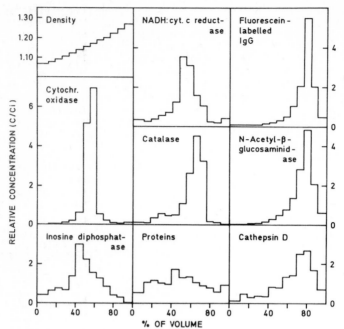

Fig. 3. Isopycnic equilibration of a post-nuclear supernatant of rat embryo fibroblasts cultivated with fluorescein-labelled IgG (100 μg/ml) present for 30 h. The experimental conditions and mode of representation are as in Fig. 1. *Recoveries of constituents varied from 92.3 to 109.7% except for cytochrome oxidase (71.3%).*

On the one hand, the rate of uptake of fluorescein-labelled globulins is reduced and the amount accumulated is only half that of the controls. On the other hand, release of micromolecular fragments into the culture fluid is nearly abolished. The overall handling of foreign proteins (uptake, lysosomal accumulation, hydrolysis and diffusion or liberation of digestion products) by lysosomes is thus largely depressed. Pre-treatment of the cells with antibodies does not significantly alter the picture: cell accumulation is reduced by half and no digestion can be detected.

The same reaction has been observed by Dingle *et al.* [17] who induced impairment of digestion of haemoglobin in macrophages with an antiserum raised against cathepsin D. A marked decrease in the transit rate of the protein (as evidenced by production of TCA-soluble haemoglobin digestion products, revealed by ^{3}H label, in the culture fluid) was noticed, and was by no means counterbalanced by the reported increase in the rate of haemoglobin accumulation [17].

Simultaneously, dramatic morphological alterations are observed in fibroblasts treated with anti-lysosome serum. They were extensively studied

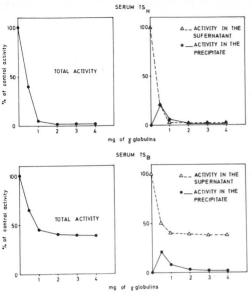

SERUM TS_H

TOTAL ACTIVITY

Δ__ ACTIVITY IN THE SUPERNATANT
●__ ACTIVITY IN THE PRECIPITATE

mg of γ·globulins

SERUM TS_B

TOTAL ACTIVITY

Δ__ ACTIVITY IN THE SUPERNATANT
●__ ACTIVITY IN THE PRECIPITATE

mg of γ·globulins

Fig. 4. Immunoprecipitation of cathepsin D of rat embryo fibroblasts. The homogenate was thawed and frozen several times in order to activate lysosomes. *General procedure:* aliquots (200 μg of protein) were treated with increasing amounts of anti-lysosome IgG at 37° for 1 h and 0° for 20 h. The reaction medium was 0.15 M aCl, 10 mM phosphate buffer pH 7.4. Precipitates were separated by centrifugation (2,000 rev/min, 30 min) and washed twice in ice-cold saline. Aliquots of the precipitates, the pooled supernatants and the original homogenates were assayed for cathepsin D through addition of acid-denatured haemoglobin buffered at pH 5 (to avoid re-dissociation of immune complexes) and incubation for 1 h at 37°.

The diagram on the *left* shows the activity of the treated homogenates prior to precipitation (as % of the activity of the untreated homogenate; reaction with unspecific IgG gave no significant inhibition). The diagram on the *right* shows the activities detected in the washed precipitates and in the pooled supernatants (as % of the activity of the untreated homogenate).

at the electron-microscope level by Van Hoof [6, 18, 19]. Dense bodies are more numerous and much larger than the normal ones. They show a pleiomorphic content, suggesting the accumulation of undigested membranous material. Quantitative morphological analysis of these cells has been performed; the results are illustrated in Fig. 7. The relative lysosomal volume increases by up to 25% of the total cell volume. Such an enlargement has only been observed with fibroblasts suffering from severe lysosomal storage [19].

Thus there is evidence that impairment of lysosomal digestion has been induced by anti-lysosome serum. Antibodies act probably first upon acid hydrolases, thus leading to a mixed storage syndrome, as antibodies are directed against several lysosomal enzymes. Switching off the endocytosis would therefore appear to be a self-defence mechanism.

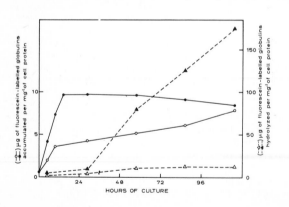

		FRACTIONS													
		1	2	3	4	5	6	7	8	9	10	11	12	13
SERUM	TSₚ														
	Pᴅ														
	S₂₁														
ENZYMES	Hydrolases								1.1	1.4	2.6	3.5	1.6		
	Cyt.-oxid.						5.1		6.9						
	Catalase							1.0	2.6	4.5	3.2				
	IDPase					1.3	3.0	2.3	1.5	1.3					
	Density	1.069	1.079	1.091	1.107	1.122	1.139	1.155	1.170	1.187	1.200	1.222	1.244	1.270	

Fig. 5. Schematic representation of immunoprecipitation lines (Ouchterlony's technique) observed between fractions from an isopycnic equilibration experiment on rat embryo fibroblasts and anti-lysosome serum (TS_p), anti-microsome serum (P_D) and anti-cell-supernatant serum (S_{21}). The distributions of enzymes throughout the gradient are indicated by their relative concentrations [12] within the fractions (values below 1 are not recorded). 'Hydrolases' distribution is the average distribution of acid phosphatase, N-acetyl-β-glucosaminidase and cathepsin D. Density values allow allow estimation of the distribution of components not assayed in this particular experiment, by comparison with the data of Fig. 1.

Fig. 6. Accumulation and digestion of fluorescein-labelled globulins (100 μg/ml) by rat embryo fibroblasts. The ordinate on *left* refers to the *solid lines* and shows the amount recovered within the cells, at various times after contact, in the presence of anti-lysosome IgG (open circles; 300 μg/ml) or unspecific IgG (closed circles, 300 μg/ml). The ordinate on *right* refers to the *broken lines* and represents the amount of fluorescein-labelled protein digested (as evidenced by detection of low molecular weight fluorescence in gel filtration chromatography [6]) by anti-lysosome IgG-treated fibroblasts (open triangles) or by unspecific IgG treated fibroblasts (closed triangles). The concentration of fluorescein-labelled globulins was about 900 μg/ml of cell protein.

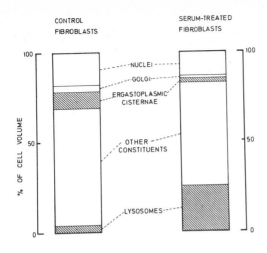

CONTROL FIBROBLASTS

SERUM-TREATED FIBROBLASTS

NUCLEI

GOLGI

ERGASTOPLASMIC CISTERNAE

OTHER CONSTITUENTS

LYSOSOMES

% OF CELL VOLUME

Fig. 7. Morphological analysis (at the electron microscope level) of rat embryo fibroblasts cultivated with anti-lysosome IgG present (300 μ g/ml for 4 days: *serum-treated fibroblasts*) or with unspecific IgG *(control fibroblasts)*. Adapted from Van Hoof [18].

It is interesting to note that Dingle *et al*. [17] also observed morphological alterations in lysosomes impaired by anticathepsin D. The appearance of the lysosomal contents was claimed to be related to accumulation of haemosiderin, a product of incomplete digestion of haemoglobin.

FATE OF ANTI-LYSOSOME SERUM

It is controversial whether anti-lysosome or anti-cathepsin D antibodies are sufficiently concentrated within secondary lysosomes as to effectively achieve intralysosomal inhibition of the enzyme. Assays performed on cell homogenates lead to ambiguous results since part of the enzyme may remain inaccessible *in vivo* in primary lysosomes. Furthermore, the conditions of the *in vitro* assays are so removed from those prevailing in lysosomes that results thus obtained serve only as an indication of what really occurs inside the cell. Dingle *et al*. [17] have also suggested that antibodies might not be distributed equally amongst all secondary lysosomes.

Direct evidence of substantial accumulation of anti-lysosome serum within secondary lysosomes is therefore needed. In the course of this study we came to the rather impressive conclusion that anti-lysosome antibodies are much more actively taken up into lysosomes than other immunoglobulins. Cells were incubated for 24 h with ^{3}H anti-lysosome IgG or ^{3}H IgG from control rabbits (henceforth designated as 'unspecific' IgG). Results of such an experiment are shown in Fig. 8. Uptake of anti-lysosome IgG was about twice that of unspecific IgG. Anti-lysosome serum was thereafter purified by immunoadsorption on the soluble lysosome fraction linked to Sepharose 4B [20]. About 20% of the total IgG from anti-lysosome serum was specifically

Fig. 8. Uptake of ^{3}H-labelled IgG (75 μ g/ml) by rat embryo fibroblasts at various times after contact. —o—: unspecific IgG; —△— : anti-lysosome IgG; —●—: anti-lysosome IgG purified by immunoadsorption.

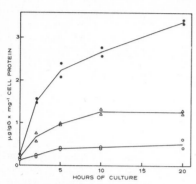

eluted (using 2% formic acid). Uptake of this purified anti-lysosome IgG was now 6 times that of unspecific IgG; the 'selective' uptake is thus dependent on the immunological properties of the IgG. Although its mechanism is still unknown, this phenomenon is obviously gratifying for the experimenter.

The localization of these antibodies was investigated by isopycnic centrifugation. The results are shown in Fig. 9. It can be observed that the distribution of the ^{3}H-labelled anti-lysosome IgG closely parallels that of the acid hydrolases and diverges substantially from the distributions of other markers. Much of the antibody is therefore likely to be associated effectively with its target-organelle. If the concentration of the anti-lysosome serum is raised to levels approaching toxicity, the intracellular content of the cells in specific anti-lysosome antibodies approaches 6% of the total cell proteins; thus, taking into account the relative lysosomal volume (maximum 25%), the intralysosomal concentration of antibodies may reach at least 25% of the lysosomal proteins. Such a concentration is very likely to provoke effectively an immunological inhibition of several enzymes, the extent of which will be studied in future experiments.

CONCLUDING REMARKS

Immunoglobulins directed towards lysosomal constituents constitute a powerful and selective tool for investigating lysosomal functions. The unique accessibility of lysosomes to external constituents allows these experiments to be undertaken on the living cell. Such studies may shed new light on the various roles of the lysosomal apparatus.

Acknowledgements

The author wishes to thank Dr. A. Trouet for invaluable help throughout this work in the immunological field, and Prof. H. Beaufay for critical collaboration in cell fractionation procedures. We also thank Dr. P. Jacques for help in final handling of the manuscript. The technical assistance of Mrs. F. Renoird was much appreciated. The author is *'Aspirant'* of the Belgian *Fonds National de la Recherche Scientifique,* and this work was supported financially

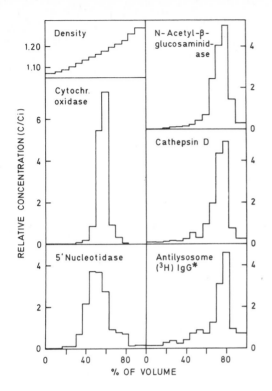

Fig. 9. Isopycnic equilibration of post-nuclear supernatant of rat embryo fibroblasts cultivated for 24 h in the presence of ^{3}H-labelled anti-lysosome IgG (30 μ g/ml). Experimental conditions were identical to those of Fig. 1. *Recoveries of constituents varied between 89.3 and 104.7%.*

by the Belgian *Fonds National de la Recherche Scientifique Médicale.*

References

1. Cinader, B. & Lafferty, K.J., *Immunology 7* (1964) 342-362.

2. Ghetie, V. & Sulica, A., *Immunochemistry 7* (1970) 175-184.

3. Fehr, K., Lospalluto, J. & Ziff, M., *J. Immunology 105* (1970) 973-983.

4. Ghetie, V. & Motas, C., *Immunochemistry 8* (1971) 89-97.

5. Weston, P.D. & Poole, A.R., *Lysosomes in Biology and Pathology,* Vol. 3 (Dingle, J.T., ed.) North-Holland, Amsterdam (1973) pp. 425-464.

6. Tulkens, P., Trouet, A. & Van Hoof, F., *Nature (Lond.) 228* (1970) 1282-1284.

7. Trouet, A., in: *Biomembranes. Cells, Organelles and Membranous Components* (Fleischer, S., Parker, L. & Estabrook, R.W., eds.), *Methods in Enzymology* (1974) *in press.*

8. Leighton, F., Poole, B., Beaufay, H., Baudhuin, P., Coffey, J.W., Fowler, S. & de Duve, C., *J. Cell Biol. 37* (1968) 482-513.

9. Trouet, A., *Caractéristiques Antigéniques des Lysosomes de Foie,* Thesis,

University of Louvain, Belgium (1969).

10. Beaufay, H. in: *Lysosomes in Biology and Pathology,* Vol. 2 (Dingle, J.T. & Fell, H.B., eds.) North-Holland, Amsterdam (1969) pp. 515-546.

11. Jacques, P., *This volume,* Art. 16.

12. Beaufay, H., Jacques, P., Baudhuin, P., Sellinger, O.Z., Berthet, J. & de Duve, C., *Biochem. J. 92* (1964) 184-205.

13. Emmelot, P., Bos, C.J., Benedetti, E.L. & Rümke, Ph., *Biochim. Biophys. Acta 90* (1964) 126-145.

14. Thinès-Sempoux,D.,in*Lysosomes in Biology and Pathology,* Vol. 3 (Dingle, J.T., ed.) North-Holland, Amsterdam (1973) pp. 278-296.

15. Novikoff, A.B. & Heus, M., *J. Biol. Chem. 238* (1963) 710-716.

16. Wattiaux-De Coninck, S. & Wattiaux, R., *Biochim. Biophys. Acta 183* (1969) 118-128.

17. Dingle, J.T., Poole, A.R., Lazarus, G.S. & Barrett, A.J., *J. Exp. Med. 137* (1973) 1124-1141.

18. Van Hoof, F., *Les Mucopolysaccharidoses en Tant que Thésaurismoses Lysosomiales,* Thesis, University of Louvain (1972).

19. Van Hoof, F. in: *Lysosomes and Storage Diseases* (Hers, H.G. & Van Hoof, F., eds.), Academic Press, New York (1973) p. 241.

20. Cuatrecasas, P., *J. Biol. Chem. 245* (1970) 3059-3065.

32 MARKER ENZYMES OF EUKARYOTIC MICRO-ORGANISMS

D. Lloyd   and   T.G. Cartledge
Dept. of Microbiology   *Dept. of Microbiology*
University College   *The University*
Newport Road   *Claverton Down*
Cardiff, CF2 ITA, U.K.   *Bath, Somerset, U.K.*

Most eukaryotic micro-organisms are devoid of many recognized marker enzymes of mammalian tissues. There is usually no lack of conventional mitochondrial or peroxisomal marker enzymes. Analysis of the enzyme complement of vacuoles is complicated by the presence of several populations of vacuoles and vesicles within the cell. The major problems appear to be the elucidation of adequate marker enzymes for endoplasmic reticulum and plasmalemma. Some of the electron microscopic evidence which claims to show separation of these two fractions is rather tenuous. For these two subcellular organelles in particular the recognized mammalian marker enzymes appear to be absent in micro-organisms.

According to the postulates of Claude, as set out by de Duve [1], a marker enzyme should satisfy two criteria.-
(a) The enzyme must belong to a single class of particles (i.e. be unilocat-ional).
(b) Its specific activity must be the same in the different particle sub-classes which may be separated by the fractionation procedures (i.e. biological homogeneity is implied).

    The former criterion is the more important in modern fractionation studies. For example cytochrome *c* oxidase is present only in inner mito-chondrial membranes. The latter criterion is not feasible in practice as it does not take into account biological heterogeneity, for example, the ageing of mitochondria which results in heterogeneous populations within the cell. In this paper we have used the term 'marker enzyme' in several places when that particular enzyme may be multi-locational within the cell.

    Virtually all of the earlier fractionation studies involved the use of rat liver tissue. A set of marker enzymes for sub-cellular organelles of this tissue is well known, as documented by Reid [2]. Unfortunately, many of the enzymes listed are not found in micro-organisms; thus, glucose-6-phosphatase, a suitable rat-liver marker enzyme for endoplasmic reticulum, is absent in many micro-organisms.

    Pioneer workers in the field, namely, Claude and the Hogeboom-Schneider team, stressed the need to establish balance sheets in which the summation of the activities of the tissue fractions is compared with that of the whole tissue. This need has also been stressed by other major contributors to the field [1]; yet several papers now cited quote results only and do not attempt to present any recovery data.

Eukaryotic micro-organisms are a very diverse group and until recently very little fractionation work on this material has been performed. In the examples shown, unless stated, the gradients were linear and of sucrose.

ILLUSTRATIVE MICRO-ORGANISMS OTHER THAN YEASTS

Tetrahymena pyriformis is a ciliate protozoan. Extracts of this organism will not reduce mammalian cytochrome *c* [3]. Thus cytochrome *c* oxidase cannot be used as a mitochondrial marker enzyme. Fractionation studies at Cardiff [4] have shown that succinate-cytochrome *c* oxidoreductase is a good mitochondrial marker. NADH-cytochrome *c* oxidoreductase is multilocational; however the antimycin A-sensitive portion is confined to the mitochondria. In sucrose gradients mitochondria have a median buoyant density around 1.20 g/ml as compared with ∿ 1.23 for peroxisomes (as indicated by sedimentable catalase activity). Zones of NADH- and NADPH-cytochrome *c* oxidoreductase activities at 1.15 g/ml indicate the presence of endoplasmic reticulum. Complex distributions of three acid hydrolases, namely, acid *p*-nitrophenyl phosphatase, DNAse and *N*-acetyl-β-glucosaminidase probably indicate both hetero- and auto-phagosomes. The sedimentable portions of acid hydrolase activities exhibit latent activity [4, 5].

The results of Müller [6] agree with those of Lloyd and co-workers. Müller used malate dehydrogenase as a marker enzyme for mitochondria. Several other acid hydrolases were assayed and confirm the existence of at least three populations of lysosomes. Müller *et al.* [7] also showed that the peroxisomes contain virtually all of the sedimentable D-amino acid oxidase, and the key enzymes of the glyoxylate cycle (isocitrate lyase, malate synthase) and also glyoxylate oxidase in addition to catalase.

Hartmannella castellanii is a soil amoeba which possesses cytochrome *c* oxidase which may be used as a mitochondrial marker [8]. Catalase is found in peroxisomes though this organelle is not well separated from mitochondria after high-speed zonal centrifugation in sucrose gradients [8,9]. The distribution of several acid hydrolases indicates the presence of more than one population of lysosomes, again probably hetero- and auto-phagosomes. Alkaline phosphatase and IDPase are located at a density of 1.12 g/ml, presumably indicating endoplasmic reticulum.

Müller [10] and Müller & Møller [11] showed that urate oxidase can also be used as a peroxisomal marker enzyme in *H. castellanii*. D-amino acid oxidases and α-hydroxy acid oxidases were not detected. Similarly the amoebae *chaos chaos* and *Amoeba proteus* have peroxisomes which contain urate oxidase but not α-hydroxy acid oxidase or D-amino acid oxidase.

Wiener & Ashworth [12] have fractionated extracts of Myxamoebae of *Dictyostelium discoideum* in linear sucrose gradients. They separated and identified mitochondria (citrate synthase), peroxisomes (catalase) and lysosomes (acid phosphatase). The fractionation of this organism is complicated, as both the densities of the organelles and the distributions of enzymes differ depending on whether the organisms are grown axenically or monoaxenically.

Trichomonas foetus is a parasitic flagellate which is found in the alimentary canal or urogenital tract of vertebrates. Its metabolism is anaerobic although certain species exhibit respiration under aerobiosis. Müller [13] reported on the following sub-cellular distributions of enzymes of homogenates of this organism. NADH and NADPH dehydrogenases, α-and β-galactosidases and over 90% of the catalase activity are non-sedimentable. Malate dehydrogenase and β-glycero-phosphatase are associated with large particles having a density of 1.24 g/ml. which have a granular matrix and are single membraned. These particles morphologically resemble microbodies (peroxisomes) of other organisms. The 10% sedimentable catalase is not associated with the microbodies and occurs in particles of density 1.22 g/ml. Several acid hydrolases associated with large particles at 1.15 to 1.20 g/ml were evident. One of these, acid phosphatase together with β-glucuronidase, was found in a second type of particle which was smaller than the former and granular in appearance. Many flattened vesicles were found in this population, which were interpreted as being fragments of the Golgi apparatus.

Polytomella caeca is classified as either a protozoan or a colourless alga. Fractionation of homogenates of *P. caeca* grown on acetate indicate that cytochrome *c* oxidase is located exclusively in the mitochondria at a density of 1.20 g/ml [14, 15]. Peroxisomes (as shown by catalase activity) have a density of 1.24 g/ml. Acid phosphatase has a complex distribution with peaks of sedimentable activity at 1.14 and 1.21 g/ml. Electron microscopy and histochemistry have shown that acid phosphatase is multilocational, i.e. in the Golgi, in the cytosol, in large vacuoles and in areas of focal degradation.

Ochromonas malhamensis is a green flagellate alga of the Division Chrysophyta. Lui *et al.* [16], fractionated extracts of this organism and identified and separated mitochondria (cytochrome *c* oxidase activity), peroxisomes (catalase) and lysosomes (acid phosphatase).

Fractionation of extracts of *Euglena gracilis* indicate that mitochondria are found at densities of 1.17 and 1.20 g/ml as indicated by the activities of succinate dehydrogenase and D-lactate dehydrogenase [17]. Graves *et al.* claim to have strong evidence of heterogeneity of mitochondria in

E. gracilis, confirming a report of Quigley et al. [18]. Over 80%
of the sedimentable activities of the glyoxylate cycle enzymes, isocitrate
lyase, malate synthase and glyoxylate reductase were found at ρ = 1.18 to
1.22 g/ml. Malate dehydrogenase and citrate synthase were also found in this
region. Graves et al. [17] conclude that these enzymes are present in organelles (glyoxysomes) which are distinct from mitochondria. Positive identification of these organelles as peroxisomes is made difficult as these workers
could not detect catalase in this organism under their particular growth
conditions.

Although chloroplasts are easily detectable by electron microscopy,
Vasconcelos et al. [19] have shown that ribulose diphosphate carboxylase is
a marker enzyme of these organelles and it has the added advantage that it is
a sensitive indicator of chloroplast integrity: minor damage to these organelles leads to detectable enzyme activity in the sample zone.

O'Sullivan & Casselton [20] published results of the fractionation of
the basidiomycete Coprinus lagopus. The wide distribution of cytochrome c
oxidase is probably indicative of mitochondrial damage during isolation. The
major peak of cytochrome c oxidase activity was different from that of malate
synthase and isocitrate lyase. In this case the peaks of the two glyoxylate
cycle enzymes were found at a density lighter than that of mitochondria.

Matile [21] isolated vacuoles from the Macroconidia of Neurospora
crassa by flotation of a homogenate through a Ficoll gradient. The vacuoles
were free of mitochondria as judged by the absence of succinate cytochrome c
oxido-reductase. These organelles contain high specific activities of
several acid hydrolases, for example, phosphodiesterase, alkaline phosphatase,
leucyl-aminopeptidase and invertase.

Cassady & Wagner [22] working with Neurospora crassa separated inner
and outer membranes of highly purified mitochondria. Centrifugation in sucrose gradients resulted in substantial separation of succinate cytochrome c
oxido-reductase (inner mitochondrial membrane) and kynurenine-3-hydroxylase
(outer mitochondrial membrane). Electron microscopic data indicated that the
outer mitochondrial membrane fractions were vesicular whereas the inner membrane fractions appeared as characteristic mitochondria but lacking an outer
membrane.

YEASTS

Bandlow & Kaudewitz [23] separated the inner and outer membranes of
yeast mitochondria (Saccharomyces sp.) after valinomycin treatment. They
showed that kynurenine-3-hydroxylase is in the outer (ρ = 1.08 g/ml) and not
inner mitochondrial membrane (ρ = 1.19 g/ml) and that monoamine oxidase (used
as an outer membrane marker for mammalian mitochondria) is not detectable in
yeast. Succinate-cytochrome c reductase and malate dehydrogenase were used

as markers for inner membranes and matrix respectively, and adenylate kinase was located in the inter-membrane space [24].

After zonal centrifugation of extracts of *Saccharomyces carlsbergensis,* cytochrome *c* oxidase and oligomycin-sensitive ATPase are located in two peaks indicating two populations of mitochondria at densities of 1.21 g/ml and 1.235 g/ml [25]. Total ATPase activity shows a complex distribution throughout the gradient. The possibility of mitochondrial heterogeneity in yeasts is also suggested by the distributions of cytochrome *c* oxidase and succinate dehydrogenase after fractionation of extracts of *Schizosaccharomyces pombe* (R.K. Poole & D. Lloyd, unpublished data).

NADH- and NADPH-cytochrome *c* oxidoreductases are also found in the mitochondrial region. Most but not all of the former is antimycin A-sensitive. Both are multimodal, in contrast to the report of Schatz & Klima [26] who claimed that NADPH-cytochrome *c* oxidoreductases is uniquely associated with the microsomal fraction.

Catalase is a conventional peroxisomal marker enzyme. Peroxisomes are not well separated from mitochondria by isopycnic centrifugation but separation by rate-zonal centrifugation may be achieved [14]. Avers and her co-workers [27] suggest key glyoxylate cycle enzymes are mainly peroxisomal.

Matile & Wiemken [28] suggested that several acid hydrolases are characteristic of the cell vacuole in *S. cerevisiae*. However Perlman & Mahler [29] found β-glucuronidase to be a soluble enzyme, and we have shown [30] that in *S. carlsbergensis* a variety of acid hydrolases are multi-locational and they are very unlikely to be associated with a single species of organelle. Recently Matile [31] found at least three types of sedimentable particles containing β-glucanase in the cell. Betata & Gascon [32] claim that the intracellular invertase of yeast is in the vacuole, whereas Holley & Kidby [33] suggest that invertase is not in the vacuoles but in separate vesicles.

One of the major problems with Saccharomyces is the separation and characterization of endoplasmic reticulum and plasmalemma. Several mammalian marker enzymes of endoplasmic reticulum are not found in *S. carlsbergensis,* e.g. glucose-6-phosphatase, IDPase, aromatic nitroreductase, aniline hydroxylase, ethylmorphine *N*-demethylase, *o*-amino-phenol-UDP glucuronate transferase, *o*-aminophenol-UDP glucose transferase and cytochrome P_{450} [25].

Klein *et al.* [34, 35] showed the presence of a ribosome fraction and a distinct microsomal membrane fraction from *S. cerevisiae*. The latter was characterized by the presence of fatty acid synthetase, acetyl Co-A synthetase and squalene oxidocyclase, though in a later communication [36] the acetyl Co-A synthetase was found to be of wide intracellular distribution.

The plasmalemma according to Matile [37] contains Mg^{++}-dependent, oligomycin-insensitive ATPase and loosely bound invertase and, according to Nurminen & Suomalainen [38] contains the ATPase mentioned above together with lipase and phospholipase.

The cell wall contains ADPase, PP_iase, saccharase and acid phosphatase [38].

Isolation of nuclei from *S. cerevisiae* [39] and *Schizosaccharomyces pombe* [40] indicate that DNA-dependent RNA polymerase is a good marker enzyme of nuclei. Duffus [40] also suggests NAD pyrophosphorylase is a nuclear enzyme.

Very few fractionation studies of anaerobically grown yeast have been reported. There is a possibility that the early reports were of semi-anaerobically grown yeasts as judged by the enzyme complements of the extracts.

From our results [14, 41] there are no organelles of peroxisomal size in anaerobically grown *S. carlsbergensis* and no catalase. In addition, cytochrome *c* oxidase, succinate dehydrogenase, succinate- or L-lactate-cytochrome *c* oxido-reductases, succinate-ferricyanide oxidoreductase and cytochromes *a*, a_3, *c* and c_1 [42] are all absent from these cells. The best 'promitochondrial' marker enzyme appears to be oligomycin-sensitive ATPase. The distribution of acid phosphatase in fractionated homogenates appears to be very similar to that for aerobically grown cells.

References

1. de Duve, C., *J. Cell. Biol. 50* (1971) 20D - 50D.

2. Reid, E., in *Separations with Zonal Rotors* (Reid, E., ed., Wolfson Bioanalytical Centre, University of Surrey, Guildford, (1971) pp. B-3.1 — B-3.6.

3. Turner, G., Lloyd, D. & Chance, B., *J. Gen. Microbiol. 65* (1971) 359-374.

4. Lloyd, D., Brightwell, R., Venables, S.E., Roach, G.I. & Turner, G., *J. Gen. Microbiol. 65* (1971) 209-223.

5. Müller, M., Baudhuin, P., de Duve, C., *J. Cell Physiol. 68* (1966) 165.

6. Müller, M., *J. Cell Biol. 52* (1972] 478.

7. Müller, M., Hogg, J.F. & de Duve, C., *J. Biol. Chem. 243* (1971)5385-5395.

8. Morgan, N.A., Howells, L., Cartledge, T.G. & Lloyd, D., in *Methodological Developments in Biochemistry* (Reid, E., ed.), *Vol. 3*, Longman, London (1973) pp. 219-232.

9. Morgan, N.A. & Griffiths, A.J., *J. Protozool. 19 Suppl.* (1972) A137.

10. Müller, M., *New York Acad. Sci.* 168 (1969) 292-301.

11. Müller, M. & Møller, K.M., *Eur. J. Biochem.* 9 (1969) 424-430.

12. Wiener, E.R. & Ashworth, J.M., *Biochem. J.* 118 (1970) 505-512.

13. Müller, M., *J. Cell. Biol.* 57 (1973) 453-474.

14. Cartledge, T.G., Cooper, R.A. & Lloyd, D., in *Separations with Zonal Rotors* (Reid, E., ed.), Wolfson Bioanalytical Centre, University of Surrey, Guildford, (1971) pp. V-4.1 − V-4.16.

15. Cooper, R.A. & Lloyd, D., *J. Gen. Microbiol.* 72 (1972) 59-70.

16. Lui, N.S.T., Roels, O.A., Trout, M.E. & Anderson, O.R., *J. Protozool.* 15 (1968) 536-542.

17. Graves, L.B., Jnr., Trelease, R.N., Grill, A. & Becker, W.M., *J. Protozool.* 19 (1972) 527-532.

18. Quigley, J.W. & Price, C.A., *Plant Physiol.* 41 (1966) 25-30.

19. Vasconcelos, A., Pollack, M., Mendiola, L.R., Hoffmann, H-P., Brown, D.H. & Price, C.A., *Plant Physiol.* 47 (1971) 217-221.

20. O'Sullivan, J. & Casselton, P.J., *J. Gen. Microbiol.* 75 (1973) 333-337.

21. Matile, Ph., *Cytobiologie 3* (1971) 324-330.

22. Cassady, W.E. & Wagner, R.P., *J. Cell. Biol.* 49 (1971) 536-541.

23. Bandlow, W. & Kaudewitz, F., *Abs. Commun. 7th Meet. Eur. Biochem. Soc.* (1971) 613.

24. Bandlow, W., *Biochim. Biophys. Acta.* 282 (1972) 105-122.

25. Cartledge, T.G. & Lloyd, D., *Biochem J.* 126 (1972) 381-393.

26. Schatz, G. & Klima, J., *Biochim. Biophys. Acta 81* (1964) 448-452.

27. Avers, C.J., *Subcell. Biochem. 1* (1971) 25-37.

28. Matile, Ph. & Wiemken, A., *Arch. für Mikrobiol.* 56 (1967) 148-153.

29. Perlman, P. & Mahler, H.R., *Arch. Biochem. Biophys. 136* (1970) 245-259.

30. Cartledge, T.G. & Lloyd, D., *Biochem. J.* 126 (1972) 755-757.

31. Matile, Ph., Cortat, M., Wiemken, A. & Frey-Wyssling, A., *Proc. Nat. Acad. Sci. 68* (1971) 636-640.

32. Betata, P. & Gascon, S., *FEBS Lett. 13* (1971) 297-300.

33. Holley, R.A. & Kidby, D.K., *Canad. J. Microbiol. 19* (1973) 113-117.

34. Klein, H.P., Volkman, C.M. & Chao, F-C., *J. Bact. 93* (1967) 1966-1971.

35. Klein, H.P., Volkman, C.M. & Weibel,'J., *J. Bact. 94* (1967) 475-481.

36. Klein, H.P. & Jahnke, L., *J. Bact. 106* (1970) 596-603.

37. Matile, Ph., *FEBS Symp. 20* (1970) 39-52.

38. Nurminen, T. & Suomalinen, H., *Biochem. J. 118* (1970) 759-763.

39. Bhargava, M.M. & Halvorson, H.O., *J. Cell. Biol. 49* (1971) 423-429.

40. Duffus, J.H., *Biochim. Biophys. Acta 195* (1969) 230-233.

41. Cartledge, T.G. & Lloyd, D., *Biochem. J. 127* (1972) 693-703.

42. Cartledge, T.G., Lloyd, D., Erecińska, M. & Chance, B., *Biochem. J. 130* (1972) 739-747.

33 USE OF ZONAL ROTORS IN THE ANALYSIS OF THE CELL-CYCLE IN YEASTS

R.K. Poole and D. Lloyd
Department of Microbiology
University College
Newport Road
Cardiff. CF2 1TA, U.K.

1. The large gradient and sample capacity of zonal rotors has been exploited to study the cell-cycle of the fission yeast Schizosaccharomyces pombe 972h⁻.
2. In one method, synchronous cultures were prepared by a scale-up of the velocity sedimentation selection method of Mitchison and Vincent [1]. Rate-zonal centrifugation of cells harvested from exponential cultures (during the phase of glucose repression) was performed in linear 10→50% (w/w) sucrose gradients, made up in defined growth medium. Small slowly-sedimenting cells were removed from the rotor and, on inoculation into fresh medium, synchronous growth was maintained for at least two full cycles. Synchrony indices [2] were 0.7 to 0.9 in typical experiments.
3. The cell cycle of this organism has also been analyzed by 'culture fractionation' [3], i.e. separation of cells from an exponential culture into classes representing successive stages in the cycle. Cells may be separated either (a) on a size basis by rate-zonal centrifugation through a shallow sucrose or Ficoll gradient, or (b) on a density basis by isopycnic-zonal centrifugation through a linear 27→33% (w/w) dextran gradient. Since age in the first ¾ of the cycle is directly proportional to cell length and inversely proportional to cell density [4], this fraction of the cycle is represented almost linearly across the radius of the rotor.
4. Measurements of activities of respiratory enzymes and of catalase and acid β-nitrophenylphosphatase in cell-free extracts of cells at different stages of the cycle revealed the expression of activity of these enzymes to oscillate during the cycle.
5. The relative merits of these three approaches to the analysis of the cell-cycle in S. pompe *are discussed.*

The cell cycle, namely the period of time between successive cell divisions, has been shown in recent years to be a series of complex biochemical events, ordered temporally with respect to each other. Research in this field is currently expanding rapidly, due to recognition that such studies provide a valuable insight into mechanisms of development and biogenesis and the control of these activities in actively growing cells.

Undoubtedly one of the greatest methodological advances in these studies has been the development of synchronously-dividing cultures, namely cultures synchronized with respect to growth and division, so that the cell population approximates to the growth behaviour of a single cell.

In general, two distinct approaches have been employed in the preparation of such cultures [5]. In the first of these, 'induction synchrony', shocks or cyclic changes of heat, light, temperature or availability of nutrients are applied to the culture to effect synchronization of a cell population. Some workers [6] have objected to such methods on the grounds that the resulting cell culture may exhibit properties partly attributable to the metabolic disturbance of the cells. The second method, 'selection synchrony', exploits the changing physical state of the cell during the cycle of growth and division. A population of cells, homogeneous with respect to size or density, is selected from an exponentially growing culture, namely a culture containing a mixture of cells in all stages in the cycle. This population of cells (which is assumed to be also homogeneous with respect to age in the cell cycle) will grow synchronously on inoculation into fresh medium.

In the method of Mitchison and Vincent [1], cells (bacteria or yeast) are centrifuged through a sucrose gradient in tubes. The small slowly-sedimenting cells are removed and used to inoculate fresh medium. Other workers [7-9] have utilized the changes in density which occur during the cell cycle of *Saccharomyces cerevisiae* to select homogeneous populations of cells after isopycnic centrifugation in dextran, renografin or Ludox gradients.

It will be apparent that since only a small proportion (5-10%) of the initial cell culture is selected and used, the yield is poor and the synchronous culture prepared is of low cell density or small volume. For this reason we have used zonal centrifugation to enable large-scale, single-step selection procedures to be employed.

A quite distinct method of cell-cycle analysis also markes use of zonal centrifugation. When it can be shown that rate or isopycnic gradient centrifugation of a mixed population of cells results in their distribution through the gradient in size or density classes representing successive stages in the cell-cycle, the whole gradient may be fractionated and cells analyzed directly. This approach has been termed 'culture fractionation' by Wells and James [3] and has been used by them and also by Sebastian *et al.* [10] to study the cell-cycle in yeasts. A similar approach was described at an earlier zonal symposium to study the life cycle of mammalian cells [11]. The technique has not previously been applied where cells may be separated on a density basis.

We now describe the preparation of large-scale synchronous cultures of yeast after size selection in a zonal rotor. The cell cycle has also been analyzed by culture fractionation by rate and isopycnic zonal centrifugation. Each method has its attendant advantages and disadvantages, as is discussed.

The advantages of using the fission yeast *Saccharomyces pombe* [12] have been exploited in this work. The cell grows as a cylinder of almost constant diameter (3-4 μm), increasing only in length during the first $\frac{3}{4}$ of the cell cycle. From single-cell studies [4] it has been shown that cell density falls during this phase. In the last $\frac{1}{4}$ of the cycle no further increase in volume occurs. Throughout this work, the haploid strain 972h⁻ was used, kindly supplied by Dr. Urs Leupold, University of Bern, Switzerland.

METHODS AND RESULTS

For separation of whole cells on a rate basis, the use of a rotor capable of operation at low speeds was found to be essential. In such experiments, the M.S.E. HS zonal rotor was used. It was loaded with gradient at 6000 rev/min in an M.S.E. 18 centrifuge at room temperature, fitted with a low-speed zonal control unit, by use of an Isco Dialagrad Programmed Gradient Pump. The gradient was linear 10→60% (w/w) sucrose and was made up in defined growth medium (which contained 1% glucose and was at pH 5.5). Cells (20-30 g wet wt.) which had been harvested from exponential cultures during the phase of glucose-repression were suspended in 5% sucrose also made up in medium, and 50-70 ml of suspension was loaded onto the rotor at 10-15 ml/min followed by an overlay (30 ml medium). Centrifugation was at 1,450 rev/min (selected on the low-speed control unit) for 8 min. This was sufficient for the band of cells to sediment approximately $\frac{1}{3}$ to $\frac{1}{2}$ the path-length of the rotor. After deceleration to 600 rev/min the rotor was unloaded by displacement with 60% (w/w) sucrose introduced to the rotor edge. The integrated field-time of the total operation, which lasted less than 30 min, varied between 900 and 1300 rad².sec⁻¹. After a brief microscopic examination of the first few fractions to be collected (to check the homogeneity of the slowly-sedimenting cells), 4 to six fractions (10 ml each) were selected and used to inoculate 4l portions of defined growth medium. In most experiments, when the conditions for satisfactory sedimentation had been established the entire gradient was not collected but only those fractions required for re-inoculation.

Re-inoculation of a homogeneous population of cells resulted in a culture exhibiting synchronous growth. After a period of elongation of the small cells (∿1-1.5 h), cell division commenced and continued until a second plateau (duration ∿1.5 h) of cell numbers was attained. This pattern was then repeated, synchronous growth being well maintained at at least 6 h, that is two synchronous doublings in cell numbers. Experiments not described indicate that a third synchronous division occurs in such cultures.

The degree of synchrony was quantitated by applying the Synchrony Index of Blumenthal and Zahler [2]. This compares the time taken for cell division to occur with the mean generation time of similar cells in exponential growth (2.7 h), and also takes into account the proportion of cells which divide at each period of division. When applied to a culture exhibiting perfect, theoretical synchrony, an index of 1.0 is obtained; when applied to an exponential (= asynchronous) culture a value of 0 is obtained. The values obtained in these experiments fall between 0.6 and 0.9, and compare favourably with those calculated from the results of other workers, who generally fail to apply such an Index.

Activities of a number of enzymes were assayed in French-Press cell-free extracts prepared from cells harvested at time intervals during the cycle. These activities were corrected for the varying degrees of efficiency of cell breakage obtained, and used to calculate the total amounts of enzyme per ml of culture. Enzyme amounts described are the means of two or more assays using different concentrations of enzyme protein.

Results from our laboratory [15] show that all enzymes studied are partly mitochondrial in their subcellular localization; cytochrome c oxidase (EC 1.9.3.1) and succinate dehydrogenase (EC 1.3.99.1) are located uniquely in this organelle. The other two enzymes, acid p-nitrophenylphosphatase (EC 3.1.3.2) and catalase (EC 1.11.1.6) are at least partially associated with membranous organelles in yeast [16].

Total amounts of all enzymes studied increased exponentially overall but differed in their patterns of expression. Cytochrome c oxidase rose to a single maximum during the period of cell elongation and fell during division. Catalase and acid p-nitrophenylphosphatase exhibited more complex patterns of variation, in which maxima were observed during phases of both cell elongation and cell division. Succinate dehydrogenase and NADH-cytochrome c oxidoreductase (EC 1.6.2.1) each showed two clear maxima of activity per cell cycle, but at different stages of the cycle.

Zonal centrifugation under the conditions described previously was thus found to be suitable for the selection of cells for establishing synchronous cultures. Separation was not improved when convex or concave sucrose gradients were employed.

Two methods of cell-cycle analysis by culture fractionation using zonal centrifugation were employed. In the first of these, cells were separated on a size basis by rate-zonal centrifugation of cells through a sucrose gradient. Conditions of fractionation were similar to those described, but a shallow aqueous 10→40% (w/w) sucrose gradient was used to achieve separation over a greater radial distance in the rotor. Centrifugation was again done in the M.S.E. HS rotor. In these experiments and in all separations where the cells were not to be used subsequently for establishing synchronous cultures, all operations were performed at 0-4° and the quantity of·cells used was reduced to 10-20 g wet wt. in a total sample volume of 30 ml. Cell length was measured microscopically using an Image Splitting Eyepiece in successive fractions removed from the rotor. Size frequency distributions of cell volume (calculated from measurements of length as previously described [4]) indicated a satisfactory separation of different size classes in the rotor. Modes of the frequency distributions increased almost linearly with distance from the rotor centre. Thus, the first $\frac{3}{4}$ of the cell cycle is represented as a linear function of rotor radius. The distribution of cell numbers, mean cell length and modes of frequency distributions of cell volumes throughout the fractions after centrifugation of cells from an exponentially growing culture through a shallow sucrose gradient are shown in Fig. 1. Cytochrome *c* oxidase showed the least variation in amount per cycle, exhibiting a single peak and trough. Two maxima were seen for all the other enzymes assayed — acid *p*-nitrophenyl-phosphatase, catalase, NADH cytochrome *c* oxidoreductase, NADH dehydrogenase (EC 1.6.99.3), malate dehydrogenase (EC 1.1.1.37) and succinate dehydrogenase. There is the possibility of a third maximum at cell division for catalase and malate dehydrogenase.

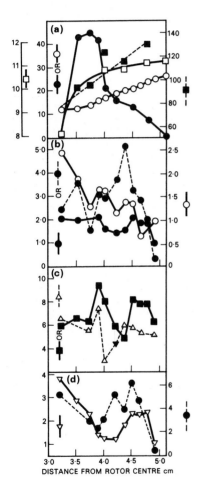

Fig. 1. Analysis of the cell cycle by rate-zonal centrifugation of cells from an exponentially growing culture of *s. pombe*. A suspension (30 ml) containing 6.7 g wet wt. of cells was loaded on the gradient in an M.S.E. HS rotor. Centrifugation was at 1,500 rev/min for 8 min (1.15×10^4 *g*-min at the sample zone; $\int \omega^2 . dt = 1.50 \times 10^4$ rad.^{2}s^{-1}).

(a) —●— Distribution of cell numbers (cells/ml of fraction)
    —□— Mean cell length, μm
    --■-- Modes of frequency distributions of cell volumes, μm^3
    --○-- Sucrose %, w/w

(b) —●— Cytochrome *c* oxidase
    —○— Catalase
    --●-- Acid *p*-nitrophenylphosphatase

(c) —■— NADH-cytochrome *c* oxidoreductase
    - -△- - NADH dehydrogenase

(d) —▽— Succinate dehydrogenase
    --●-- Malate dehydrogenase.

Amounts of enzymes are expressed as units /10^7 cells *(in Fig. 2 also)*.
Figs. 1-3 are based on Figs. in ref. [18].

        Zonal centrifugation in isokinetic Ficoll gradients or in convex sucrose gradients did not improve separation.

        In the second method of culture fractionation, cells were separated on a density basis by zonal centrifugation through a dextran gradient. The dextran used was Sigma chemical grade, of average mol. wt. 40,000. The gradient was linear

27→33% (w/v) dextran (540 ml) and was generated and loaded with the Isco Dialagrad pump. Loading was at 2,500 rev/min in the M.S.E. B-XIV zonal rotor running in an M.S.E. Superspeed 40 centrifuge. As in the previous experiment, the operating temperature was 4°. Cells were suspended in 10% (w/v) dextran to a total volume of 30 ml and the overlay was 30 ml distilled water. Centrifugation at 35,000 rev/min for 50 min was sufficient for cells to sediment to their isopycnic densities (integrated field-time 4.9×10^{10} rad.^{2}/sec). After deceleration and collection of fractions, cells were collected by centrifugation at 37,000 g for 30 min after dilution of the gradient medium with water.

Microscopic examination of cell volumes in these zonal fractions revealed a reciprocal relationship between cell volume and density, confirming cell cycle described previously [4]. In this case, the first $\frac{3}{4}$ of the cell cycle is inversely related to distance from the rotor centre. The distribution of cell numbers, mean cell length, and modes of frequency distributions of cell volumes throughout the fractions after centrifugation of cells from an exponentially growing culture to their equilibrium densities in a dextran gradient, are shown in Fig. 2. Here the abscissa has been reversed so that the cell cycle is still represented from left to right. Patterns of enzyme activities during the cycle are very similar to those observed in the previous experiment. Cytochrome c oxidase again shows a single peak and trough. Two maxima were again evident for all the other enzymes studied (Fig. 2). Again there is the possibility of multiple fluctuations of catalase and malate dehydrogenase.

DISCUSSION

Thus, three independent methods of cell-cycle analysis have been employed, and all show that the enzymes investigated in *s.pombee* manifest periodic patterns of expression as 'peaks'. The results obtained can be conveniently compared and summarized as cell cycle maps (Fig. 3). Here, the cell cycle is represented as a linear scale extending from 0 (the time of cell division)

Legend to Fig. 3 *(opposite, on right)*.
Scheme to represent timings of maxima of enzyme amounts during the cell cycle of *s. pombe*. In (a) the abscissa represents (as a linear scale, 0→1.0) the cell cycle taken as the period between the mid-points of the first and second doublings in cell numbers in synchronous cultures. (b): Results of analysis of the cell cycle by rate-zonal centrifugation, assuming the cell cycle to be represented linearly with distance from the rotor centre, and using the single maximum of cytochrome c oxidase at 0.67 of a cell cycle as a reference point for timings of maxima of other enzymes. (c): Results of analyses of the cell cycle by isopycnic-zonal centrifugations, assuming the cell cycle to be represented linearly as a reciprocal function of distance from the rotor centre, and again normalizing the positions of maxima of enzyme amounts with respect to the position of the maximum of cytochrome c oxidase.   *[Continued opposite, bottom left*

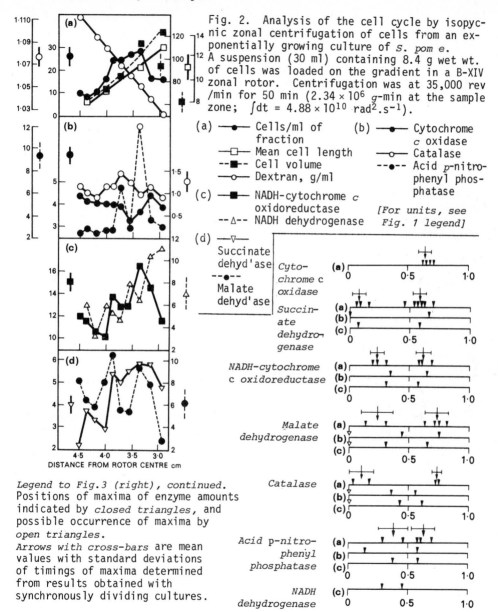

Fig. 2. Analysis of the cell cycle by isopycnic zonal centrifugation of cells from an exponentially growing culture of *S. pom e*. A suspension (30 ml) containing 8.4 g wet wt. of cells was loaded on the gradient in a B-XIV zonal rotor. Centrifugation was at 35,000 rev /min for 50 min (2.34×10^6 *g*-min at the sample zone; $\int dt = 4.88 \times 10^{10}$ rad^2.s^{-1}).

(a) —●— Cells/ml of fraction
—□— Mean cell length
--■-- Cell volume
—○— Dextran, g/ml

(c) —■— NADH-cytochrome *c* oxidoreductase
--△-- NADH dehydrogenase

(b) —●— Cytochrome *c* oxidase
—○— Catalase
--●-- Acid *p*-nitrophenyl phosphatase

[For units, see Fig. 1 legend]

(d) —▽— Succinate dehyd'ase
--●-- Malate dehyd'ase

DISTANCE FROM ROTOR CENTRE cm

Legend to Fig.3 (right), continued.
Positions of maxima of enzyme amounts indicated by *closed triangles*, and possible occurrence of maxima by *open triangles*.
Arrows with cross-bars are mean values with standard deviations of timings of maxima determined from results obtained with synchronously dividing cultures.

to 1.0 (timing of the next cell division). The single maximum of cytochrome
c oxidase is consistently observed at 0.67 of a cell cycle in synchronous
cultures, and has been used as a reference point for timings of enzyme maxima
observed after culture fractionation. The good agreement with the three
approaches indicates that all three represent valid procedures for studying
the cell cycle of this organism.

Since volume and density change only during the first ¾ of the cell
cycle in *S. pombe*, the last quarter of the cycle cannot be resolved by culture fractionation methods, but can be studied when synchronous cultures are
prepared. The latter procedure also has the advantage that results may be
confirmed in successive synchronous cycles. On the other hand, the quantity
of cells that can be obtained is limited, and it may be technically difficult
to sample sufficiently rapidly from cultures of microorganisms with short
generation times, e.g. bacteria. Representation of the cell cycle across a
gradient in a zonal rotor eliminates these deficiencies and greatly increases
the yield of cells for analysis. We have not investigated the maximum loading capacity of gradients used for rate and isopycnic separations.

In our experience, separation of cells by isopycnic zonal centrifugation yields the most homogeneous populations with respect to size, but the
expense of suitable gradient media at high concentrations may prove prohibitive for some purposes.

We have found that neither sucrose, dextran nor Ficoll cause flocula-
tion of cells in these separations, in contrast to the results with avian
erythrocytes [17]. We have not investigated the use of Ludox and other
gradient media. When cell-types liable to floculate are used, however,
efficient dispersal of clumps is necessary prior to zonal centrifugation,
particularly when the cell cycle is to be represented across a gradient.

In summary, zonal centrifugation provides a powerful tool for the
separation of homogeneous populations of intact cells. Choice of conditions
to be adopted for separation of a particular class of cell obviously requires
a knowledge of the physical nature of the cell-type in question. With the
fission yeast *S. pombe*, both rate and isopycnic zonal centrifugation procedures are suitable for studying the cell cycle. In the work reported here,
such studies have shown that the amounts of several enzymes associated with
membranous organelles oscillate during the cell cycle. If enzyme amounts
reflect concentrations of enzyme proteins, then the observation that different enzymes of a particular membrane system (e.g. cytochrome c oxidase and
succinate dehydrogenase of the inner mitochondrial membrane) are not expressed at identical times in the cell cycle, suggests that the stoichio-
metry of individual enzyme components in a membrane may vary at different
stages of the cell cycle. Elsewhere [18] we present more fully the results
obtained by the present approaches.

Acknowledgement

Figs. 1-3 appear by courtesy of the *Biochemical Journal,* being based on Figs. 2 & 3 and Scheme 1 in ref. [18].

References

1. Mitchison, J.M. & Vincent, W.S., *Nature (Lond.) 205* (1965) 987-989.

2. Blumenthal, L.K. & Zahler, S.A., *Science (Wash.) 135* (1962) 724.

3. Wells, J.R. & James, T.W., *Exp. Cell Res. 75* (1972) 465-474.

4. Mitchison, J.M., *Exp. Cell Res. 13* (1957) 244-262.

5. James, T.W., in *Cell Synchrony* (Cameron, I.L. & Padilla, G.M., eds.), Academic Press, London (1966) pp. 1-13.

6. Maaloe, O., in *The Bacteria* (Gunsalas, I.C. & Stanier, R.Y., eds.), Vol. 4, Academic Press, New York (1962), pp. 1-32.

7. Wiemken, A., von Meyenburg, H.K. & Matile, Ph., *Acta Fac. Med. Univ. Brun. 37* (1970) 47-52.

8. Hartwell, L.H., *J. Bacteriol. 104* (1970) 1280-1285.

9. Shulman, R.W., Hartwell, L.H. & Warner, J.R., *J. Mol. Biol. 73* (1973) 513-25.

10. Sebastian, J., Carter, B.L.A. & Halvorson, H.O., *J. Bacteriol. 108* (1971) 1045-1050.

11. Warmsley, A.M.H., Phillips, B. & Pasternak, C.A., *Biochem. J. 120* (1970) 683-688.

12. Mitchison, J.M., *Methods Cell Physiol. 4* (1970) 131-165.

13. Campbell, A., *Bacteriol. Rev. 21* (1957) 263-272.

14. Poole, R.K., Lloyd, D. & Kemp, R.B., *J. Gen. Microbiol. 77* (1973) 2⅟9-220.

15. Poole, R.K. & Lloyd, D., unpublished results.

16. Cartledge, T.G. & Lloyd, D., *Biochem. J. 126* (1972) 381-393.

17. Mathias, A.P., Ridge, D. & Trezone, N. St. G., *Biochem. J. 111* (1969) 583-591.

18. Poole, R.K. & Lloyd, D., *Biochem. J. 136* (1973) 195-207.

34 MARKER ENZYMES OF PLANT CELL ORGANELLES

B. Halliwell*
Botany School
Oxford University
South Parks Road
Oxford, OX1 3RA, U.K.

Fractionation of leaf homogenates by density-gradient centrifugation in swing-out or zonal rotors produces good separation of peroxisomes (marked by catalase) from mitochondria (marked by cytochrome oxidase or matrix enzymes). Damaged chloroplasts (marked by chlorophyll) band at a lower density than mitochondria, but intact chloroplasts (marked by chlorophyll plus a stromal enzyme such as $NADP^+$-dependent triosephosphate dehydrogenase) contaminate the mitochondrial band. When these techniques are applied to young plant tissues, there is often heavy contamination of the peroxisome band by proplastids or protein bodies. The importance of assessing contamination of fractions by unmarked organelles and of the accurate assay of enzyme activities in all fractions is stressed. The application of density-gradient techniques to the subcellular location of enzymes in plant tissues is illustrated by the behaviour of nitrite reductase, phenolase, NADH-cytochrome c reductase and 'formate oxidase' on density gradients.

The use of density-gradient centrifugation, coupled with the development of 'marker' enzymes each characteristic of a single organelle, has permitted enormous advances in our understanding of the subcellular location of enzymes in animal cells [e.g. 1, 2]. These techniques are now beginning to be widely applied to plant systems. Perhaps the two most successful recent examples are the isolation of the glyoxysome from castor-bean endosperm [3] and the peroxisome from spinach leaves [4]. The glyoxysome is an organelle containing the enzymes for β-oxidation of fatty acids, together with those of the glyoxylate cycle. The peroxisome is a related particle, containing enzymes involved in photorespiration. Indeed, such 'microbodies' can be isolated from a wide range of plant tissues [5−8]. They have in common a high equilibrium density on sucrose gradients (1.24-1.26 g/cm^3 for leaf peroxisomes) and a high content of catalase. Because of this high density, they may be readily separated from mitochondria and chloroplasts on sucrose gradients. When a homogenate of spinach leaves is subjected to differential centrifugation (25 min at 500 g followed by 20 min at 6,000 g) and the re-suspended 6,000 g pellet resolved on a simple linear sucrose gradient, the results obtained are as shown in Fig. 1. Note that the peak of catalase is

* *Present address: Dept. of Biological Sciences, Portsmouth Polytechnic.*

well-separated from that of cytochrome oxidase, a mitochondrial marker enzyme, and also from chlorophyll, indicative of chloroplasts. If an enzyme X behaves in the same way on the gradient as one of these marker enzymes, this is good evidence that X is associated with that organelle. Catalase is usually used as a marker for peroxisomes. Glycollate oxidase, NAD^+-glyoxylate reductase, glutamate-glyoxylate aminotransferase and serine-glyoxylate aminotransferase behave in exactly the same way on density gradients of spinach leaf fractions, and are thus considered to be exclusively located in the peroxisomes [5, 9, 10]. Cytochrome oxidase is the most commonly-used mitochondrial marker. Fumarase, citrogenase and other enzymes of the mitochondrial matrix show a similar distribution [11]. Little work has been done on marker enzymes for the outer mitochondrial membrane of plant tissues.

    Successful determination of the subcellular location of an enzyme by density-gradient centrifugation requires that certain elementary precautions be taken. Obviously, the assay used must accurately measure the activity of all fractions from the gradient; yet the precaution of calculating the recovery of the activity applied initially to the gradient is frequently neglected. Secondly, the technique is only valid to the extent that a given marker enzyme satisfactorily diagnoses the presence of the appropriate organelle. Thirdly, if some other organelle bands at a similar density to the one in question, then mistakes can be made. Consideration is now given to how these basic principles apply to plant systems.

PROBLEMS ENCOUNTERED WITH PLANT SYSTEMS

During the differential centrifugation procedure used to prepare the 6,000 g pellet for the gradient shown in Fig. 1, most chloroplasts lose their outer membranes, causing the release of stromal enzymes such as ribulose-diphosphate carboxylase and $NADP^+$-glyceraldehyde-3-phosphate dehydrogenase. These enzymes are thus found mainly in the supernatant fraction, although their true location within intact chloroplasts has been shown by non-aqueous fractionation techniques [12]. Thus most of the chloroplasts applied to the gradient in Fig. 1 are 'broken', i.e. have lost their stromal enzymes, and they band at a fairly low density, well separated from mitochondria and peroxisomes. However, by using various rapid centrifugation procedures to obtain particulate fractions for analysis on density gradients, or by applying the leaf homogenate directly to a tube or zonal gradient, some intact chloroplasts will enter the gradient. This is clearly seen in Fig. 2, taken from ref. [13]. A spinach-leaf homogenate was centrifuged at 1,000 g for 5 min and the supernatant then centrifuged at 3,000 g for 15 min. The pellets were gently resuspended and then separately applied to 40-80% linear sucrose gradients and centrifuged to equilibrium. The '3,000 g pellet' gradient shows a clear separation of the chlorophyll and mitochondrial peaks, whereas that obtained for the 1,000 g pellet shows two peaks of chlorophyll, one overlapping into the mitochondrial band. This small peak is characterized by the presence of $NADP^+$-glyceraldehyde-3-phosphate dehydrogenase, and therefore represents intact chloroplasts, banding at a higher density than broken chloroplasts and contaminating the mitochondrial band. Therefore, if intact chloroplasts have entered a density-gradient, they must be 'marked' by a stromal enzyme, since they behave differently from broken chloroplasts. On the basis of zonal centrifugation on a sucrose gradient of a tobacco leaf homogenate, it has been claimed that nitrite reductase is located in plant peroxisomes [14]. However, the separation of organelles achieved was poor, and the catalase peak contained some chlorophyll. No marker was included for intact chloroplasts. Recent work, using such a marker, has shown that nitrite reductase is most probably a chloroplast stromal protein [15]. It would be interesting to obtain preparations of chloroplast outer membranes and examine them for enzymic activities.

PROFILE IN RELATION TO AGE OF PLANT

Care must also be taken in applying density-gradient/marker enzyme techniques to tissues from young plants. Fig.3, taken from ref. [16], shows the distribution of enzymes obtained when a homogenate of sunflower cotyledons is resolved on a sucrose gradient. With plants germinated in the dark, DOPA oxidase and triose-phosphate isomerase activities are associated with the catalase peak. Indeed, it has been claimed on the basis of experiments of this type that phenolase activity (assayed using 4-methylcatechol as a substrate) is located in peroxisomes of potato tubers [17]. However, after 2 days' exposure of the plants to light, these activities in sunflower cotyledons have begun to move to lower densities (Fig. 3). Phenolase activity of mature leaves is known to be associated almost exclusively with chloroplasts [18-20]. Microscopic examination shows that, in the etiolated tissue, the microbody band is heavily contaminated with proplastids, the precursors of chloroplasts, this accounting for the presence of phenolase and triose-phosphate isomerase in this region of the gradient. Also, storage proteins in seeds are located in particles surrounded by a single membrane. In sunflower cotyledons, these 'protein bodies' have an equilibrium density of 1.26-1.36 g/ml on sucrose gradients, and so heavily contaminate the peroxisome band [21].

PLANT MICROSOMES

Some work has been done on specific marker enzymes for plant microsomes. Crude microsomal fractions may be obtained by differential centrifugation of homogenates: these fractions are usually enriched in NADPH-cytochrome c reductase and antimycin-insensitive NADH-cytochrome c reductase [22]. Such crude fractions contain various biochemical activities, including enzymes for the synthesis of phospholipids [23] and for the N-demethylation of N-methylphenylurea herbicides [24, 25]. Normally, centrifugation at high speeds is necessary to sediment the microsomes, so that fractions obtained by centrifugation at, say, 6,000 g for 20 min, will contain few of these particles. However, if the total homogenate, rather than a single

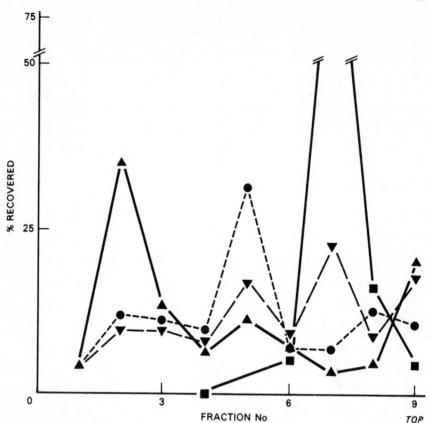

Fig. 1. Subcellular location of enzyme activities in spinach leaves.
A pellet sedimented at 6,000 *g* by differential centrifugation was resolved
on a simple linear sucrose gradient (1.3 - 2.0 M). Symbols used:
——▲——, catalase (recovery of activity = 97%); --●-- ·cytochrome oxidase
(93%); ——■——, chlorophyll; --▼- -, formate oxidase, pH 5 (75%).

Fig. 2 *(opposite)*. Sucrose-density gradient elution profiles (from Rocha
& Ting [13]). *Upper,* 1000 *g* pellet: fraction 27 = broken chloroplasts;
18 = intact chloroplasts and an occasional mitochondrion- 10 = microbodies.
Lower, 3000 *g* pellet: fraction 23 = broken chloroplasts; 16 = mitochondria; 8 = microbodies.
G.O. = glycollate oxidase; MDH = malate dehydrogenase; cyt *c* = cytochrome *c*
oxidase; Chl = chlorophyll; TDH = NADP$^+$-glyceraldehyde-3-phosphate dehydrogenase.   Enzyme activities are expressed as % of activity recovered.

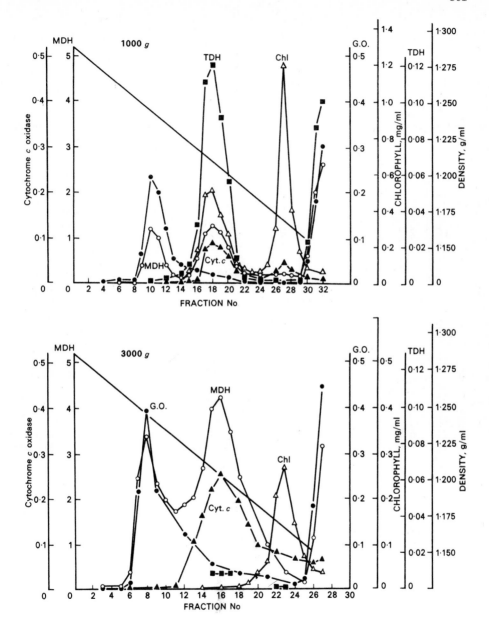

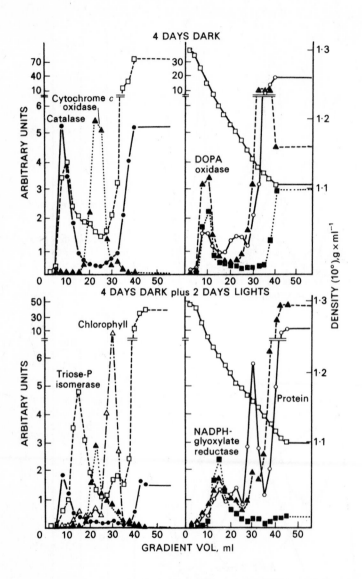

Fig. 3.
Distribution of
cell organelles
from cotyledons
of sunflower seed-
lings germinated
for 4 days in the
dark *(top)* or for
an additional 2
days in the light
(bottom) after
isopycnic centri-
fugation in suc-
rose density
gradients (from
ref. [16]).
Symbols used:
——●——, catalase
(microbodies), 1 u.
= 2 m-moles ml^{-1}
min^{-1}; ——▲——,
cytochrome *c* oxid-
ase (mitochondria),
1 u. = 1 µmole
min^{-1} ml^{-1};
–·–△–·–, chloro-
phyll (broken
chloroplasts), 1 u.
= 100 µg ml^{-1};
——□——, triose-P-
isomerase (whole
plastids), 1 u. =
2µ moles min^{-1} ml^{-1};
——▲——, DOPA oxidase
(whole plastids),
1 u = 0.2 absorbance
units min^{-1} ml^{-1};
——■——, NADPH gly-
oxylate reductase
(whole plastids),
1 u. = 50 n-moles
min^{-1} ml^{-1};
——○——, protein,
1 u. = 1 mg ml^{-1};
——□——, density.

particulate fraction from it, is applied to a gradient, one must consider how the microsomes will behave. Fig. 4, taken from ref. [26], shows the distribution of NADH-cytochrome c reductase activity on a sucrose gradient of a homogenate of sunflower cotyledons. Two peaks are observed. The first overlaps with cytochrome oxidase and is largely (but not completely) anti-mycin-sensitive; it is presumably due to mitochondria. The second is insensitive to antimycin, and presumably represents the microsomes. Simi-larly, Lord *et al.* [27] have recently isolated, by sucrose density-gradient centrifugation, a 'smooth-membrane fraction' from castor bean endosperm. This fraction banded at a density of 1.12 g/ml and contained NADPH-cyto-chrome c reductase, antimycin A-insensitive NADH-cytochrome c reductase and phosphorylcholine-glyceride transferase. Glucose-6-phosphatase was not present, but spectral studies indicated the presence of cytochromes b_5 and P_{450} in this fraction. Antimycin-insensitive NADH-cytochrome c reductase can also be used as a marker for microsomes in spinach-leaf fractions [26], but its presence in outer mitochondrial membranes may tend to confuse the results obtained.

It is thus clear that cross-contamination of fractions may be a serious problem in the subcellular location of plant enzymes by density-gradient techniques and must be checked for.

HYDROGEN PEROXIDE FORMATION

Finally, I present an interesting result recently obtained in my own research on formate oxidation in spinach leaves. Crude particulate fractions obtained by differential centrifugation of spinach-leaf homogenates readily oxidise formate to CO_2. At pH 5, this activity is almost entirely due to the peroxidatic action of catalase [28]. However, when a 6,000 g precipitate was resolved on a sucrose density-gradient, activity was everywhere *except in* the catalase peak (Fig. 1). Also, only 75% of the activity of formate oxi-dation applied to the gradient was recovered. A detailed investigation revealed that H_2O_2 is generated in chloroplasts and, to a lesser extent, in mitochondria, so enabling peroxidatic action to occur in these fractions because of the contaminating catalase. Isolated peroxisomes oxidize little formate because they do not generate H_2O_2 under these conditions. It is interesting that the clue to such a complex situation was merely a 25% loss of the activity applied to the density gradient.

Acknowledgments

For use of a Figure from each of refs. [13], [16] & [26], as Figs. 2, 3 & 4, respectively, grateful acknowledgment is made to the authors and publishers concerned.

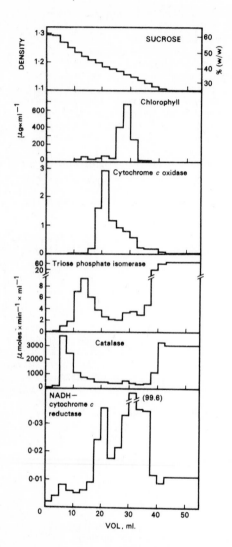

Fig. 4. Distribution of subcellular organelles from sunflower cotyledons on a sucrose density gradient (from ref. [26]). Cotyledons were obtained from seedlings which had been germinated for 4 days in the dark and 2 days in the light. The organelles represented are the microbodies (catalase), whole chloroplasts (triose phosphate isomerase), mitochondria (cytochrome c oxidase), broken chloroplasts (chlorophyll), and microsomes (NADH-cytochrome c reductase, pH 7.3).

References

1. Baudhuin, P., Beaufay, H. & de Duve, C., *J. Cell Biol.* 26 (1965) 219-243.

2. de Duve, C., *Physiol. Rev.* 46 (1966) 323-357.

3. Breidenbach, R.W. & Beevers, H., *Biochem. Biophys. Res. Commun.*, 27 (1967) 462-469.

4. Tolbert, N.E., Oeser, A., Kisaki, T., Hageman, R.H. & Yamazaki, R.K., *J. Biol. Chem.* 243 (1968) 5179-5184.

5. Tolbert, N.E., *Annu. Rev. Plant Physiol.* 22 (1971) 45-74.

6. Huang, A.H.C. & Beevers, H., *Plant Physiol* 48 (1971) 637-641.

7. Parish, R.W., *Planta* 104 (1972) 247-251.

8. Parish, R.W., *Zeit für Pflanzenphysiol.* 67 (1972) 430-442.

9. Tolbert, N.E., Yamazaki, R.K. & Oeser, A., *J. Biol. Chem.* 245 (1970) 5129-5136.

10. Rehfeld, D.W. & Tolbert, N.E., *J. Biol. Chem.* 247 (1972) 4803-4811.

11. Yamazaki, R.K. & Tolbert, N.E., *J. Biol. Chem.* 245 (1970) 5137-5144.

12. Heber, U., Pon, N.G. & Heber, M., *Plant Physiol.* 38 (1963) 355-360.

13. Rocha, V. & Ting, I.P., *Archs. Biochem. Biophys.* 140 (1970) 398-407.

14. Lips, S.H. & Avissar, Y., *Eur. J. Biochem.* 29 (1972) 20-24.

15. Dalling, M.J., Tolbert, N.E. & Hageman, R.H., *Biochim. Biophys. Acta* 283 (1972) 505-512.

16. Schnarrenberger, C., Oeser, A. & Tolbert, N.E., *Plant Physiol.* 50 (1972) 55-59.

17. Ruis, H., *Hoppe-Seyler's Z. Physiol. Chem.* 352 (1971) 1105-1111.

18. Arnon, D.I., *Plant Physiol.* 24 (1949) 1-15.

19. Parish, R.W., *Eur. J. Biochem.* 31 (1972) 446-455.

20. Bartlett, D.J., Poulton, J.E. & Butt, V.S., *FEBS Lett.* 23 (1972) 265-267.

21. Schnarrenberger, C., Oeser, A. & Tolbert, N.E., *Planta* 104 (1972) 185-194.

22. Martin, E.M. & Morton, R.K., *Biochem. J.* 62 (1956) 696-704.

23. Marshall, M.O. & Kates, M., *Biochim. Biophys. Acta* 260 (1972) 558-565.

24. Frear, D.S., *Science (Washington, D.C.)* 162 (1968) 674-675.

25. Frear, D.S., Swanson, H.R. & Tanaka, F.S., *Phytochemistry* 8 (1969) 2157-2169.

26. Donaldson, R.P., Tolbert, N.E. & Schnarrenberger, C., *Archs. Biochem. Biophys. 152* (1972) 199-215.

27. Lord, J.M., Kagawa, T., Moore, T.S. & Beevers, H., *J. Cell Biol. 57* (1973) 659-667.

28. Halliwell, B., *Biochem. J. 138* (1974) 77-85.

35 SEPARATION AND CHARACTERIZATION OF PLANT ZYMOGEN BODIES

Y. Shain, E. Cohen and Y. Ben-Shaul
Laboratory for Electron Microscopy
Tel-Aviv University
Tel-Aviv, Israel

Plant zymogen bodies have been isolated from pea seeds by means of separation on a Ficoll density gradient. These organelles contain an inactive form of amylopectin 1,6-glucosidase which can be activated by limited proteolysis [1]. These organelles possess a single limiting membrane. An acid nucleotide phosphatase with a preference for ATP was also shown to exist in these bodies [2]. This enzyme is membrane-bound and could be located cytochemically. The ATPase activity is present within the single limiting membrane and could be localized by a new method combining both cytochemical and freeze-etch techniques. Thus it can be shown that only a portion of the 9.8 nm diameter particles appearing on freeze-fractured membrane surfaces appear to possess ATPase activity. These zymogen bodies also show β-galactosidic and proteolytic activity.

Plant cells contain numerous types of subcellular organelles possessing a single limiting membrane. These organelles have been called lysosomes [3], spherosomes [4], peroxisomes [5], glyoxysomes [6], microbodies [7], aleurone vacuoles [8], depending on their appearance in electron micrographs or according to their enzymatic complement.

This has, at times, resulted in considerable terminological duplicity. For instance, the enzymatic content of what was known to be an aleurone vacuole or spherosome is similar to that of a lysosome [9,10]; the glyoxysome is a more specialized form of the peroxisome [11].

A new type of subcellular organelle in plants, a zymogen body, was isolated from pea seeds. It was shown to contain an inactive form of the starch-debranching enzyme (amylopectin 1,6-glucosidase) which can be activated by limited proteolysis [1, 12]. Although zymogen bodies are known to exist in animal tissues [13], this was the first report of such an organelle in plants. From the structural and enzymatic data presented below we can conclude that these zymogen bodies are a distinct class of subcellular plant organelles.

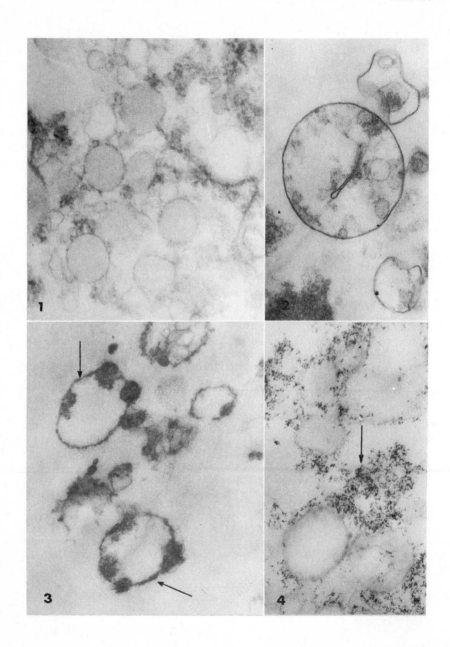

Isolation

Zymogen bodies can be prepared from pea seeds, imbibed for $4\frac{1}{2}$ h at 26°
as previously described [2]. They are purified on a stepwise 10-30%
(w/v) Ficoll density gradient which is prepared in 0.1 M tris-maleate
buffer pH 6.6 containing 0.1 mM lauric acid and 0.5 M sorbitol. The
inclusion of lauric acid in the grinding and density gradient medium
results in more intact preparations which were not sensitive to cold
shocks (transfer from room temperature to an ice-bath). In the stepwise Ficoll gradient the zymogen bodies appear at the top and within
the 10% Ficoll band. With a continuous 0-50% (w/v) Ficoll gradient
they equilibrate at a density of 1·03. If, however, these particles
are recovered from this gradient and ran in a continuous 0-60% (w/v)
sucrose density gradient they equilibrate at a density of 1·17.

Ultrastructure

Zymogen bodies isolated on a stepwise Ficoll gradient were pelleted
(20,000 × g, 10 min). The pellet was then fixed with 5% (v/v) glutaraldehyde in 0.1 M phosphate buffer pH 7.3 (4°, 1 h) and post-fixed with
2% (w/v) osmium tetraoxide in the same buffer (4°, 15 min). The fixed
particles were dehydrated and embedded in Epon 812. Sections were cut
with glass knives and collected on Formvar-coated, carbon-reinforced
grids. They were stained overnight with a saturated solution of uranyl
acetate in 30% ethanol and post-stained with lead citrate for 5 min.

The zymogen body fraction consists mainly of spherical bodies with
a diameter of 0.8 - 1.5μm (Fig.1) which are limited by a single membrane
80-100Å wide (Fig.2). There was some (5-10%) contamination by a smaller
particle with a diameter of less than 0·8μm.

Enzymatic content

Besides the inactive starch-debranching enzyme, additional enzymatic
activities were detected within these zymogen bodies. These included an
acid nucleotide phosphatase with a preference for ATP [2]. However, no

---

Legends to Figs. 1 - 4 (opposite)

Fig. 1. Zymogen body fraction isolated from Ficoll density gradient.
[× 11,000

Fig. 2. Isolated zymogen body with single limiting membrane. × 25,000

Fig. 3. Localization of ATPase activity (arrows) on zymogen body
membrane. × 17,000

Fig. 4. Non-specific lead deposits (arrow) in control mixture
without ATP. × 13,500

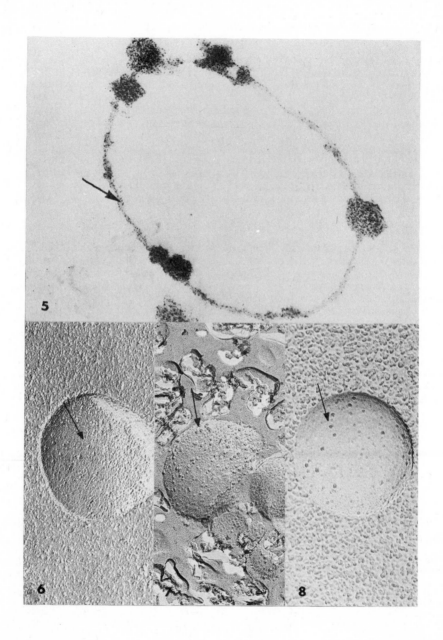

RNase, α-galactosidase or succinic dehydrogenase activities could be
found in this preparation.

Localization of the ATPase activity

The ATPase activity could be localized within the zymogen body by a modification of the Wachstein-Meisel method using a lower lead nitrate concentration (0.5 mM) [2]. The reaction product of the ATPase is restricted
to the single limiting membrane (Figs. 3, 5). Controls containing no ATP
showed non-specific lead deposits (Fig.4). It thus appears that this
acidic ATPase activity is located on or within the membrane.

   ATPase activities have been histochemically demonstrated in
electron microscopic sections of various sub-cellular organelles [14,15].
On the other hand, various particles on surfaces of freeze-etched membranes have been designated as sites for ATPase activity [16,17].
However, these designations were not based on histochemical observations
and it was not possible to determine whether all of the particles seen on
these membranes possess ATPase activities. Since the reaction product of
ATPase activity histochemically yields small lead phosphate deposits
(Fig.5.)it should be possible to observe a change in size of the freeze-
fractured membrane particles related to their ATPase activities. Zymogen
bodies from pea seeds were chosen as objects for this combined histochemical and freeze-etch study, as membrane-bound ATPase activity could be
demonstrated in electron microscope sections of these organelles.

   The zymogen bodies were incubated with 10 mM ATP, 50 mM tris-
maleate pH 5.4 and 0.5 mM lead nitrate at 37° for 2 h; the zymogen bodies
were recovered by centrifugation (20,000 g × 10 min) and then were either
freeze-etched or deep-etched. In the case of freeze-etching 0.5 ml of
27% glycerol was added and the suspension was left overnight. After
centrifugation the recovered pellet was resuspended in a drop of 27%
glycerol, transferred to gold discs and then rapidly frozen in liquid
Freon 22 and transferred again into liquid nitrogen. The frozen specimens were freeze-etched according to the method of Moor & Muhlethaler [18].

Legends to Figs. 5 - 8 (opposite)

Fig. 5. Localization of ATPase activity on zymogen body - fine lead
deposits on membrane (arrow).   × 71,000

Fig. 6. Freeze-fractured zymogen body incubated in cytochemical control
medium (without ATP), showing smaller particles on the fractured surfaces
(arrow).   × 51,000

Figs. 7 & 8. Freeze-fractured zymogen bodies incubated in ATP-lead-
nitrate medium, showing enlarged particles on the fractured surface
(arrow). Smaller particles can also be seen. × 51,000

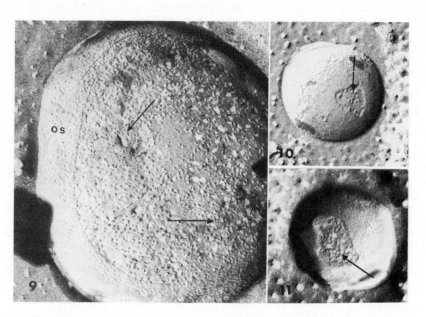

Fig. 9. Deep-etched zymogen body incubated in ATP-lead-nitrate medium, showing enlarged particles within the membrane (arrows). Smaller particles can also be seen. × 83,000

　　　os- outer surface.

Figs. 10 & 11. Deep-etched zymogen bodies (apices) incubated in cyto-chemical control medium (without ATP) showing smaller particles within the membrane (arrows). × 83,000

In some cases the pellet containing the zymogen bodies was resuspended in a drop of distilled water (deep-etching) and treated as described for freeze-etching.

　　　Control preparations of freeze-etched zymogen bodies, which were not incubated in the histochemical medium, show particles with a diameter of 11.2 ± 0.1 nm (Fig.6; based on measurements of at least 150 particles). Controls which were incubated with lead nitrate only show similar particles with a diameter of 9.2 ± 0.05 nm.

　　　However, after incubation in the ATP-lead-nitrate medium, larger particles - 23.3 ± 0.09 nm in diameter - are seen on the freeze-etched surfaces (Figs. 7, 8). The small particles (9.9 nm) can also be recognized on these same surfaces indicating that not all particles located on the membrane demonstrate ATPase activity. This method may therefore help us to distinguish between particles which participate in a given enzymatic reaction, and those which do not, even though both classes of particles appear similar in size prior to the reaction.

Particles appearing on deep-etched membranes after the histochemical reaction are 25 ± 0.07 nm in diameter (Fig.9). Occasionally these lead deposits penetrate the membrane and appear as hillocks on its outer surface. Once again, sites not participating in this histochemical reaction have the same size — 10 nm — **as** those of the control (Figs.10, 11). These deep-etch studies show that the ATPase activity is located within the zymogen body membrane and not on its surface.

Changes in the size and/or shape of the suspected sites of enzymatic reaction due to attachment of large substrates, or reaction products, to these sites can be detected by freeze- or deep-etching and therefore may provide a method for the localization of the enzymatic reaction without the use of chemical fixatives or stains. We have demonstrated the utility of such method for the location of ATPase activity. It remains to be seen whether other enzymatic reactions could also be localized on membranes with a similar degree of success.

References

1. Shain, Y. and Mayer, A.M., *Physiol. Plant 21* (1968) 765-76.

2. Cohen, E., Shain Y., Ben-Shaul, Y., and Mayer, A.M., *Can. J. Bot. 49* (1971) 2053-2057.

3. Matile, Ph., *Planta 79* (1968) 181-196.

4. Mishra, A.K. and Colvin, J.R., *Can.J. Bot. 48* (1970) 1477-1480.

5. Tolbert, N.E. and Yamazaki, R.K., *Ann. N.Y. Acad. Sci.168* (1969) 325-341.

6. Cooper, T.G. and Beevers, H., *J. Biol. Chem. 244* (1969) 3507-3513.

7. Frederick, S.E., Newcomb, E.H., Vigil, E.L. and Wergin, P., *Planta 81* (1968) 229-252.

8. Perner, E., *Planta 67* (1965) 324-325.

9. Matile, Ph., *Z. Pflanzenphysiol. 58* (1968) 365-368.

10. Matile Ph. and Spichiger, J., *z. Pflanzenphysiol. 58* (1968) 277-280.

11. de Duve, C.H., *Ann. N.Y. Acad. Sci. 168* (1969) 369-381.

12. Mayer, A.M. and Shain, Y., *Science (Wash.D.C.) 162* (1968) 1283-1284.

13. Schram, M., *Annu. Rev. Biochem. 36* (1967) 307-320.

14. Grossman, I.W. and Heitkamp, D.H., *J. Histochem. Cytochem. 16* (1968) 645-653.

15. Sabnis, D.D., Gordon, M. and Galston, A.W., *Plant Physiol.45* (1970) 25-32.

16. Arntzen, C.Y., Dilley, R.A. & Crane, F.L., *J. Cell Biol. 43* (1969) 16-31.

17. Park, R.B. and Pfeifhofer, A.O., *J. Cell Sci. 5* (1969) 299-312.

18. Moor, H. and Muhlethaler, K., *J. Cell Biol. 17* (1963) 609-628.

36 THE USE OF SUBCELLULAR FRACTIONS OF LYMPHOCYTES IN THE PREPARATION OF ANTILYMPHOCYTIC SERA

H. Zola, Betty Mosedale & D. Thomas
Biological Division
Wellcome Research Laboratories
Langley Court, Beckenham,
Kent BR3 3BS, U.K.

Antilymphocytic sera owe their immunosuppressive properties to antibodies against constituents of the lymphocyte membrane. Antisera raised against membranes or against solubilized components may have advantages in specificity and potency over antisera raised using whole cells. Methods used in the isolation of particulate and soluble antigenic fractions, and the properties of the antisera, are described and discussed.

The immunosuppressive properties of antilymphocyte sera are well known, as are the undesirable features of many of those sera (review, [1]). There are advantages in using isolated membranes rather than whole cells as immunogen [2, 3]. Using membranes from mouse lymphocytes a vigorous immunization schedule could be used in the rabbit without producing toxic sera, whereas whole cells produced less potent or toxic sera. This has been confirmed in several laboratories, including our own [4], in the rabbit anti-mouse system. However, direct translation to the horse anti-human situation, clearly the one that interests us most, has not been so successful [4, 5], and there remains a need for further studies. We now summarize the techniques we have used for the isolation of subcellular forms of antigen, both particulate and soluble, and describe the properties of some of the antisera produced.

PREPARATION OF LYMPHOCYTE MEMBRANES

Cell lysis

Initially [4] we compared six methods of lysing lymphocytes. Hypotonic lysis using a Dounce homogenizer and lysis in the 'nitrogen bomb' were found to be the most promising methods. The two lysates were equally suitable for subsequent fractionation, and the resultant membrane fractions gave rise to equally potent antisera, although electron microscopy showed that they differed morphologically. Nitrogen bomb lysis produced a membrane fraction consisting mainly of microsomes, whilst hypotonic lysis produced large membrane fragments. Antisera have also been raised against antigens prepared by the methods of other investigators [6, 7] and will be discussed below.

The procedure used for hypotonic lysis has evolved gradually since the earlier publication [4] and the current practice is as follows: a single-cell suspension is centrifuged at 500 g for 10 min and the cells are re-suspended in cold hypotonic buffer (0.01 M NaCl/0.01 M Tris, pH 7.4/1.5 mM MgCl$_2$; 10 ml for 2.5×10^9 cells). The suspension is kept at 4° for 30 min. Subsequent treatment depends upon the ease of lysis of the cells. Some cells, particularly some leukaemic lymphocytes and occasionally thymocytes, lyse on contact with the hypotonic buffer. These often leave a gel of nucleoprotein which is dispersed by brief treatment in a Silverson Homogenizer (Silverson Machines Ltd., London) with a micro head. Cells which are not lysed by suspending in buffer are treated with a moderately tight-fitting Dounce homogenizer. If 100 strokes do not produce 80-100% cell lysis, assessed microscopically, the Silverson homogenizer is used. Human peripheral and cultured lymphocytes tend to resist lysis in the Dounce homogenizer. An alternative way of treating resistant cells is to freeze the suspension at -20° overnight. This usually results in complete lysis.

Fractionation of lysates

Density gradient fractionation has generally given better separation of enzymic and antigenic markers than differential centrifugation [4], although this can in some instances produce a very adequate separation of antigenic markers. Membrane fractions are routinely isolated by centrifugation through sucrose gradients [4] or on sucrose steps derived by simplification of the gradient.

PREPARATION OF SOLUBLE ANTIGENS

Six different methods for the extraction of mouse soluble lymphocyte-specific antigens (ALS receptors) have been compared [8] and the extracts assayed by a fluorescence inhibition test [8,9]. The extract giving the highest specific activity was obtained by hypotonic lysis using the procedure described above for membrane preparation. The soluble fraction was obtained by centrifuging the lysate at 100,000 g for 30 min [10]. When the membrane fraction is not needed the soluble fraction can be isolated by filtration through successive membrane filters of decreasing porosity, finishing with a 0.22 μm filter. Some of the properties of the extracts and a partial purification of the fraction which reacts with ALS have been described recently [8].

ANTISERA RAISED AGAINST MURINE SUBCELLULAR ANTIGENS

Rabbits were immunized with a priming dose in Complete Freund's Adjuvant i.m. or in the footpad followed by i.v. boosters on days 21, 28 and 35. Antigen dose was 750 μg membrane protein or 5 mg soluble antigen protein for the priming injection, followed by boosters of 250 μg of membrane protein or 2.5 mg soluble antigen protein. Bleeds were taken on days 21, 35 and 42. Table 1 shows the properties of antisera raised against fractions from a nitrogen bomb lysate of mouse thymocytes. A clear separation of immunosuppressive

Table 1. Properties of rabbit antisera raised by hyperimmunization with density gradient fractions from 'nitrogen bomb' lysate of mouse thymocytes. In Tables 1 & 2, survival was by the mouse-skin homo-graft test (groups of 6 - 8 mice); the stated error function being the standard deviation about the mean survival time; controls receiving normal rabbit serum showed a median survival of 14 ± 1 days. *NT* in a titre column denotes 'not tested'.

Fraction	Median survival time	Immuno-fluorescence titre
Dense	20 ± 2	1/16
particulate	19.5± 2	1/16
Membrane	40.5 ±8	1/32
	28.5 ±9	*NT*
	25 ± 9	1/64
Low density	14 ±3	*NT*
particulate	15 ±2	*NT*
	16.5 ±3	1/4
	13.5 ±3	1/2
Soluble	28.5 ±4	1/64
	23.6 ±5	*NT*

activity is observed, the antisera to the membrane fraction being superior to antisera against the other fractions. Table 2 shows the properties of antisera raised against membrane and soluble fractions from hypotonic lysates of mouse thymocytes. Most of the antisera are highly immunosupp-ressive and have low haemaggluti-nation titres. In general the soluble fractions lead to less potent antisera than membrane antigens. This is probably due to the lower immunogenicity of soluble molecules rather than low antigen content of the extracts, since in hypotonic lysis 90% of the antigen goes into solution [8].

Fluorescent antibody studies, using fluorescein-conjugated anti-rabbit immunoglobulin showed that antibodies raised against membranes or soluble antigen concentrated on the plasma membrane but also reacted with intracellular components.

ANTISERA RAISED AGAINST HUMAN SUBCELLULAR ANTIGENS

Table 3 shows the *in vitro* properties of rabbit antisera prepared against human lymphocyte antigens. The first set of sera were prepared in our laboratory using human thymocyte antigen prepared by Drs. D.Allan and M.J. Crumpton of the National Institute of Medical Research, Mill Hill [6]. The remaining antisera were prepared using antigen solubilized from thymocytes and peripheral lymphocytes by hypotonic lysis. The *in vitro* titres against lymphocytes are exceptionally high in most of the sera. Studies of horse anti-human lymphocyte globulin (11) have shown that preparations with an immunofluorescence titre in excess of 1/128 and a rosette inhibition titre in excess of 1/4000 are usually immunosuppressive in speciosa monkeys. On this basis the sera and the globulin in Table 3 would be expected to show strong

Table 2. Properties of rabbit antisera to membrane and soluble
fractions from hypotonic lysates of mouse thymocytes

Antigen	Median survival time	Immuno-fluorescence titre	Rosette inhibition titre	Haemagglutination titre
Membrane				
546	30.4 ± 4	1/256	1/16000	1/192
547	35 ± 6	1/64	1/64000	1/192
548	26 ± 6	1/64	1/128000	1/48
549	42.5 ± 9	1/64	1/64000	1/96
574	32.5 ± 4	1/64	1/8000	1/4
575	20.0 ± 4	1/64	1/2000	1/32
Soluble				
379	28 ± 4	1/128	1/8000	1/72
433	27 ± 7	1/96	*NT*	1/48
434	24 ± 4	1/8	*NT*	1/18
477	14 ± 1	1/2	1/2000	1/24
478	16 ± 3	1/2	1/2000	1/24

immunosuppression. One possible problem associated with antigen prepared
from thymus is the appearance in the antiserum of antibodies which cross-
react with the glomerular basement membrane of the kidney and are thus neph-
rotoxic [12]. A preliminary test carried out in rats as described by Feldkamp-Vroom and Balner [13] failed to detect such antibody in one of the antisera described in Table 3 (personal communication from Dr. J.G. Woodrooffe,
Pathology Department, Wellcome Research Laboratories). Antisera raised
against peripheral cells may contain antiplatelet antibodies; but no such
antibody could be detected in the last two sera of Table 3 [10]. The sera
all had low haemagglutination titres.

The distribution of receptor sites for the ALS in human lymphocytes
was examined using fluorescein- and ferritin-labelled antibody techniques.
The indirect fluroescence technique, using fluorescein-conjugated anti-rabbit
I_gG, showed strong membrane fluorescence, with a ring of weaker staining
around the nucleus. Higher resolution was achieved by electron microscopy
using ferritin-labelled antilymphocyte globulin (the globulin described in
Table 3). As can be seen in Fig. 1, the ferritin concentrated on the plasma

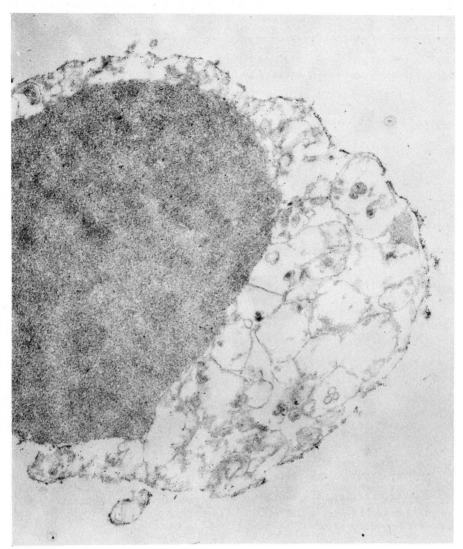

Fig. 1. Electron micrograph of human peripheral lymphocyte treated with ferritin-labelled rabbit anti-(thymocyte membrane) globulin. The properties of the globulin are described in Table 3. × 35,000.

Table 3. *In vitro* titres of antisera against human plasma membrane and soluble antigens

Antigen	Haemagglutination titre	Immunofluorescence titre	Rosette inhibition titre
	1/24	1/16	1/16000
	1/24	1/32	1/4000
Thymocyte plasma membrane [ref. 6]	1/16	1/128	1/32000
	1/16	1/128	1/16000
	1/8	1/64	1/24000
	1/16	1/100	1/24000
(I_gG fraction)	1/2	1/256	1/64000
Thymocyte soluble antigen (hypotonic lysis)	1/24	1/128	1/8000
	1/24	1/128	1/8000
Peripheral lymphocyte soluble antigen (hypotonic lysis)	1/192	1/32	1/32000
	1/24	1/256	1/16000
	1/12	1/256	1/8000

membrane but could also be found associated with intracellular membrane elements and the nucleus. The presence of receptor sites for ALS inside the cell confirms quantitative studies with mouse lymphocytes which indicated that only 20% of the receptor activity is present on the outer membrane of the cell [8].

Two simplified approaches to the use of membrane antigens have been attempted using human and mouse lymphocytes. In the first, acetone-dried thymus antigen was prepared as described by Zweibaum & Bouhou [7]. Antisera raised using this form of antigen had activity *in vivo* (in the mouse system) and *in vitro*, but were considerably less active than antisera raised with membrane prepared by hypotonic lysis and density gradient fractionation.

The second approach was to use antisera to isolate the antigens from crude lysates in the form of antigen-antibody complexes. The antisera were raised against membranes and were absorbed to remove haemagglutinating antibodies. When such antisera were mixed with crude lysates, antigen-antibody complexes formed and were isolated by differential centrifugation. The complexes were injected into rabbits and gave rise to potent ALS [14].

DISCUSSION

In an attempt to produce more potent and more specific ALS several laboratories have examined the use of subcellular fractions, mainly membranes or soluble antigens, as immunogens. The results obtained have been encouraging, but at present large-scale production of ALS in horses is still generally carried out using whole cells. This is principally because there is an element of experiment in changing antigens, and experiments involving horses are costly. An example of the problems which can occur is the finding that antisera raised in horses using human peripheral lymphocyte membranes contain anti-platelet activity while rabbit antisera raised against the same antigens do not.

Concerning the nature of the antigens, their ease of solubilization by hypotonic lysis suggests that the antigens are not of the hydrophobic class of protein normally found as structural constituents of membranes. It is not necessary to use detergent, chaotropic ions, hydrogen-bond scavengers, or other dissociating agents to solubilize the antigens. In view of the ease with which the antigens are released from the membrane, it is probable that they are continuously shed from the membrane and regenerated. Our results show that ALS receptor activity is not confined to the outer membrane of the lymphocyte. The significance of this is not clear at present, and interpretation depends on the specificity of the antiserum used in detecting the antigens. Thus it is possible that the antigens on the surface are not identical with those found inside the cell and that a more specific antiserum would show this. The separation of non-antigenic from antigenic particles by isopycnic centrifugation suggests that if antigen occurs on intracellular organelles as well as the plasma membrane, these antigen-bearing organelles have a similar density to the plasma membrane, and probably form part of the endoplasmic reticulum. Ferritin-labelled antilymphocyte globulin can indeed be seen bound to internal, as well as external membrane elements.

Acknowledgments

The authors are indebted to Mrs. L. Rahr for haemagglutination and rosette inhibition titres, to Dr. R.O. Thomson for ferritin-labelling and Mr. J. Short for electron microscopy. Mrs. S. Barnes and Miss S. Couper provided expert technical assistance.

References

1. James K., in *Handbook of Experimental Immunology* (D. Weir, ed.), 2nd edn.,Blackwell, Oxford (1973) pp. 31.1-31.27.

2. Moynihan, P.C., Janes, M.D., Grogan, J.B. & Hardy, J.D., *Surg. Forum 18* (1967) 235-237.

3. Lance, E.M., Ford, P.J. & Ruszkiewicz, M., *Immunol. 15* (1968) 571-580.

4. Zola, H., Mosedale, B.M. & Thomas, D., *Transplantation 9* (1970) 259-272.

5. Woiwod, A.J., Courtenay, J.S., Edwards, D.C., Epps, H.B.G., Knight, R.R., Mosedale,Betty, Phillips, A.W., Rahr, L., Thomas, D., Woodroffe, J.G. & Zola H., *Transplantation 10* (1970) 173-186.

6. Allan, D. & Crumpton, M.J., *Biochim. Biophys. Acta 274* (1972) 22-27.

7. Zweibaum, A. & Bouhou, E., *Transplantation 13* (1972) 622-624.

8. Zola, H. & Thomas, D., in *Protides of the Biological Fluids* (H. Peters, ed.), Vol. 21, Pergamon, Oxford (1973), in press.

9. Thomas, D., Edwards, D.C. & Zola, H., *J. Immunol. Meth.,* in press.

10. Zola, H., Thomas, D., Mosedale Betty and Courtenay, J.S., *Transplantation 12* (1971) 49-53.

11. Edwards, D.C., Thomas, D., Mosedale, B., Woodrooffe, J.G. & Zola, H., in *Behring Institute Mitteilungen 51* (Seiler, F.R. & Schwick, H.G., eds.), Behringwerke AG, Marburg (1972) 28-33.

12. Taylor, H.E., *Transpl. Proc. II* (1970) 413-416.

13. Feldkamp-Vroom, Th. M. & Balner, H., *Eur. J. Immunol. 2* (1972) 166-173.

14. Zola, H., Thomas, D. & Mosedale, Betty, *Experientia 28* (1972) 192-193.

37 CHARACTERIZATION OF MEMBRANE PROTEINS

A.H. Maddy
Department of Zoology
University of Edinburgh
West Mains Road
Edinburgh EH9 3JT, U.K.

It is unlikely that any one method or set of conditions will prove to be adequate for the isolation of proteins from membranes. Indeed, limited confidence can be placed on results obtained by only one solvent system. Techniques for the characterization of the isolated proteins, now considered, include centrifugation, electrophoresis, especially gel electrophoresis in the presence of SDS, N-terminal analysis and amino acid analysis.

This paper is concerned with the study of membrane proteins as isolated molecules and with those methods that are available to determine the chemical nature of these proteins. The central problem in this approach is the fractionation of membrane proteins into their various molecular species free from each other and from lipids. Once this has been achieved, hopefully there only remains the relatively straightforward elucidation of the structures of the different proteins by classical techniques.

APPROACHES TO SOLUBILIZATION, FRACTIONATION AND CHARACTERIZATION

There is an abundance of methods which 'isolate' and 'fractionate' membrane proteins, but an equal dearth of evidence on the precise molecular status of the various 'fractions'. Rarely can it be concluded that any given fraction is a distinct molecular species rather than a stable multimolecular aggregate with perhaps no biological significance. This state of affairs is largely a consequence of the complexities of the integration of proteins into membranes. Most, if not all, membranes contain aqueous and non-aqueous domains which arise by the coming together of the amphipathic constituents. The protein molecules, which themselves contain both hydrophilic and hydrophobic regions, are built into, and contribute towards, this larger superstructure composed of hydrophilic and hydrophobic regions. Consequently in most cases the native dispositions of the proteins require the integrity of the supermolecular organisation, and proteins cannot be isolated from the complex, no matter how mild the procedure, without a conformational rearrangement.

Fig. 1 illustrates schematically the possible relationships between proteins and a lipid bilayer and the effects of transferring such proteins into an aqueous medium, i.e. 'solubilization'. In all cases except 1 the operation will probably result in the rearrangement of the hydrophobic facets

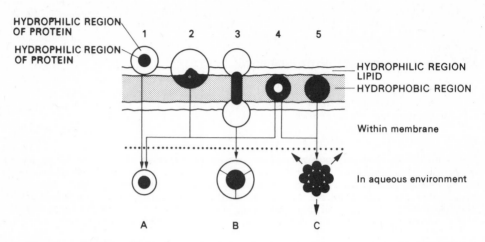

Fig. 1. A schematic representation of the possible interrelationships between proteins and a lipid bilayer and of the consequences of transferring the proteins into a wholly aqueous environment.

of the proteins and produce either states A, B, or C. The difficulties begin when B and C occur, and in any given experiment it is often difficult to demonstrate which of the various alternatives have been produced. (The diagram assumes that there is a complete separation of protein and lipid. Incomplete separation aggravates the problem.)

A wide range of reagents can effect the transition from membrane to water-soluble complex [1]. The great variety of the reagents strongly suggests that they do not have a common mechanism and implies that lipids and proteins are held together to form membranes by several types of forces. Thus one group of reagents - high or low ionic strength, chelating agents, low pH - modifies electrostatic conditions and releases proteins which are bound into the membrane predominantly by electrostatic bonds. It is unusual for such treatments to release more than 50% of the proteins. Other reagents - e.g. butanol, 2-chloroethanol, pyridine, chloroform/methanol - which completely disrupt membrane structure and which must modify the hydrophobic interactions of the membrane - are more likely to give rise to complexes of types B and C. A third approach involves the use of detergents and seeks, in essence, to replace the native interactions between the molecules of the membrane with interactions between the membrane molecules and detergent in the hope that these new complexes are less intractable than the intact membrane. Irrespective of which method is employed the final result is a 'solution' of protein. This usually means no more than that the proteins are not sedimented by e.g. 2×10^8 g-min. and while it frequently does not even

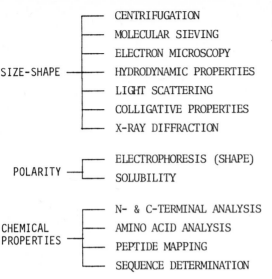

SIZE-SHAPE
- CENTRIFUGATION
- MOLECULAR SIEVING
- ELECTRON MICROSCOPY
- HYDRODYNAMIC PROPERTIES
- LIGHT SCATTERING
- COLLIGATIVE PROPERTIES
- X-RAY DIFFRACTION

POLARITY
- ELECTROPHORESIS (SHAPE)
- SOLUBILITY

CHEMICAL PROPERTIES
- N- & C-TERMINAL ANALYSIS
- AMINO ACID ANALYSIS
- PEPTIDE MAPPING
- SEQUENCE DETERMINATION

Fig. 2. Techniques for the analysis of membrane proteins.

approach being a solution in any strict sense of the word the process has at least converted the proteins to a state where they can, with caution, be analyzed by some of the traditional methods of protein chemistry.

Let us now consider what properties of proteins can be used to characterize the components of these various solutions, and which of the traditional techniques of protein chemistry can, and have, been used on membrane proteins [Fig. 2].-

a) The first set of methods depend upon the size and shape of the particles; but primarily they give information pertaining to the particle in solution, and its molecular nature is of only secondary significance. Thus, in most cases, these methods will not indicate whether a particle consists of just one polypeptide chain or is an aggregate of several.

b) The term 'polarity' used for the second group is intended to cover not only the electrical charge on a protein but also the protein's hydrophobicity. Electrophoretic techniques will depend chiefly on the electrostatic nature of the protein but will also be markedly affected by the size and shape of the protein, especially when the electrophoresis takes place through a sieving medium such as polyacrylamide. The term 'solubility' does not only refer to a protein's solubility in aqueous solutions and to aqueous chromatographic methods, but also covers its behaviour in non-aqueous solvents. Techniques based on non-aqueous solvents which could be of considerable potential value have, for several reasons, not been exploited but may be expected to be developed in the near future.

c) Direct chemical analyses of the protein fractions yield the most unequivocal information, especially if they can be carried right through to the stage of amino-acid sequencing to show that each fraction contains but one unique sequence. (Sequence is of course not the end of the story, for it leaves unsolved the three-dimensional arrangement of the proteins which ultimately must be known when the proteins are part of the membrane complex, i.e.

not as pure protein crystals.)

At whatever level the question is considered, the need for the application of several techniques and cross-reference between them is evident. Harris [2] describes work leading from the electron microscope to other techniques, and I now illustrate how centrifugation and electrophoresis can be combined. For reasons that need not concern us here, we wished to identify within an extract prepared by a butanol treatment, containing virtually all the proteins of the erythrocyte membrane, those proteins which could be extracted separately from the membrane by dilute solutions of EDTA [3]. The EDTA extract of ghosts consists chiefly of protein with relatively low sedimentation coefficient. Similar material is present in the butanol extract, but here it is accompanied by faster sedimenting proteins and the complex schlieren pattern cannot be unequivocally analyzed. Nevertheless we were able to confirm the identity of the slowly sedimenting molecules of the two extracts by supplementing the sedimentation data with electrophoretic evidence. This was made possible by repeating the centrifugation on a sucrose gradient and so isolating the various components of the butanol extract, each being examined by polyacrylamide gel electrophoresis. Prior electrophoresis of the EDTA extract in a Tris/glycine buffer had established its characteristic band pattern, and by demonstrating a similar pattern in the fractions at the top of the sucrose gradient from the butanol extract we confirmed that the slowly sedimenting proteins in the butanol extract were indeed those proteins of the EDTA extract.

Thus it is possible to recognise a given set of proteins by their sedimentation properties and by the way they fractionate electrophoretically, but it certainly cannot be concluded that because a mixture splits up into a given number of bands this represents the number of protein species in the mixture. One really must know the molecular nature of each band. Are there as many, or more, proteins in the mixture as there are bands in its electropherogram? Does each band consist of one protein, or are some of them aggregates of several proteins ? Can one protein give rise to several bands by aggregating with itself ? Questions of this type can be answered indirectly by comparing the mixture under several different electrophoretic conditions, but it is preferable to isolate and analyze the individual bands. It certainly helps to establish the molecular homogeneity of the components of a mixture if it can be shown that the same number of components are present under several different conditions, and that if a particular component is isolated from one ststem it runs true in the others.

ELECTROPHORETIC EXAMINATION

The most widely used technique is electrophoresis in the presence of SDS. This is claimed to be much more than merely an alternative to any other buffer system, for it is said to have the advantage that after reduction of disulphide bridges the detergent destroys all non-covalent interactions between and within protein molecules and that the mobilities of the detergent-protein complexes are simply related to the molecular weights of the polypeptides. Thus it is claimed that the method not only demonstrates unequivocally the polypeptide content of a mixture (except for the coincident mobility of chains of the same size) but also permits a rapid and simple estimation of their molecular weights. The result of SDS electrophoresis of the EDTA extract is well known [e.g. 4,5]. The extract contains a group of proteins between 200,000 and 250,000: there are two major bands but this group is more complex than a simple doublet. - Lenard [6] describes 5 bands which we also see. Clearly separated from these is a faster band or pair of bands with a mobility corresponding to a molecular weight of around 40,000. This fractionation can be compared with the fractionation of the same extract on a Tris/glycine gel by adapting the latter to a preparative scale and then analyzing each individual fraction on a SDS gel. (The large-scale fractionation is effected on a 20 x 20 x 0.4 cm slab of polyacrylamide which is cut into slices after the run. Each slice is homogenized and the protein extracted). These experiments show that band A and B in the Tris/glycine system consists of the 200,000 doublet, band C contains only the faster component of the doublet, and the fast moving bands in Tris/glycine are the 40,000 components of the SDS gels [Fig. 3]. These results are consistent with estimations of the sizes of bands A, B, and C in Tris/glycine by the Fergusson procedure, i.e. the estimation of molecular weight by measuring the effect of polyacrylamide concentration of protein mobility [7]. It is found that the weight of band C falls between that of aldolase (160,000) and fibrinogen (310,000); band B is bigger than 310,000 and Band A even bigger than B.

Should the matter now be considered to be more or less closed ? Band C is indeed a polypeptide chain of about 200,000 daltons; bands A and B are complexes of polypeptides of this size and are dissociated into their constituent chains by the detergent: the high-mobility bands are not related to these large polypeptides and their co-extraction is merely a coincidence. These would not be unreasonable conclusions if the claims made for the SDS method are valid for membrane proteins. There is increasing evidence to show that they are not.

The validity of the SDS method for the characterization of the molecular constitution of proteins depends on two principal conditions being fulfilled. First, polypeptide chains must bind a constant amount of SDS per unit chain length to form SDS-protein complexes whose structures are simply related to the size of the polypeptides. Although these conditions would appear to be fulfilled by many proteins [8,9], the amount of SDS bound can vary [10] and can be affected by amino-acid content. Thus anomalous behaviour has been reported for maleylated proteins [11], for highly charged proteins [12] and for glycoproteins [13]. The second condition which must be met is that the interaction of the detergent with the polypeptides must eliminate all other non-covalent interactions of the proteins. This has been

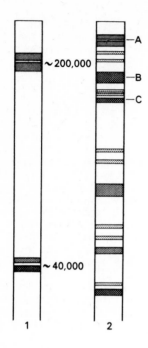

Fig. 3. The fractionation of the EDTA
extract of ox erythrocyte ghosts by polyacrylamide gel electrophoresis -
1. in the presence of SDS.
2. in a Tris-glycine buffer.
Bands A *and* B *contain both components of
the 200,000 doublet, and band* C *only the
faster component.*

demonstrated to be the case with soluble
proteins whose properties and behaviour
have been well characterized by alternative
techniques. However, membrane proteins
are known to form intractable complexes
both with themselves and with each other.
The extrapolation of results from well
characterized soluble proteins to membrane
proteins ignores this complication. Indeed
there are reports in the literature which
tend to suggest that SDS alone does not
overcome all the non-covalent forces between membrane proteins. Lenard [6] found
that in addition to SDS, EDTA was also
essential for all the erythrocyte ghost
protein to enter the gels, and the presence
of calcium ions during haemolysis can cause
formation of stable aggregates [14]. Organic solvents can cause an aggregation of
90,000 peptide which cannot be reversed by
SDS [4,15].

    Against this background it is not too surprising to come across evi-
dence suggesting the '200,000' proteins, or at least some of them, are aggregates of the '40,000' units [16]. For reasons that are not clear, during passage of solution of the smaller molecules isolated by preparative electrophoresis down a G-200 Sephadex column, some are converted to '200,000' components.
When the '40,000' molecules are passed down the column in a dilute phosphate
buffer *in the absence of detergent* most of the material is excluded, but a
small amount is retarded and this retarded protein is found by SDS electrophoresis to be '200,000' although no protein of this size is present in the
sample applied to the column. The retention of large protein by the column
indicates that it is not functioning as a simple molecular sieve, but is in
this instance interacting with and consequently aggregating some protein. (When
the 200,000 fraction isolated by preparative electrophoresis is passed down a
G-200 column it is all excluded: the exclusion of '40,000' molecules by G-200

is not necessarily anomalous as this molecular weight is derived for the protein only after detergent treatment, not in the phosphate buffer used for the chromatography.) As some '200,000' protein aggregates which are not dissociated by SDS can exist, it is possible that the '200,000' proteins in the original membrane extract are similar complexes and not very long polypeptide chains. The identity of the artificially produced large complexes with those extracted from the membrane has not been established, and the evidence can only be circumstantial until the molecules have been more fully characterized by direct chemical analysis. (As covalent cross-links between polypeptide chains of EDTA-soluble membrane proteins have recently been observed [17], aggregation could arise from the formation of such bonds; but it does seem improbable that the chromatography could bring about their formation.)

CHEMICAL ANALYSIS

The simplest direct chemical analysis which can be applied to a protein solution is identification of its N-terminal residues (C-terminal analysis of a complex mixture is more difficult). The N-terminal analysis will estimate the minimum number of polypeptide chains present. Qualitative analysis underestimates the heterogeneity if more than one chain has the same terminal amino acid and if any chains have blocked termini. Quantitative analysis of a complex mixture runs into the problem of estimating the percentage recovery of the labelled derivative from each terminus. Dansylation in the most effective method in our hands, as it is sufficiently sensitive to be used on the small samples of protein obtained from electrophoretic gels and can also be used in the presence of SDS. Dansylation of the '200,000' and '40,000' fractions of the EDTA extract reveals N-terminal tyrosine, lysine, glycine, threonine, serine, aspartic and glutamic acids. The multiplicity of N-terminal residues illustrates the complexity of both samples, and the identity of the two maps is consistent with one being an aggregate of the other.

Logically, peptide mapping and amino acid analysis follow determination of terminal residues, especially if the solution is homogeneous, but heterogeneity undermines their value. Comparison of the peptide maps of different proteins from *Micrococcus lysodeikticus* is very interesting in that the most prevalent peptides of all three proteins are common to all three maps [18]. The proteins are an ATPase, and NADH dehydrogenase and a component characterized by its high electrophoretic mobility on polyacrylamide gels (this characteristic is shared by our '40,000' fraction. Fukui and Salton [18] conclude that membrane proteins have certain common regions which anchor them into the membrane, and considers the alternative explanation that the results are due to cross-contamination to be unlikely. A third explanation of general interest needs to be explored, namely that the peptides are only functionally common, in the sense used by Fukui and Salton, but are structurally different and composed of different apolar amino acids. The functional requirement could be met by various combinations of hydrophobic side chains, and the apparent identity of peptides could be a consequence of the inability of chromatographic solvents to resolve peptides differing only in this respect.

The futility of the amino acid analysis of protein solutions of indeterminate heterogeneity should have been fairly well established by the comparison of the published analyses of several structural proteins with that of the postribosomal supernatant of *E. coli* [19]. By this criterion the latter was also a structural protein ! Analysis of a homogeneous protein is quite a different matter, although the figures in themselves are not all that remarkable, for it is the distribution of amino acids along a sequence which is significant, not its overall composition. A number of reports indicate the segregation of hydrophobic residues into certain segments which are presumably those regions of the protein surrounded by the hydrocarbon chains of the lipids. This is almost certainly the case in the human erythrocyte membrane glycoprotein, which is the only membrane protein to have been fully sequenced [20]. The protein probably traverses the membrane, and it has a central hydrophobic sequence that is presumably coincident with the lipid bilayer in the membrane.

The determination of the amino-acid sequence of one membrane protein shows what is possible and what, among other things, must be done for the others. Membrane proteins are not mystical objects; they are in principle amenable to analysis by the classical techniques of protein chemistry. Nevertheless their interactions achieve a degree of complexity which remains a challenge to those who study them and it is improbable that this challenge will be adequately answered by any one miraculous reagent or technique.

Acknowledgement

I am indebted to the Medical Research Council for financial support for the research work mentioned in this article and to my collaborators M.J. Dunn and W.McBay.

1. Maddy, A.H., *Subcell. Biochem. 1* (1972) 293-301.

2. Harris, J.H., *this volume,* Art. 38.

3. Maddy, A.H., Dunn, M.J. & Kelly, P.G., *Biochim. Biophys. Acta 288* (1972) 263-276.

4. Fairbanks, G., Steck, T.L. & Wallach, D.F.H., *Biochemistry 10* (1971) 2606-2616.

5. Hoogeveen, J. Th., Juliano, R., Coleman, J. & Rothsterin, A., *J. Membr. Biol. 3* (1970) 156-172.

6. Lenard, J., *Biochemistry 5* (1970) 1129-1132.

7. Rodbard, R. & Crambach. A., *Anal. Biochem. 40* (1971) 95-134.

8. Webern, K. & Osborn, M., *J. Biol. Chem. 244* (1969) 4406-4412.

9. Reynolds, J.A. & Tanferd, C., *J. Biol. Chem. 245* (1970) 5161-5165.

10. Nelson, C.A., *J. Biol. Chem. 246* (1971) 3895-3901.

11. Tung, J.S. & Knight, C.A., *Biochem. Biophys. Res. Commun. 42* (1971) 1117-1121.

12. Williams, J.G. & Gratzer, W.B., *J. Chromatog. 57* (1971) 121-125.

13. Bretscher, M.S., *Nat. New Biol. 231* (197) 229-232.

14. Triplett, R.B., Wingate, J.M. & Carraway, K.L., *Biochem. Biophys. Res. Commun. 49* (1972) 1014-1020.

15. Tanner, M.J.A. & Boxer, D.M., *Biochem. J. 129* (1972) 333-347.

16. Dunn, M.J. & Maddy, A.H., *FEBS Lett. 36* (1973) 79-82.

17. Birckbichler, P.J., Dowben, R.M., Matacic, S. & Loewy, A.G., *Biochim. Biophys. Acta 291* (1972) 149-155.

18. Fukui, Y. & Salton, M.R.J., *Biochim. Biophys. Acta 288* (1972) 65-72.

19. Rosenberg, S. & Guidotti, G., in *The Red Cell Membrane* (Jamieson, G.A. & Greenwalt, T.J., eds.), Lippincott, Philadelphia (1969) 93-109.

20. Segrest, J.P., Kahane, I., Jackson, R.L. & Marchesi, V.T. *Arch. Biochem. Biophys. 155* (1973) 167-183.

38 THE PURIFICATION OF SOME MEMBRANE-ASSOCIATED PROTEINS FROM ERYTHROCYTE 'GHOSTS'

James R. Harris
Department of Physiology
University of St. Andrews
Bute Medical Buildings
St. Andrews, Fife, U.K.

The proteins under investigation are released from mammalian erythrocyte 'ghosts' when intact haemoglobin-free membranes are made to undergo fragmentation in a low ionic strength solution. This fragmentation is produced by overnight dialysis against distilled water followed by freezing and thawing. The initial extract is obtained as the supernatant following the centrifugal pelleting of the membrane fragments. This extract contains several proteins, two of which have been characterized by electron microscopic and electrophoretic studies. For further studies which require larger quantities of purified material, a scheme has been developed to separate the extracted proteins, with monitoring by analytical polyacrylamide gel electrophoresis, Ouchterlony immunodiffusion and electron microscopy. A preliminary fractionation by column chromatography using Biogel P300 removes traces of haemoglobin, if present. The fractions under the excluded peak from the P300 column contain a mixture of high molecular weight proteins. These fractions are pooled for the next stage of the purification. Sucrose density-gradient centrifugation very effectively separates the larger of the two proteins (the hollow cylinder protein, 22.5 S; m.wt. ∿ 900,000). Preparative electrophoresis on polyacrylamide gel has been employed to purify the second protein (the torus protein, 9.0 S; m.wt. ∿ 225,000) from the heterogeneous region taken from the top of the gradient. The overall scheme is discussed and proposed characterization studies are briefly described.

Electron microscopic studies and preliminary biochemical investigations have been performed on a class of high molecular weight multi-subunit proteins that are released when haemoglobin-free mammalian erythrocyte 'ghosts' undergo a fragmentation which is brought about by distilled water dialysis followed by freeze-thawing [1, 2, 3]. From electron microscopic observations of the protein molecules escaping from lesions in intact 'ghosts' and of molecules situated inside stromalytic tubules [4], it has been suggested that the proteins may be associated with the inner surface of the erythrocyte membrane. Additional support for this interpretation comes from the immuno-ferritin studies of Howe and Bächi [5] who have shown that the ferritin antibody directed against the torus protein will bind to the inner surface of erythrocyte ghosts. The experiments of Hoogeveen *et al.* [6] using a fluorescent label for the outer surface of the erythrocyte membrane indicated the absence of fluorescence on the proteins released from labelled 'ghosts' by low ionic strength solutions, thus suggesting that these proteins came from the inner surface.

Recently, work has commenced on the characterization of the hollow cylinder and torus proteins. These studies need milligram quantities of the purified proteins, and to this end a more comprehensive purification scheme has been developed for the separation of the various proteins present in the initial extract obtained from the erythrocyte 'ghosts'.

MATERIALS AND METHODS

Preparation of erythrocyte 'ghosts'

Out-dated human blood (group O, +ve) was obtained from the Blood Transfusion Service at Victoria Hospital, Kirkcaldy. Fresh, defibrinated cattle blood was obtained direct from the St. Andrews slaughter-house. The methods employed for washing the erythrocytes and for producing their 'ghosts' by osmotic haemolysis and repeated washing are as previously published [2, 3], except that the Beckman J-21 centrifuge with the JA-10 (3 1 capacity) angle rotor was used throughout. Up to 1 1 of packed erythrocytes can be converted to haemoglobin-free erythrocyte 'ghosts' within one working day, as the J-21 centrifuge has extremely rapid acceleration and deceleration.

Preparation of the initial protein extract

Intact erythrocyte 'ghosts' are made to undergo fragmentation by overnight dialysis followed by freeze-thawing [1, 2, 3]. The membrane fragments are then pelleted at 21,000 rev/min (48,000 × g) for 1 h and the supernatant taken as the crude extract. Prior to any of the fractionation procedures the protein extract is concentrated by rotary film evaporation at 35° and pervaporation, and dialyzed against 5mM Tris-HCl buffer (pH 8.0) at 4. The solution is then clarified by centrifugation at 50,000 rev/min (165,000 × g) for 10 min and the small pellet discarded.

Electron microscopy

Negatively stained specimens of the various protein solutions are prepared on carbon-coated grids (Type 400) using 2.0% sodium phosphotungstate (pH 7.0). The 'single-drop' technique is used to apply the protein and stain to the grid [7]. Specimens are routinely studied at electron optical magnifications in excess of 50,000 diameters. The Phillips EM 300 and EM 301, and also the AEI EM 6B, have been used to study the specimens. Photographic recording has been done on Ilford EM6 and Special Lantern Contrasty glass plates.

Analytical polyacrylamide gel electrophoresis

A simple system employing 7.0% polyacrylamide gels in 50 mM Tris-HCl buffer (pH 8.0) has been used. Gels 6.0 cm in length and 0.5 cm in diameter are prepared in the tubes provided with the Quickfit Instrumentation analytical polacrylamide gel electrophoresis modules. A solution of 7.0% acrylamide (BDH) and 0.1% *N,N'*-methylene *bis*acrylamide (BDH) in 50 mM Tris-HCl buffer (pH 8.0) is polymerized by the addition of *N,N,N',N'*-tetramethylethylene-diamine to 0.1% and ammonium persulphate to 0.05%.

Gels are left at least 24 h at room temperature before use. In both the electrode compartments of the electrophoresis module, 50 mM Tris-HCl buffer (pH 8.0) is used. The gels are pre-run at 6 mA per tube until a sample of bromophenol blue has passed through. The electrode buffers are then replaced before applying the protein samples. Approximately 50 μl quantities of protein solution mixed with an equal quantity of 10% glycerol in 5 mM Tris-HCl buffer (pH 8.0) containing 0.01% bromophenol blue, are applied to each tube. The current is passed initially at 1 mA per tube for 5 min and then increased to 5 mA per tube. The completion of the electro-phoretic separation is indicated by the emergence of the tracker dye from the bottom of the tubes. Gels are stained with Coomassie Blue dye in 10% acetic acid-20% methanol-70% water.

Immunodiffusion

Immunodiffusion plates are prepared from 1% agar solution in phosphate-buffered saline containing 0.1% sodium azide on microscope slides and 5 cm plastic Petri dishes. Outer-wells are cut in the conventional hexagonal pattern around a central well. Protein samples are put in the outer wells and antiserum prepared by injecting rabbits with human and cattle erythro-cyte ghosts is put in the central well. Precipitin lines form over a period of 48 h at 4°. The gel is then exhaustively washed in phosphate-buffered saline and stained with Coomassie Blue dye.

RESULTS

Fractionating the protein extract from erythrocyte 'ghosts'

The complexity of the initial protein extract is indicated by the stained bands of protein revealed on analytical polyacrylamide gels (Fig. 1) and also by the multiple precipitin lines produced on the immunodiffusion plates. In the electron microscope the extract is seen (Fig. 2) to contain the hollow cylinder and the single ring or torus protein, along with smaller particles.

Column chromatography

This is the first step of the purification scheme, and is included only if significant quantities of haemoglobin have been carried over from the 'ghosts'

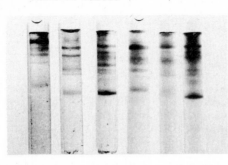

Fig. 1. Analytical polyacrylamide gels, showing the electrophoretic separation of the proteins in extracts made from several separate human erythrocyte 'ghost' preparations. *Note the overall similarity of the pattern of bands on the gels.*

into the protein extract and require to be removed before separating the other proteins. Gel filtration chromatography on Biogel P300 (Biorad Laboratories) with 10 mM Tris-HCl buffer (pH 8.0) has been used to separate haemoglobin as a retarded fraction, whereas the higher molecular weight proteins are eluted in the excluded peak (Fig. 3). Analytical polyacrlamide gels corresponding to the Biogel P300 separation are shown in Fig. 4. It is important to note that all the fractions under the excluded peak are heterogeneous; these excluded fractions are pooled for further separation.

The second step in the purification scheme is separation of the proteins by rate-dependant sedimentation through a sucrose density gradient. This technique is excellent for separating the high molecular weight hollow cylinder protein from the other proteins. Linear 0.25 M to 1.0 M sucrose gradients in 10 mM Tris-HCl buffer (pH 8.0) are used in the 40 ml tubes of the Beckman S.W. 27.1 rotor. Centrifugation is performed at 27,000 rev/min for 24 h. A typical sedimentation profile of the protein in the tube following centrifugation is shown in Fig. 5, and the corresponding analytical polyacrylamide gels in Fig. 6. The fractions under the small fast moving peak are taken as the partially purified hollow cylinder protein. Higher up the gradient the fractions are all heterogeneous and these are pooled for separation by the third stage, preparative polyacrylamide gel electrophoresis.

The partially purified hollow cylinder protein can be readily obtained in a more highly purified state by concentrating the pooled fractions from the first sucrose gradient run and reapplying it as the sample for a second identical sucrose gradient. A typical sedimentation profile is shown in Fig. 7, and an analytical polyacrylamide gel from the pooled fractions under the main protein peak is shown in Fig. 8. Electron microscopy of this material (Fig. 9) reveals that there is only one molecular species present and that the background is very clean, supporting the homogeneity indicated by Fig. 8. In addition, immunodiffusion analysis supports this claim for homogeneity, as shown in Fig. 10.

The more slowly sedimenting proteins on the first sucrose gradient run cannot be purified by applying them to a second gradient; accordingly,

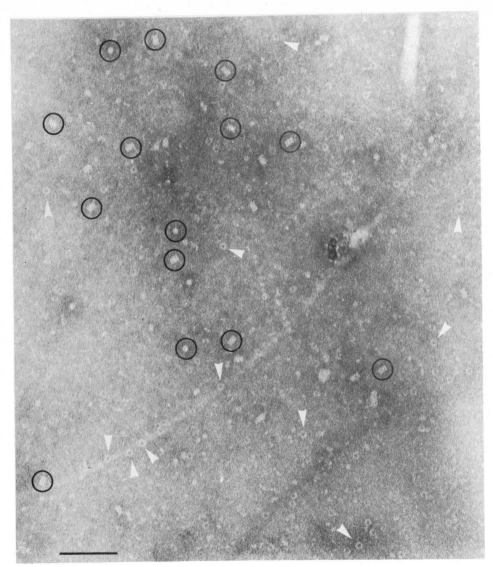

Fig. 2. An electron micrograph of the initial protein extract from human erythrocyte 'ghosts', negatively stained with Na phosphotungstate. White arrows indicate the torus protein and black circles the hollow cylinder protein. *The line represents 0.1 μm.*

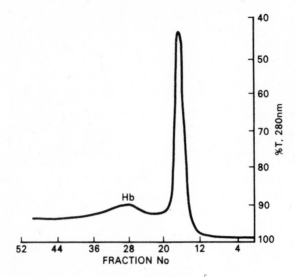

Fig. 3. The 280 nm absorption profile given by the effluent on passing the human erythrocyte 'ghost' extract through a Biogel P300 column.

Fig. 4 *(below, left)*. The analytical polyacrylamide gels corresponding to the P300 fractions indicated in Fig. 3.

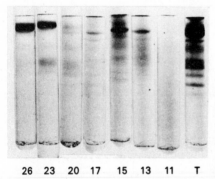

as mentioned above, preparative polyacrylamide gel electrophoresis has been employed

The apparatus manufactured by Quickfit Instrumentation Ltd. has been used, with the same gel and buffer system that was employed for the analytical polyacrylamide gel electrophoresis. The gels (30 ml) are prepared 24 h before use. The apparatus is pre-run to remove traces of contaminating material which absorbs at 280 nm, prior to the application of the protein sample (~2 ml), which is dialyzed against 5 mM Tris-HCl buffer (pH 8.0) before application and is mixed with a small quantity of the bromophenol blue tracker dye. A typical 280 nm absorption profile obtained from the preparative electrophoretic separation of a sample containing approximately 30 mg of protein is shown in Fig.11, and the corresponding analytical gels are shown in Fig. 12. By electron microscopy (Fig.13) Fractions 13 to 15 were shown to contain the torus protein. Immunodiffusion analysis supports the electrophoretic and electron microscopic assessment of the homogeneity of the torus protein (Fig. 14). For comparison purposes, analytical gels of the purified hollow cylinder and torus proteins are shown beside the initial extract (Fig. 15).

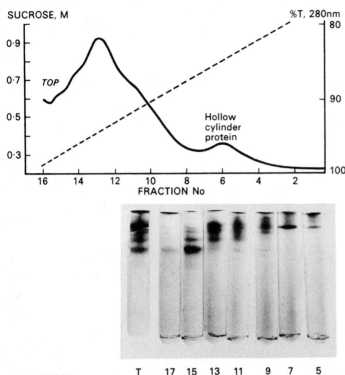

Fig. 5. The 280 nm adsorption profile given by the Biogel P300-excluded proteins on a 0.25 M to 1.0 M sucrose density gradient after centrifugation for 24 h at 27,000 rev/min.

Fig. 6. The analytical polyacrylamide gels corresponding with the density gradient fractions shown in Fig. 5.

DISCUSSION

The purification scheme presented above for the separation of the loosely associated erythrocyte membrane proteins release by freeze-thawing following distilled water dialysis successfully isolated the two proteins under consideration, along with several others, for which no morphological characteristics of enzymic functions have yet been assigned. The scheme is in fact a further development of that described previously [3] in which sucrose density gradient centrifugation was the main technique employed. Preparative polyacrylamide gel electrophoresis was recently used successfully by the author to purify tumour-specific antigen from rat hepatoma [8] by the same method as described above, which was therefore transferred directly to the erythrocyte membrane proteins. In an early publication [2] it was mentioned that by slicing analytical polyacrylamide gels the torus protein could be detected by electron microscopy, so the preparative electrophoretic system naturally presented a hopeful separating method.

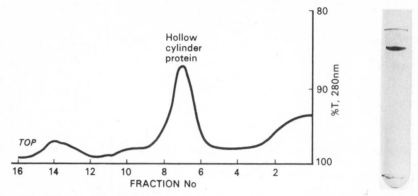

Fig. 7.   The 280 nm absorption profile given by the partially purified human erythrocyte 'ghost' hollow cylinder protein on sedimenting through a $0.25 \rightarrow 1.0\,M$ sucrose density gradient for 24 h at 27,000 rev/min.

Fig. 8.   An analytical polyacrylamide gel corresponding to the main peak shown in Fig. 7.

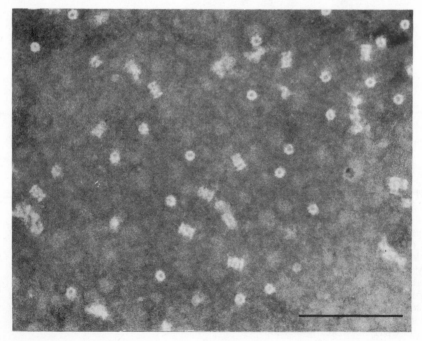

Fig. 9.
Legend opposite.

The step-by-step separation of the hollow cylinder and torus proteins, though time-consuming, does enable highly purified fractions to be obtained more easily than if a one-step separation with preparative electrophoresis is attempted. The very slow electrophoretic migration of the hollow cylinder is not compatible with the easy use of the preparative electrophoresis, and the fact that the migration of haemoglobin is only slightly greater than that of the hollow cylinder would be a further complication in this instance.

Having obtained reasonable quantities of the purified hollow cylinder and torus proteins, one can now characterize them by the many standard bio-chemical and physical procedures currently available. Preliminary results obtained by several approaches will later be presented elsewhere in detail. In brief, molecular weight studies are in progress using meniscus-depletion sedimentation equilibrium in the analytical ultracentrifuge and sodium dode-cylsulphate gel electrophoresis. These studies supplement the preliminary

---

Fig. 9 *(left, below).* Electron micrograph of the highly purified human ery-throcyte 'ghost' hollow cylinder protein obtained from a second sucrose den-sity gradient run, as in Fig. 7. Negatively stained with phosphotungstate.
[*The line represents 0.1* µm.

Fig. 10 *(below).* Ouchterlony immunodiffusion plates using the human eryth-rocyte 'ghost' extract and the hollow cylinder protein at various stages of purification.

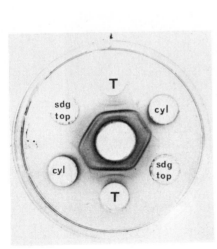
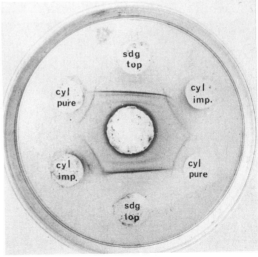

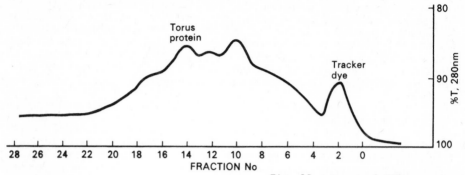

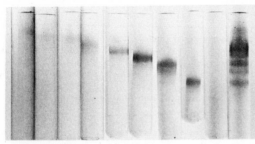

Fig. 11 *(above)*. A 280 nm
absorption profile given by
the eluate from the preparative polyacrylamide gel electrophoresis apparatus after
application of the proteins
in the pooled sucrose density
gradient top fractions as
the sample.

Fig. 12 *(left)* &   *LEGENDS*
Fig. 13 *(below)*   *OPPOSITE*

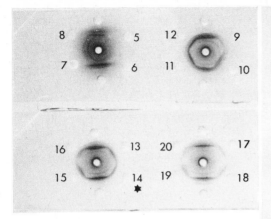

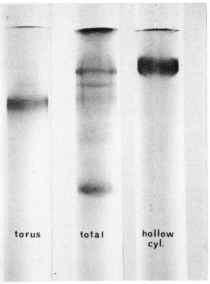

torus   total   hollow cyl.

Fig. 14. Ouchterlony immunodiffusion plates of the preparative electrophoresis purification of the torus protein, corresponding with Fig. 11.

Fig. 15. Analytical polyacrylamide gels comparing the purified hollow cylinder and torus proteins with the initial protein extract from human erythrocyte 'ghosts'.

sedimentation-coefficient determinations already performed [3], which are also due to be amplified. Studies on the amino acid composition of the hollow cylinder protein from human and cattle erythrocyte 'ghosts' has so far indicated a close similarity of composition. In both cases the composition is compatible with that of other associating proteins. *N*-terminal analysis is to be undertaken in the near future. Isoelectric point determination is also of interest, and is relevant to the amino acid composition, as is a study of the absorption spectra of the proteins under different conditions.

Fig. 12 *(opposite, centre)*. Analytical polyacrylamide gels corresponding to the preparative electrophoretic separation shown in Fig. 11.

Fig. 13 *(opposite, below)*. Electron micrograph of fraction no. 14 from the preparative electrophoretic separation shown in Fig. 11.

The line represents 0.1 µm.

Antigenic analysis is to be performed using the native and dissociated proteins to see if the subunits are antigenically identical or different. Antigenic cross-reactivity of the proteins and their subunits from different mammalian species is also a point of interest. In conjunction with a further high-resolution electron microscopic study of the two proteins, the interaction of antibody will be investigated using monospecific antiserum directly, and following ferritin labelling. This latter complex will be used in an attempt to repeat and extend the studies of Howe and Bächi [5],i.e. to label the hollow cylinder and torus proteins when still associated with the inner surface of the erythrocyte membrane. High-resolution negative-contrast staining studies may elucidate the subunit interactions and the dissociation/ association properties of the proteins and also their crystallization properties. Finally, enzyme screening has been commenced in an attempt to assign some function(s) to the proteins under investigation, so far with negative results.

References

1. Harris, J.R., *Biochim. Biophys. Acta 150* (1968) 534-537.

2. Harris, J.R., *Biochim. Biophys. Acta 188* (1969) 31-42.

3. Harris, J.R., *Biochim. Biophys. Acta 229* (1971) 761-770.

4. Harris, J.R., *J. Ultrastruct. Res. 36* (1971) 587-594.

5. Howe, C. & Bächi, T., *Exp. Cell Res. 76* (1973) 321-332.

6. Hoogeveen, J.Th., Juliano, R., Coleman, J. & Rothstein, A., *J. Memb. Biol. 3* (1970) 156-172.

7. Harris, J.R. & Agutter, P., *J. Ultrastruct. Res. 33* (1970) 219-232.

8. Harris, J.R., Price, M.R. & Baldwin, R.W., *Biochim. Biophys. Acta 311* (1973) 600-614.

39 NOTES AND COMMENTS

Editor's explanation.- *The material in this Section is intended to reinforce the usefulness of the book. It consists largely of abbreviated presentations and of discussion remarks from the 1973 Symposium which was the basis for this book. The subject categories below correspond to those in the CONTENTS list.*

STRATEGY AND EQUIPMENT (Articles 1-9): *SUPPLEMENTARY MATERIAL*

QUANTITATIVE IMMUNOLOGICAL METHODS FOR BIOCHEMICAL STUDY OF THE LOCALIZATION, TRANSLOCATION, SECRETION AND TURNOVER OF PROTEINS WITHIN TISSUES

A.R. Poole

Strangeways Research Laboratory, Cambridge CB1 4RN, U.K.

In addition to the immunohistochemical techniques described in Article 1 for the localization of proteins, it is possible to use antisera in rocket immunoelectrophoresis [1], radial immunodiffusion [2] or radioimmunoassay [3] to determine quantitatively the concentration of a given component within a tissue homogenate, a subcellular fraction derived therefrom, or a body fluid.

The synthesis, turnover [5], intracellular translocation and secretion of proteins can also be examined by pulse-labelling tissues with suitable radiochemicals, followed by immunoprecipitation of labelled and unlabelled antigen. Details of these methods and examples of their applications are provided in the references cited.

1. Weeke, B., in *A Manual of Quantitative Immunoelectrophoresis. Methods and Applications* (Axelsen, N.H., Kroll, J. & Weeke, B., eds.), Universitetsforloget, Oslo (1973) pp. 37-46.

2. Mancini, G., Carbonara, A.O. & Heremans, J.J., *Immunochemistry 2* (1965) 235-254.

3. Smith, F.R., Raz, A.& Goodman, D.S., *J. Clin. Invest. 49* (1970) 1754-1761.

4. Muto, Y., Smith, J.E., Milch, P.O. & Goodman, D.S., *J. Biol. Chem. 247* (1972) 2542-2550.

5. Granner, D.K., Hayashi, S.I., Thompson, E.B. & Tomkins, G.M., *J. Mol. Biol. 35* (1968) 291-301.

The foregoing Note was submitted at the Editor's invitation. The questions put to Dr. Poole at the Symposium, as summarized below, refer largely to the work described in Article 1. ‒
Would the cell particles in which haemoglobin and anti-cathepsin D become localized all be secondary lysosomes (G. Sachs) ? - Yes, by definition.

G. Yagil (Rehovot) pointed out that the anti-cathepsin D serum might be valuable in deciding a widely debated question — whether normal cellular protein turnover involves lysosomal cathepsins or some unknown cytoplasmic proteolytic system. He wondered whether the effect of the antiserum on the rate of ^{3}H-leucine release had been examined, with HeLa or other cultured cells.

How specific is the Laurell technique, and what happens to the height of a rocket if a cross-reacting or denatured antigen is present in the sample (J. Steensgaard) ? - If the electrophoretic mobility of a cross-reacting antigen is similar, the height is proportionally increased; if the mobility is different, then several peaks may be discernible. If an excess of antibody is not present, the height will be decreased with cross-reaction.

Questions by O. Touster, *answered in the negative.*- Have you encountered any antibody to a lysosomal enzyme which would cross-react with any other lysosomal enzyme ? Have you tried to exploit the sensitivity of your fluorimetric assays to estimate how many primary lysosomes fuse with a pinocytotic vacuole ?

Various points concerning centrifugation and gradients (Arts. 3-7)

Answering J.O. Molloy who asked about the durability of steeply discontinuous gradients (Art. 4), G.B. Cline said that they do indeed degrade quite quickly but can have up to 2 h of useful life; steps may of course be used to make fully continuous gradients, and would largely be smoothed out in very long runs.

Replying to N.L. Webb, J. Steensgaard confirmed that their basic equations (Art. 7) indeed account for radial dilution.

Nomogram for centrifugal data: *see* Rasmussen, H.N., *Eur. J. Biochem. 34* (1973) 502-505.

Tables due to C.R. McEwen, for computation of particle sedimentation coefficients, have been amended and expanded as *Sedimentation Tables for Use with Shallow Sucrose Density Gradients* by B.D. Young in an Appendix to an Application Information leaflet (A6/6/72; M.S.E. Ltd., Crawley, Sussex, U.K.), in which Anna Hell deals with centrifugation of small DNA fragments.

Attention is drawn to the article *A Plea for a General Treatment of the Principles Underlying Centrifugal Separation of All Classes of Particles* [Porteous, J.W., *Biochem. Education 1* (1973) 75-81], where the limitations of the Svedberg and related equations are stressed, particularly from an instructional viewpoint, and equations readily applicable to organelles are given. T.B. Christensen & Eva Ruud (Biochem. Dept., Oslo Univ.) have submitted for publication a simple method for determining S values in a linear sucrose gradient; S is derived from the final position by a Δ-distance/Δ-time relationship.

Use of the B-XXIX rotor for rate-zonal edge unloading: *see* Brown, P.H.,*et al., Arch. Biochem. Biophys. 155* (1973) 9-18.

Importance of temperature control in gradient separation of osmotically active organelles: *see* Pollack, J.K. & Morton, M., *Biochem. Biophys. Res. Commun. 52* (1973) 620-626. (Dip in temperature prone to occur with Ti rotors.)

DENSITY GRADIENT BEHAVIOUR, DESIGN AND CONSTRUCTION

J.A.T.P. Meuwissen
Liver Physiopathology Laboratory, Dept. of Medical Research,
Rega Institute, Katholieke Universiteit te Leuven, Belgium

Experimental studies and theoretical considerations concerning the zone capacity problem have recently led the author to a plausible physical explanation of the phenomenon known as anomalous zone spreading [1,2]. Qualitatively, these studies have demonstrated that in practice zone instability is not related to density inversion, but is due to a hydrodynamic phenomenon which arises from the physical impossibility of constructing convection-free three-component systems in which opposed concentration gradients of solutes are arbitrarily chosen. From comparative studies on block zones and inverted zones, at 1 *g* and in a zonal rotor, it also became apparent that hydrodynamic instability is not the same as 'droplet-sedimentation' [3] or 'streamer formation' [4]. The latter form of instability is not observed in a zonal rotor nor in inverted sample zones, even at 1 *g*, although hydrodynamic instability can persist. A quantitative expression describing the limiting supporting capacity at each distance point of a given density gradient was derived and proved useful in designing optimal separation conditions. This equation takes the form:

$$\frac{dc_1}{dc_2} \geq - \frac{\overline{v}_2 \cdot D_1}{\overline{v}_1 \; D_2} \qquad (\text{Eqn.}1)$$

where dc_i denotes the concentration derivative, $\overline{v}_i$ the partial specific volume, and D_i the experimental diffusion coefficient of the respective solutes. Subscripts 1 and 2 denote the sample solute and the gradient solute respectively. This equation is in fair agreement with experimental stability limits [1,2,5].

Guided by these principles, optimization of gradient profiles and sample application procedures was attempted in order to attain maximal separation efficiency, i.e. to enable the maximal amount of sample material to be separated with maximal resolution. In general, it was found that sigmoidal density gradients with application of the zone at the inflection point give optimal results for the separation of complex mixtures. The stability principles mentioned are equally applicable to other gradient profiles, such as isokinetic [6], isometric [7] or equivolumetric [8] gradients, but with some sacrifice of efficiency.

In order to achieve these goals it was necessary to re-think current procedures for gradient formation and sample application. The need to form inverted-gradient sample zones and, moreover, the pursuit of analytical use of density-gradient methodology called for very accurate construction and withdrawal of density gradients. Commercially available density-gradient formers to not meet these requirements, because pumping accuracy and/or the feasibility of forming inverted gradients is deficient or absent. Accordingly a new gradient-forming device* was designed, constructed and tested. The

* *prototype; patents applied for*

following principles and features were incorporated.-
1. Accurate flow rates were obtained by using syringe pumps driven by stepper motors (coefficient of variation of flow rates: ±0.025%).
2. By combining individual syringe pumps into a continuous infusion-withdrawal pump-set, virtually unlimited volume capacity could be attained (the same relative degree of accuracy for gradients from 5 to 1600 ml).
3. Gradients are formed with the heavy end first, which is a prerequisite for sample application as an inverted zone and has other advantages, especially for the formation of gradients in zonal rotors.
4. All solutions are held in thermostatted containers and are driven by water, the syringes thus acting only on water.
5. By using electronically controlled valves, the gradient-former has been largely mechanized. A 7-bit word switch-register controls the state of the device so that human intervention is restricted to action upon a push-button at pre-calculated time points, depending on the gradient profile selected.
6. The time-consuming procedure of calculating the time points at which action should be taken during gradient formation has been shortened by using a program in conversational BASIC run on the laboratory computer. The computer asks (a) some essential data concerning the physical dimensions of the particular gradient system to be used, and (b) the flow rates of the pumps of the gradient-former. It then returns the time points when action should be taken.
7. For gradient recovery and analysis, the same automated pump-set is used, and on-line monitors continuously determine some physical parameters of the gradient column, e.g. the extinction at a selected wavelength by use of a flow-through spectrophotometer. Fractional collection of the gradient on a time base finishes the complete cycle.

1. Meuwissen, J.A.T.P., in *Methodological Developments in Biochemistry,* Vol. 3 (Reid, E., ed.), Longman, London (1973) pp. 29-44.

2. Meuwissen, J.A.T.P., in *European Symposium of Zonal Centrifugation* (Chermann, J.C., ed.) *[Spectra (2000) 4],* Editions Cité Nouvelle, Paris (1973) p. 21.

3. Brakke, M.K., *Arch. Biochem. Biophys. 55* (1955) 175-190.

4. Anderson, N.G., in *Physical Techniques in Biological Research,* Vol. 3 (Oster, G. & Pollister, A.W., eds.), Academic Press, New York (1956) pp. 299-352.

5. Meuwissen, J.A.T.P. & Heirwegh, K.P.M., *Biochem. Biophys. Res. Commun. 41* (1970) 675-681.

6. Steensgaard J., *Eur. J. Biochem. 16* (1970) 66-70.

7. Spragg, S.P., Morrod, R.D. & Rankin, C.T. Jr., *Sep. Sci. 4* (1969) 467-481.

8. Pollack, M.S. & Price, C.A., *Anal. Biochem. 42* (1971) 38-47.

For questions relating to this contribution, see p. 411.- Ed.

THE USE OF A REORIENTING ZONAL ROTOR FOR CELL MEMBRANE ISOLATION

Edward A. Smuckler, Matt. Riddle, Marlene Koplitz & John Glomset
Departments of Pathology and Medicine,
Washington University, Seattle, Washington 98195, U.S.A.

Zonal centrifugation utilising the 'A' type zonal rotor permits the simultaneous isolation of the several cellular constituents of rodent liver, although the technique is used frequently as the first step in isolation of plasma membranes [1]. We were interested in determining the subcellular location of a neutral cholesterol esterase that has been identified in sera [2,3,4], the supposition being that it would be located on the cell surface. One of us (E.A.S.) had the opportunity to use the techniques described by Evans [1], while visiting the National Institute for Medical Research, Mill Hill, U.K. Upon return to the U.S.A. we had access to a Sorvall reorienting zonal rotor and adapted the Evans procedure for this use.

Rat liver was homogenized in 1 mM $NaHCO_3$, filtered through nylon bolting cloth, and centrifuged at 1,000 g for 10 min. The supernatant was discarded, and the pellet suspended in the bicarbonate buffer. A Sorvall reorienting rotor was used in an RC2-B centrifuge. The rotor was carefully balanced, and brought to a speed above that showing precession, generally about 2,000 rev/min. We found that the time period of loading the rotor was very important, that the more time used for forming the gradient the better the final isolation. To load on 1.04 litres we utilized about 30-40 min, the rotor was accelerated to about 4,000 rev/min, and the sample was then layered on the preformed gradient. Centrifugation was carried out at 4,000 rev/min for 40 min. We found that the period of reorientation of the rotor was critical. The rotor was permitted to come to rest without braking; scrupulous attention was given prior to the run to assure a proper balance of the rotor so that there was minimal precession and shudder. The gradient was unloaded by pumping; we found that the rapidity of emptying decreased diffusion artifacts. The several fractions obtained were analyzed for enzymes as indicated in Fig. 1; in addition, the sucrose concentration was measured, and fractions showing activity peaks were fixed for ultrastructural analysis.

We were able to separate an enriched cell membrane fraction with little contamination by the endoplasmic reticulum (Fig. 1). Our enzymic assays indicate also that there was a separation of the mitochondria and the lysosomal fraction, although there was cross-contamination noted morphologically. These data also show that the acid cholesterol-ester hydrolase [2] is present in the plasma membrane fraction, and that the neutral esterase runs with a fraction rich in succinate dehydrogenase [4]. These experiments indicated that indeed the reorienting zonal rotor can be employed for the separation of the several cytoplasmic constituents; however, there are some specific safeguards and difficulties involved in its use. Unlike the 'A' zonal rotor, there is not the ease of visually following the separation, and indeed in

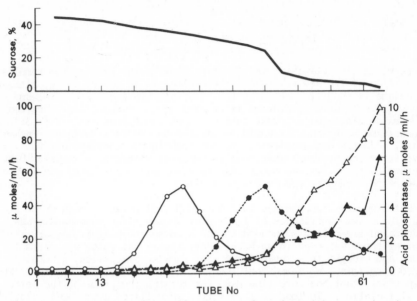

Fig. 1. Preparation of a cell membrane fraction with a reorienting zonal rotor. o——o, 5'-Nucleotidase; ∆---∆ glucose-6-phosphatase; ●···●, succinate dehydrogenase; ▲-·-·▲, acid phosphatase.

this type of rotor there is no ready means of assessment of the success of the run until the rotor has been emptied. We also found that the care taken in assuring the best balance was very important during periods of change of rotation speed. Minor imbalances were associated with precession and shuddering that resulted in less clean zone separation. This was especially true in deceleration. Nevertheless, with this extra care we were able to achieve satisfactory separations using the Sorvall products.

The work was supported in part by USPHS grants AM-1000 and CA 13600.

1. Evans, W.H., *Biochem J. 116* (1970) 833-842.

2. Riddle, M.C. and J.A. Glomset, *Fed. Proc. 32* (1973) 363abs.

3. Riddle, M.C., J.A. Glomset and E.A. Smuckler, *in preparation*.

4. Stein, O., Stein, Y., Goodman, D.S. & Fidge, N.H., *J. Cell Biol. 43* (1969) 410-431.

Editor's Note.- The authors of the preceding Note (submitted by invitation) were not present at the Symposium where Dr. Sheeler gave his paper on a re-orienting zonal rotor. Questions asked there are given on the next page.

A reorienting zonal rotor   (P. Sheeler *et al.*, Art. 5)

Questions put to Dr. Sheeler at the Symposium were answered as follows.
Must the SZ-14 rotor be filled to its capacity, 1,400 ml, for use ? (D.R. Headon).— In our experience the rotor should contain at least 1,000 ml. If a much smaller gradient volume is desired, then an overlay and/or cushion could be added to increase the total volume to 1,000 ml or more.
In isopycnic batch runs, can the sample be loaded statically ? (N.L. Webb).— Yes: the sample is loaded first, to the floor of the rotor bowl, and is then floated upward during introduction of the density gradient. However, gradient loading takes about 20 min, so the sample is exposed to the rotor temperature for this extra time. Also (*remark by* J. Steensgaard; Dr. Sheeler agreed) with static loading of the sample there is a risk of 'droplet sedimentation'; it is thus preferable to load the sample dynamically.
Does the SZ-14 rotor require a special control to provide slow acceleration and deceleration ? (G.B. Cline).— Yes: for the Sorvall RC2-B centrifuge as commonly used there is a 'rate controller' accessory for <1,000 rev/min.
Are face seals used in the SZ-14 in continuous-flow operation ? (K. Grinstead).— No.

Further points on gradients (arising from contribution by Meuwissen - p.407)

In answer to J.O. Molloy, who asked what type of discontinuity was regarded as being bad for resolution, J.A.T.P. Meuwissen said that the discontinuity which exists at the front edge of a block-loaded sample must be eliminated, but that the trailing edge is also important. *Answers to* G.B. Cline: albumin was indeed used in all the experiments, but tissue homogenates have recently been used, with good separation;   the albumin is not specially cleaned up to remove polymers and aggregates. *Comment* (G.B. Cline): with a syringe and about 2 ft of tubing one can easily get an inverted sample zone.

A SIMPLIFIED ASSAY FOR SUCCINATE DEHYDROGENASE

T.D. Prospero
Wolfson Bioanalytical Centre   *[Present address: Shell*
Univ. of Surrey, Guildford GU2 5XH, U.K.   Research, Sittingbourne]

As a convenient and seemingly satisfactory mitochondrial marker, succinate dehydrogenase activity (EC 1.3.99.1) may be assayed using the synthetic acceptor 2-(*p*-iodophenyl)-3-(*p*-nitrophenyl)-5-phenyltetrazolium chloride, 'INT' [1]. On reduction, INT yields a red water-insoluble formazan which, after stopping the reaction with trichloroacetic acid, may be extracted into ethyl acetate and read photometrically. The following modification eliminates the extraction step.

To 0.25 ml of 0.3 M phosphate buffer, pH 7.4, containing 2 mg/ml INT (a saturated solution at 37°) are added 0.5 ml of the tissue sample (e.g. a homogenate equivalent to 2 mg liver/ml) and 0.25 ml of pH 7.4 0.3 M Na succinate or (blanks) malonate. After mixing, and incubation at 37° for 10-20 min, the reaction is stopped by the addition of 6 ml of a mixture of 96% (v/v) ethanol : ethyl acetate : 10% (w/v) trichloroacetic acid [130 : 200 : 20 by vol.]. The tubes are stoppered or covered with a piece of polythene sheet (not 'Parafilm') and shaken. After centrifugation to remove denatured protein, the absorbance of the (single-phase) supernatant is measured at 490 nm. Activity may be calculated in terms of μmoles INT reduced by use of the molar extinction coefficient for reduced INT, viz. 20.1×10^3 [1]. The assay is linear with enzyme concentration to an E_{490} of at least 2. The incubation should not, however, exceed 20 min, as the assay ceases to be linear.

The above mixture is suitable for sucrose concentrations up to 1 M; for higher concentrations of sucrose, or other gradient materials such as Ficoll, it may be necessary to modify the procedure slightly to ensure a single phase. Either the proportion of ethanol, or the volume of mixture added, or both, may be increased. Care should be taken in increasing the proportion of ethanol, as high concentrations alter the extinction coefficient. In general the trichloroacetic acid has to be added to the ethanol : ethyl acetate on the day of use, as the ethyl acetate would otherwise hydrolyze on storage. A further point arises if an automatic flow-cell unit is used in the photometry: the tubing should be resistant to ethyl acetate, suitably 'Solvoflex' (Technicon) or, better, PTFE.

1. Pennington, R.J., *Biochem. J. 80* (1961) 649-654.

Triton interference in phosphate assay (P.J. Jacques - pers. comm. to E. Reid) In phosphatase assays with Triton X-100 present as a releasing agent, phosphate measurement may be impaired by turbidity. The following measures may obviate this.— After TCA addition, wait 1 h before removing the protein; the Triton concentration should be such that $\not> 0.2$ mg is present in the aliquot taken from the TCA supernatant; with tissue fractions low in protein, 10 mg of bovine serum albumin should be added to the assay mixture before TCA addition.

INFORMATIONAL MACROMOLECULES AND NUCLEOPROTEIN PARTICLES
(Articles 10-14): *SUPPLEMENTARY MATERIAL*

Question by R. Williamson to R.H. Hinton (cf. Art. 10): do you find any separation of messenger RNP from ribosomal RNP, as found in CsCl gradients with formaldehyde-fixed material ? — No; in Metrizamide they have both banded at the same density in studies with pulse-labelled mRNP-enriched fractions.

Refs. noted by Editor:
Levin, D. & Hutchison, F., *J. Mol. Biol. 73* (1973) 455-478. — Handling very large DNA molecules on sucrose gradients.
Wilt, F.H., Anderim, M. & Ekenburg, E., *Biochemistry 12* (1973) 959-966. — Nuclear RNP particles fractionated in Cs_2SO_4.
Ormerod, M.G. & Lehmann, A.R., *Biochim. Biophys. Acta 247* (1971) 369-372. — Artefacts in sedimentation of high mol. wt. DNA on sucrose gradients.

ISOLATION OF FREE AND MEMBRANE-BOUND POLYSOMES

T. Hallinan and D. Lowe

Dept. of Biochemistry, Royal Free Hospital School of Medicine London WC1 1BP, U.K.

Free polysomes (FP) and bound polysomes, i.e. rough endoplasmic reticulum (rough e.r.), are separable on discontinuous sucrose gradients, constructed so that FP traverse the densest step to form a pellet, whereas the less dense rough e.r. remain layered over the densest step [e.g. 1,2]. In an alternative method of separation, the rough e.r. being prepared as very large fragments by Dounce homogenization, differences between FP and rough e.r. in sedimentation rate through 0.25 M sucrose are utilized [3]. We favour the former (isopycnic) technique. There is some disagreement on what is the lowest concentration of sucrose that can be used in the separatory step whilst avoiding the risk that the densest rough e.r. may penetrate it and contaminate the FP; 1.6 M sucrose has been used, but at least 1.8 M seems necessary for a clean centrifugal separation, and an even denser sucrose step may be required for flotation separation.

In addition to the well-recognized possibility of contamination of FP by ferritin and glycogen [4], substantial contamination by agranular membranes (possibly co-sedimenting with glycogen) can also be met [5]. Finally it should be noted that it is difficult to reproducibly isolate FP and rough e.r. in good yield from livers of non-fasted animals [6]. Methodological improvement is still required here.

1. Webb, T.E., Blobel, G. & Potter, V.R., *Cancer Res.* 24 (1964) 1229-1237.

2. Bloemendal, H., Bont, W.S., DeVries, M. & Benedetti, E.L., *Biochem. J.* *103* (1967) 177-182.

3. Loeb, J.N., Howell, R. & Tomkins, G.M., *J. Biol. Chem.* 242 (1967) 2069-2074.

4. Reid, E., in *Enzyme Cytology* (Roodyn, D.B., ed.) Academic Press (1967), pp. 321-406.

5. Murty, C.N. & Hallinan, T., *Biochem. J.* *112* (1969) 269-274.

6. Lowe, D., Reid, E. & Hallinan, T., *FEBS Lett.* 6 (1970) 114-116.

ELEMENTS FROM LIVER (Articles 15-24): *SUPPLEMENTARY MATERIAL*

Lysosomes: isolation and membrane studies (Arts. 16-20)

Refs. contributed by the Editor.-
Aas, M. *(Inst. of Clinical Biochemistry, Rikshospitalet, Oslo), Abs. 9th Int. Congr. Biochem.* (Stockholm, 1973), p. 31. - Use of Metrizamide (cf. Arts. 9, 10 & 12) to separate mitochondria and lysosomes.
Davies, M., *Biochem. J. 136* (1973) 57-65 [cf. Note by Jacques, p. 415]. - Lysosomal fate of injected albumin and Trypan Blue as affected by Triton WR-1339.

Answering questions on disrupted lysosome preparations (Art. 19), M. Dobrota agreed that electron-microscopic characterization was desirable: preliminary morphological results as mentioned in the article need to be supplemented by cytochemical staining for acid phosphatase. Concerning acid phosphatase distribution (B. Arborgh), he summarized as follows:

> *Proportion of homogenate activity in lysosome-rich region from HS run:* ~16%
>   "   "   "   "   " *'Region 3' from B-XIV run:* ~4.0%;
> *or, in terms of total activity taken from the HS run and loaded into the*
>   *B-XIV, proportion recovered in 'Region 3' is* ~23%;
>   "   "   " *'Region 4' is* ~26%;
>   "   "   *'Soluble' (sample) Region is* ~45%.

The suggestion (D. Thinès-Sempoux) that 'Region 3' might contain matrix is a distinct possibility, as the density (1.175 g/ml) in this region is rather high for pure membranes and could reflect protein attachment to the fragments. Results bearing on a Symposium query (O. Touster), concerning further biochemical characterization of the membrane fraction, are included in the Article, but the centrifugation results (Figs. 3 & 4) are not at this stage easy to interpret; the only clue to enzyme specificity is that in this region the distribution patterns (with or without Triton X-100) are identical for β-glycerophosphatase and p-nitrophenylphosphatase.

Remark by O. Touster (cf. Art. 23): we have reported the presence of three acid phosphatases in rat-liver lysosomes, separable from each other on DEAE-cellulose columns; one is a specific nucleotidase.

Editor's Note.- In a forthcoming review article [Hinton, R.H. & Reid, E., in *Mammalian Cell Membranes,* Vol. 1 (Jamieson, G.A. & Robinson, D.M., eds.), Butterworth/Van Nostrand, in press], the general question of membrane-associated enzymes is considered, and a paper relevant to acid phosphatase in lysosomal membranes is cited [Sloat, B.F. & Allen, J.M., *Ann. N.Y. Acad. Sci. 166* (1969) 574-600.

Suggestion arising from R. Henning's Symposium presentation (Art. 20): might trihexosyl ceramide be useful as a lysosomal membrane marker ? - With further work, this hope has been abandoned (*see* concluding pages of Arts. 20 & 21). O. Touster: lysosomal membrane preparations could be contaminated by membranous and other intracellular material from secondary lysosomes.

ISOLATION OF LIVER LYSOSOMES BY AID OF IRON OR SILVER LOADING

Bengt Arborgh, Jan L.E. Ericsson & Hans Glaumann
Pathology Dept., Sabbatsberg Hospital,
Karolinska Institutet, Stockholm, Sweden

Rats may be injected with iron (in a sorbitol-citric acid complex) or with silver (colloidal silver iodide), which accumulate in liver lysosomes and thereby increase their equilibrium density in sucrose. Thus, as reported for iron [1], a purified fraction of secondary lysosomes may be isolated. With iron-loaded lysosomes and, for comparison, silver-loaded lysosomes, studies have been made of chemical composition and enzymology. An attempt has been made to determine the half-life of secondary lysosomes by studying the turnover of cholesterol.

1. *Authors as above, FEBS Lett. 32* (1973) 190-194.

PROBLEMS IN APPLICATION OF STANDARD ANALYTICAL CENTRIFUGATION METHODS TO THE LOCALIZATION OF EXOGENOUS COMPOUNDS IN RAT LIVER ORGANELLES*

Pierre J. Jacques
Lab. de Chimie Physiologique, Univ. Catholique de Louvain, Belgium

In tissues for which methods of homogenization and of analytical centrifugation have been satisfactorily developed, it has become easy to accurately localize compounds belonging to the cell's own machinery, at least where they show a single location. One only has to compare the distribution pattern of the substance of unknown location with that of marker constituents characterizing the various organelles, and to carefully avoid the possibility of technical or graphical artifacts. In contrast, serious difficulties were generally encountered when the same methodology was applied, in the same tissue homogenates, for the localization of compounds of exogenous origin.

These difficulties result from two types of event. One consists of a virtually systematic change in the distribution pattern of the exogenous compound as time elapses between its administration to the animal and the tissue homogenization. The other is less frequent and consists of a more or less specific adsorption of the exogenous compound, from the cytosol on to the organelles. Reviews on application of centrifugal methodology to the identification or purification of lysosomes [1,2] completely ignore these problems. The present attempt to remedy this is prompted by the importance of endocytotic uptake of (exogenous) lysosomotropic drugs for cellular therapy [3-7].

STATEMENT OF THE PROBLEMS

In all the experiments reported in this section, a single agent was injected into the animal, as a single intravenous dose at time zero. After a suitable delay the liver was perfused *in situ*, by means of an ice-cold isotonic solution, in order to avoid contamination of the homogenate by the amount of agent contained in residual blood plasma. The homogenate was then fractionated by differential pelleting in 0.25 M sucrose or mannitol and, in some cases, one

* *Space limitations entailed editorial compression at end of article*

of the fractions so obtained was subfractionated by isopycnic equilibration in a suitable medium. Finally, the concentrations of the agent, of total proteins and of reference enzymes for various organelles were determined in each fraction, to allow plotting of the distribution patterns. Evidently the assay of the agent has to be strictly quantitative in each fraction, as is also requisite for the measurement of marker constituents [8].

In vitro fixation

Since such experiments aim at localizing the drug after its *in vivo* uptake from the body fluids, it is necessary to check systematically the distribution of the agent after it has been added *in vitro* to the liver homogenate from untreated animals. Possible interpretive artifacts can thereby be detected in the case of dyes [9-11] or other compounds which can adsorb directly on to organelles or even penetrate into them *in vitro*. In all the examples discussed below, 85-98% of the agent added *in vitro* was recovered in the particle-free supernatant. The only exception was iodoinsulin; but even in this doubly unfavourable case, where *in vitro* fixation was both important and specific for one particulate fraction (microsomes), the amount fixed *in vitro* remained small enough to be neglected in the interpretation of the distribution profiles of iodoinsulin after its *in vivo* administration.

Changing distribution patterns

Firstly the liver homogenates were submitted to differential pelleting. Here the 5-fraction technique designed by de Duve *et al.* [12] proved definitely superior, although it does not allow lysosomes and peroxisomes to be distinguished. When this distinction was desired, the two fractions (M and L) containing 65-70% of the large cytoplasmic organelles (LCO) and of the captured drug were combined and, in a second step, subfractionated in systems suitable for isopycnic equilibration, e.g. a gradient of sucrose in D_2O [13, 14] or of glycogen in 0.5 M sucrose/H_2O [14, 15].

Differential pelleting

The distribution profile of diverse exogenous compounds varied, after differential pelleting, with the time interval separating the injection of a single dose from the processing of the liver. At some point the pattern matched that of one of four cytoplasmic entities which radically differ from one another on the structural and functional levels (Table 1).

Only in the case of biodegradable compounds supplied in a continuous manner at the same rate as they are destroyed within the cell can one hope that, after a reasonable time for equilibration, a stable distribution pattern will appear, as seems to occur with serum neutral protease [27] and with injected homologous serum albumin and transferrin [28]. This might also occur when cells are exposed to agents for days in culture [29]. In all cases reported in Table 1, the distribution patterns for marker enzymes were unaffected by the agent used, whatever the time after administration to the animal.

The information provided by Table 1 is misleading in two respects.

Table 1. Similarity in distribution patterns between exogenous compounds and marker enzymes, after differential pelleting of rat liver homogenates. The tabulated values indicate the time (h) or range of time after injection at which similar distribution patterns were observed for the agent and the indicated marker enzyme(s). *Differential pelleting was performed in 0.25 M sucrose (0.25 M mannitol in the case of invertase), according to [12], after perfusion of the liver.*

Exogenous compound *(single injection at time zero)*	*M a r k e r e n z y m e*				
	Glucose-6-phosphatase ('microsomes')	β-Glycero-phosphatase (lysosomes)	Urate oxidase (peroxisomes)	Cytochrome oxidase (mitochondria)	*Ref.*
Horseradish peroxidase	-	0.05 to 0.25	0.05 to 0.25	3 to 6	16-19
Triton WR-1339	0.16	17 to 144	17 to 144	-	20,21
Yeast invertase	-	0.25 to 1	0.25 to 1	13 to 480	19,22
Suramin	-	6	6	-	23,24
Iodoinsulin	0.08	1	1	-	25,26

Firstly, it should be recalled that the distribution patterns "à la de Duve" [12], on which the comparisons of Table 1 are based, are only normalized patterns; they do not take into account the absolute tissue levels of the agent and the marker enzymes. Secondly, the change in distribution for the agent is continuous with time; 'transition patterns' were thus observed for times intermediate between those tabulated. In order to describe the phenomenon in a general manner, one can say that whatever the initial pattern (microsomal type or lysosomal/peroxisomal type) shown by the agent, the latter invariably and progressively shifts from lighter fractions (P or L) towards heavier ones (L and/or M); in other words, the average sedimentation coefficient of the agent-containing organelles increases with time.

Isopycnic equilibration

Several exogenous substances exhibit, in density gradients also, a distribution pattern which changes with time after injection. This does not necessarily cause a problem. For instance, in a gradient of sucrose in H_2O, *Triton WR-1339* equilibrates at fairly high densities a short time after injection and, four days later, its median equilibrium density has decreased by more than 0.10 g.ml^{-1}. But the distribution of lysosomes is specifically altered and faithfully follows, in the course of time, that of the lysosomotropic detergent. Thus, far from raising a problem, the distribution shift of Triton WR11339 in fact solved an important one; indeed, by this very means, liver lysosomes were isolated for the first time, in a state of high purity [20, 21]. Quite different is the case of exogenous compounds which do not alter the behaviour of known organelles. The most thoroughly studied example is that of *horseradish peroxidase* [16-19].

Table 2. Shift in median equilibration density of horseradish peroxidase in two gradient systems. Data from Jacques [19].

Time after peroxidase injection, *h*	GRADIENT	
	Sucrose/H_2O [ref. 13]	Glycogen in 0.25 M sucrose /H_2O [ref. 15]
0.06	-	1.096
0.25	1.207	-
0.5	-	1.114
2	-	>1.115
3	1.225	-
3.5	-	1.114
6	1.240	-
15	1.235	-

Table 2 shows the mean equilibrium densities of the peroxidase in two different systems of density gradients, in which mitochondria, peroxisomes and lysosomes equilibrate more or less at the same levels [13,15,19]. With glycogen in isotonic sucrose, the median density of peroxidase increased slightly with time; but this increase was small and ceased, at the latest, after 0.5 h. With the hypertonic sucrose gradient, the median density was found to resemble that of the large organelles (LCO), 0.25 h after injection; then it increased appreciably until the sixth hour, when the distribution was very different from that of LCO.

Interpretation of distribution results

Whatever the difficulties that may be encountered in interpreting the results of analytical centrifugations when applied to the localization of exogenous substances which do not alter the properties of known organelles in these systems, we do not imply that these techniques are inadequate in all cases. Such an impression would be absolutely unjustified. Thus, the simple gradient of hypertonic sucrose in H_2O, whilst hardly allowing the three main types of LCO to be distinguished from one another, is perfectly sufficient for the accurate localization of exogenous compounds such as Triton WR-1339 [20, 21], dextran [30, 31], Macrocyclon [24] and metal colloids [32] which specifically change the properties of the organelles within which they become stored (e.g. secondary lysosomes and, presumably also, heterophagosomes and post-lysosomes). Moreover, that gradient system enables a substantial percentage of the vacuoles containing these substances to be obtained in high purity.

Yet with most other compounds the interpretation of results can be very laborious and require extensive additional experimentation. Whilst horseradish peroxidase is now considered as an example [16-19], it must be stressed that each agent leads to species problems. The superimposability of peroxidase and cytochrome oxidase at 3-6 h after injection (Table 1) does not connote identity with mitochondria even at that time, in view of the equilibration results of Table 2. Similarly, the results of Table 1 do not imply a lysosomal or peroxisomal locus for peroxidase at 0.25 h; indeed, other isopycnic equilibration studies [19] showed matching between peroxidase and lysosomal but not peroxisomal markers with glycogen in 0.5 M sucrose/H_2O, and conversely with sucrose in D_2O, the analytical gradient systems giving sharp distinctions of the LCO.

Table 3. *In vitro* disruption of cytomembranes by digitonin *(as mg/g tissue)* in rat liver homogenates#

Cytomembrane & marker enzyme(s)	Release, %*		
	10	50	90
Lysosomes: β-glycerophosphatase, β-*N*-acetylaminodeoxyglucosidase, deoxyribonuclease, β-galactosidase *(25†)*	0.8	2.4	4.7
P-a$_2$ (a microsomal subfraction [35]; plasma membrane elements): 5'-nucleotidase *(7)*	1	3.5	6
Peroxisomes: catalase *(3)*	11.5	19.5	25
Mitochondria (external): polyadenylase, sulphite cytochrome *c* reductase *(12)*	17.8	33.3	63.5
Endoplasmic ret.: inosine diphosphatase *(8)*	32	51	62
Mitochondria (intL): glut. dehydrogenase *(3)*	85	116	150

*ratio of free to total activity *of latent marker enzymes*
#*Mrs. G. Huybrechts-Godin adapted the technique to homogenates*
†*No. of experiments*

It is, then, emphasized that similar distribution patterns for exogenous compounds and constitutive markers for organelles do not always - in fact rarely do - suggest the precise intracellular location of the agent. Yet even in a complex case like peroxidase in liver, biochemical centrifugations are by no means useless. In fact, a detailed study [19] involving supplementary experiments (kinetics of peroxidase uptake and intracellular decay after various doses, distribution study after repeated injections), elimination of graphical artifacts *(see above)*, and also stepwise analysis of peroxidase (marker concept [8]) and then of peroxidase-granule properties *in vitro* in relation to *in vivo* events, provided much information on the nature and properties of the granules without recourse to the morphological counterpart of the whole study.

SEARCHING FOR BETTER ADAPTED METHODS

A first approach consisted of applying the above standard techniques after treating the animal with an additional compound *(foregoing Section)* capable of specifically modifying the behaviour of organelles involved in the intracellular transport, storage and disposal of exogenous material. This approach as first applied to peroxidase in liver [34] is now widely exploited, but has its own limitations and, worse, can lead to artifacts. One such compound is the lysosomotropic non-lysosomolytic detergent Triton WR-1339, which has its own *limitations:* its presence in both lysosomes and peroxidase-containing granules establishes merely a functional relationship (e.g. ability to sequestrate Triton), not necessarily identity, and only 15-25% of the agent at the best can be localized in Triton-containing vacuoles: where is the bulk of the intracellular agent? *Artifacts* can result from the pharmacological action of Triton, relatively inert though it is, e.g. it increases lipaemia, depresses hepatic and renal endocytotic uptake of peroxidase, increases autophagy in hepatocytes, is cytotoxic to cultured cells, and profoundly alters properties of secondary lysosomes. Should these artifacts qualitatively as distinct from quantitatively alter the intracellular pathway of some agents, danger arises.

Thus, approaches avoiding *in vivo* treatment by cytopharmacological agents seem definitely safer, e.g. (Table 3; cf. [16-19]) differential release of markers from individual LCO *in vitro,* suitably by digitonin, in relation to the concentration of added agent. Besides obviating *in vivo* treatment and tissue manipulation other than homogenization, this procedure may overcome the two limitations described above for the Triton technique. Particularly desirable are new approaches that help purify a substantial proportion of the agent-containing organelles, on the basis of more specific properties than size or equilibrium density in the various media thus far exploited.

References [editorially abbreviated; 'UCL' = Université Catholique de Louvain]

1. Beaufay, in *Lysosomes, 2* (Dingle & Fell, eds.) N. Holland (1969) p.515.
2. Beaufay, in *Lysosomes: a Laboratory Handbook* (Dingle, ed.) (1972) p. 1.
3. Jacques, *as for 1.,* p. 395.
4. Jacques, = ref.18 in Art. 16.
5. Jacques, = ref.19 in Art. 16.
6. de Duve & Trouet, in *Non-specific Factors Influencing Host Resistance* (Braun & Ungar, eds.) Karger (1973) 153.
7. Jacques & Demoulin-Brahy, in *Macrophage Activation* (Wagner, ed.), Excerpta Medica, *in press.*
8. Jacques, *this Vol.,* Art. 16.
9. Koenig, *as for 1.,* p. 111.
10. Dingle & Barrett, *Proc. Roy. Soc., Ser. B, 173* (1969) 85.
11. Allison & Young, *as for 1.,* p. 600.
12. de Duve *et al.,* = ref. 2 in Art. 16.
13. Beaufay *et al.,* = ref. 6 in Art. 16.
14. de Duve *et al.,B.B.A. 40* (1960) 186.
15. Beaufay *et al.,B.J. 92* (1964) 184.
16. Jacques, *Arch. Int. Physiol. Biochim. 68* (1960) 683.
17. Jacques, *Memoir,* UCL (1960).
18. Jacques, *as for* 16., *71* (1963) 306.
19. Jacques, = ref. 12 in Art. 16.
20. Wattiaux *et al.,* in *Lysosomes* (de Reuck & Cameron, eds.) Churchill (1963) p. 176.

21. Wattiaux, *Etude Exp. de la Surcharge des Lysosomes,* Duculot, Gembloux (1966).
22. Jacques & Bruns, *2nd FEBS Mtg. Abstr.* (1965) 26.
23. Smeesters & Jacques, *Excerpta Med., Int. Congr. Ser. 166* (1968) 82.
24. Smeesters, *Memoir,* UCL (1968).
25. Jacques & Wattiaux-de Coninck, *6th FEBS Mtg. Abstr.* (1969) 279.
26. Alsteens, *Memoir,* UCL (1971).
27. Bartholeyns, *Doct. Thesis* UCL (1973).
28. Gordon & Jacques, in *Labelled Proteins in Tracer Studies* (Donato *et al.,* eds.) Euratom 2950 Bruxelles (1966) 127.
29. Tulkens, *this Vol.,* Art. 31.
30. Thinès-Sempoux *et al., 3rd FEBS Mtg. Abstr.* (1966) 213.
31. *Idem, Doct. Thesis,* UCL (1968).
32. Arborgh *et al., this Vol.,* 415.
33. de Duve, in *Enzyme Cytology* (Roodyn, ed.) Acad. Press (1967) 1.
34. Jacques & Wattiaux, *6th Int.Congr. Biochem., Abstr.* (1964) 654.
35. Amar-Costesec *et al.,* in *Microsomes and Drug Oxidations* (Gilette *et al.,* eds.) Acad. Press (1969) 41.

Answering O. Touster, P.J. Jacques *(cf. above)* said there was no experimental evidence bearing on the possibility that where a protein is readily hydrolyzed by lysosomal proteases without prior denaturation, the synthesis of other proteins might be accelerated, perhaps by promotion of acidic conditions within the lysosomes. One might assume acceleration in Kwashiorkor, where amino acids, especially the supply of essential amino acids, are probably the rate-limiting step in protein synthesis. Concerning possible promotion of acidic conditions, it may be recalled that protein digestion in the stomach increases the pH of gastric contents. The question is amenable to experiment.

Various hepatocellular elements *[see later for membranes]*

NON-ENZYMIC LIPID PEROXIDE FORMATION IN TISSUES, AND ITS POSSIBLE ROLE IN INFLAMMATION[*]

Coimbatore R. Krishna Murti
*Central Drug Research Institute,
Lucknow-226001, India*

It is not yet fully clear how lipid peroxides are formed and how they bring about tissue injury, although under well-defined conditions tissue homogenates and subcellular organelles can be shown to produce lipid peroxides in large amount [1-5]. Work in our laboratories since 1960 has revealed that the substrates which are transformed into lipid peroxides are located in the mitochondrial, lysosomal and microsomal fractions, and the pro-oxidant catalytic factor in the organelle-free cytosol fraction [S.K. Sharma & S.C. Sharma, unpublished observations]. Some salient points leading to a possible correlation between *in vitro* lipid peroxidation in liver and the extent of carrageenin-induced experimental inflammation in mice [6] are now presented.

Methods

Since sucrose interferes with the malondialdehyde assay, 150 mM KCl was used as the medium throughout. Stock albino rats (100-150 g) or albino mice (20-25 g) were fasted overnight with water *ad lib*. They were decapitated, and tissues were excised and well washed with chilled medium. Homogenates (10% w/v) prepared with a Potter-Elvehjem homogenizer (1-2 min at top speed) were centrifuged (always at 4°): 15 min at 3,000 g (MSE angle rotor), then (supernatants; Janetzki K-24 centrifuge, angle rotor 904) 20 min at 9,000 g, the resulting pellet being washed by re-dispersing in 10 vol and re-centrifuging. The mitochondria-free supernatant was centrifuged for 60 min at 105,000 g (Spinco Model L, rotor 40), and the pellet similarly washed. The washed pellets were suspended in the medium (5 ml for 4-5 g tissue equivalent).

To assay lipid peroxide formation, aliquots of freshly prepared homogenates or fractions with or without pro-oxidant factor and other additions as desired were shaken at 37° in 25 ml Erlenmeyer flasks (100 horizontal strokes/min, amplitude 10 cm). At intervals 1 ml portions were transferred to 1.5 ml 10% (w/v) trichloracetic acid, well mixed, and centrifuged to give a clear supernatant, 2 ml of which was mixed with 2 ml 0.67% thiobarbituric acid (BDH, recrystallized). The mixture, adjusted to pH 7.4 with N NaOH, was heated for 10 min at 100°, cooled, made up to 5 ml, and read at 535 nm [7].

Results and Discussion

Pre-incubation almost abolished the ability to oxidize succinate aerobically with brain homogenate or mitochondrial fraction (Fig. 1). With the former but not the latter there was considerable production of lipid peroxides during the 3 h, which could be inhibited almost completely by the addition of rutin (rhamnoside of hesperidin). With a rat-liver lysosome-rich fraction, pre-incubation gave extensive peroxide formation and released acid phosphatase into the medium.

[*] *Communication No. 1867 from the Central Drug Research Institute, Lucknow*

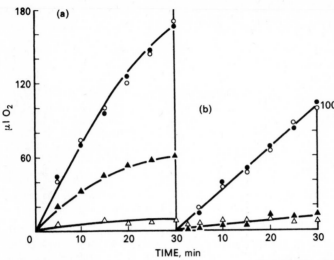

Fig. 1. Succinate oxidation by homogenate *(left)* and mitochondria *(right)* from rat brain as affected by pre-incubation (37°, 3 h). Data from ref. [17].

o——o, not pre-incubated

●——●, *ditto*, + rutin

△——△, pre-incubated

▲——▲, *ditto*, + rutin

Rutin promptly prevented the peroxidation but not the release [8].

Table 1. Effect of particle-free supernatant ('sup'; pro-oxidant fraction) on lipid peroxidation by rat mitochondrial fraction.

Substrate source (mit. fr'n)	Additions (R = rutin, 20 µg/ml)	Peroxidation: malonyldialdehyde, µmoles /mg protein 0 h	3 h
Brain	Brain sup	22	285
"	" " + R	5	16
"	*None*	20	60
None	Brain sup	0	0
Liver	Liver sup	20	240
"	" " + R	5	16
"	*None*	15	55
None	Liver sup	0	0
Kidney	Kidney sup	12	52
"	" " + R	3	6
"	*None*	10	22
None	Kidney sup	0	0

Evidently (a) lipid peroxidation alone cannot account for the succinoxidase loss in the mitochondrial fraction or the acid phosphatase leakage from the lysosomal fraction by membrane damage, and (b) the lipid peroxidation is maximal in whole homogenates and minimal in particulate subcellular fractions.

More extensive experiments with different subcellular fractions and their combinations [9] established that if any fraction was incubated alone the peroxidation was marginal, but when any particulate fraction was incubated with the cytosol fraction there ensued rapid and profuse peroxidation. Table 1 shows some of the findings. Rutin was consistently a potent anti-oxidant in such reconstituted systems as well as in homogenates.

Results so far obtained on the nature of the pro-oxidant factor indicate that it is thermostable, is of relatively low molecular weight, diffuses out of cellophan bags on prolonged dialysis, is not related to haematin, contains no Fe^{3+}

or ascorbate, and is insensitive to trypsin. The dialyzed fraction resolves into multiple ninhydrin spots by TLC on silica plates. Before resolution it absorbs maximally at 265-270 nm and gives a blue fluorescence. Pro-oxidant activity is lost by lyophilization, or storage for even 24 h below 10°. Acid hydrolysis and paper chromatography revealed glycine, glutamic acid, methionine, serine, and a few uncharacterized spots.

We were prompted to ascertain the correlation, if any, between lipid peroxide formation and inflammation from the following known facts: (a) many anti-inflammatory drugs are reported to be potent cellular anti-oxidants, and (b) erythaemia induced by irradiation elicits an inflammatory response [10]. Evidence has been adduced [e.g. 6] that in carrageenin-induced inflammation in albino mice, maximal oedema in the legs occurred within 3 h, and liver homogenates from these mice showed maximal peroxidation within this period. It appears that carrageenin triggered some as yet uncharacterized metabolic reaction in the liver, with consequent stimulation of *in vitro* peroxidation. Since the increased output of lipid peroxides implies lysosome damage [11], it is likely that carrageenin elicits a local reaction in the paw which is signalled to the liver, where in turn lysosomal activation occurs possibly as a defensive measure.

We conclude that besides the extensively studied microsomal enzymic formation of lipid peroxides, NADP-mediated, lipid peroxidation *in vitro* can also occur by a non-enzymic mechanism requiring the interaction of lipid substrates contributed by the particulate fractions and a thermostable low-molecular weight pro-oxidant factor located in the cytosol fraction. Rutin is a powerful inhibitor of lipid peroxide formation. The loss of functional activity of subcellular organelles on pre-incubation or *in vitro* 'ageing' appears to be unrelated to production of lipid peroxides. A correlation between the inflammatory response to carrageenin and *in vitro* lipid peroxide formation in mouse liver has emerged, the significance of which awaits exploration.

Acknowledgements

I thank my junior colleagues and graduate students Srikant Sharma, Satish Sharma and Hasan Mukthar as active participants in the above work, Dr. M.L. Dhar (Director of the Institute) for his interest, and the Council of Scientific and Industrial Research for sponsoring a Symposium visit to the U.K.

1. Carpenter, M.P., Kitabchi, A.E., McCay, P.B. & Caputto, R., *J. Biol. Chem.* *234* (1959) 2814-2818.

2. Tappel, A.L. & Zalkin, H., *Nature (Lond.) 185* (1960) 35.

3. Hunter, F.E., Gebicki, J.M., Hoffsten, P.E., Weinstein, J. & Scott, A., *J. Biol. Chem. 238* (1963) 828-835.

4. Ghosal, A.K. & Recknagel, R.O., *Life Sci. 4* (1965) 1521-1530.

5. Sharma, S.K. & Krishna Murti, C.R., *Ind. J. Exp. Biol. 1* (1963) 5-11.

6. Sharma, S.C., Mukhtar, Hasan, Sharma, S.K. & Krishna Murti, C.R., *Biochem. Pharmacol. 21* (1972) 1210-1214.

7. Utley, H.G., Bernheim, F. & Hochstein, P., *Arch. Biochem. Biophys. 118* (1967) 29-32.

        [2029.

8. Sharma, S.K. & Krishna Murti, C.R., *Biochem. Pharmacol. 15* (1966) 2025-

9. Sharma, S.K. & Krishna Murti, C.R., *J. Neurochem. 15* (1968) 147-149.

10. Whitehouse, M.W., in *Progress in Drug Research, 8* (Jucker, E., ed.) (1965) 321-429.

11. Wills, E.D. & Wilkinson, A.E., *Biochem. J. 99* (1966) 657-666.

Further refs. on hepatocellular elements contributed by the Editor.-

Blyth, C.A., *et al., Biochim. Biophys. Acta 32* (1973) 57-62. - Steroid binding with plasma-membrane preparations.
Prospero, T.D., *et al., Biochem. J. 132* (1973) 449-458. - Plasma-membrane alkaline ribonuclease and phosphodiesterase (perhaps identical).
Shakoori, A.R., Romen, W., Oelschläger, W., Schlatterer, B. & Siebert, G., *Hoppe-Seyl. Zeit. Physiol. Chemie 353* (1972) 1735-1748. - Isolation of nucleoli, with a zonal-rotor step in organic solvents (cf. *this series*, Vol. 3, pp. 157-158).

Cytomembranes and relevant lysosomology; marker constituents (Arts. 21-24)

Remark by O. Touster on glycosyl transferase studies by D.J. Morré *et al.* (Art. 21): whilst the cell fractions are well prepared, one has to look out for complications in the enzyme assays, which can be influenced by detergents and other variables. — Dr. Morré agrees, and comments thus: As for any enzyme, assay conditions should be determined for each new tissue and/or cell fraction. We have examined in some detail the lactose synthetase from mammary gland and the UDP-galactose : *N*-acetylglucosamine galactosyltransferase and UDP-galactose : *N*-acetylgalactosaminyl-(*N*-acetylneuraminyl)galactosylglucosyl-ceramide galactosyltransferase from rat liver. The others have been only spot-checked, relying on the more detailed studies of H. Schachter (Univ. of Toronto) and S. Basu (Univ. of Notre Dame). With the glycosphingolipid glycosyltransferase purified 50-100 fold, the activity can be enhanced 5-6 times using phospholipids as solubilizing agents, as compared with synthetic detergents [F.A. Wilkinson, M.S. Thesis]; but specificity and relative activity seem not greatly affected. The activities have relatively broad pH optima and a requirement for metal, usually Mn^{2+}. Some detergents inhibit.

Re Art. 22.- Of the glycerol-labelled microsomal phospholipids, a small proportion (perhaps 5% or even 10%) is attributable to the VLDL precursors entrained in the intracisternal space. — H. Glaumann, *in answer to* T. Hallinan, *who then commented:* this agrees fairly well with the proportion of choline-labelled phospholipids in intracisternal secretion products *(cf. a Note which follows).*

NOTES: membrane markers

PHOSPHOLIPID BASES AS MEMBRANE MARKERS

T. Hallinan
Department of Biochemistry,
Royal Free Hospital School of Medicine, London WC1N 1BP, U.K.

Several radioactive phospholipid bases, injected into intact animals *in vivo,* are incorporated efficiently, and with a high degree of specificity into membrane phosphatides [1-3]. Cultured cells can similarly be labelled [4,5]. With rat liver microsomes, 30-40 min after injection of [^{14}C-] or [^{3}H-]choline, up to 95% of the radioactivity in a crude acid-washed precipitate is in *membrane* phosphatides (not more than 2% in intracisternal secretion products) and 98% of the phospholipid labelled is lecithin (PC), which is a macro-components of membranes of the endoplasmic reticulum (e.r.), making up about 10% of their dry weight. Smooth and rough e.r. membranes are virtually equilabelled; hence membrane distribution among microsome subfractions separated by a wide variety of techniques (including detergent treatments which would inhibit or labilize marker enzymes), can be accurately and very sensitively monitored, simply by counting crude acid-washed fractions [2].

With longer labelling periods, uniform labelling of PC in all organelles appears to occur, and if allowance is made for small differences in their PC content, choline can be used to monitor the distribution of all of the cellular membranes. Phospholipase C quantitatively solubilizes radioactive choline-phosphate off membranes at a rate identical with the rate of hydrolysis of lecithin, chemically determined. Hence the labelling technique can be very conveniently used to quantitate the extent of de(phospho)-lipidation of membranes, when studying the role of phospholipids in membrane structure [2] or function [3] (phospholipid dependence of enzymes). Experience to date suggests that ethanolamine and inositol can also be used exactly like choline to 'mark' membranes.

1. Nagley, P. & Hallinan, T., *Biochim. Biophys. Acta 163* (1968) 218-225.

2. Hallinan, T., Nagley, P., Murty, C.N., Bennett, J. & Grant, J.H., *Biochim. Biophys. Acta 173* (1969) 554-563.

3. Cater, B.R., Poulter, Jane & Hallinan, T., *FEBS Lett. 10* (1970) 346-348.

4. Plagemann, P.G.W., *Arch. Biochem. Biophys. 128* (1968) 70-87.

5. Pasternak, C.A. & Bergeron, J.J.M., *Biochem. J. 119* (1970) 473-480.

Comment by G. Siebert.- It would be interesting to know how the nuclear envelope, or at least its outer layer, fits into the concept of choline incorporation being a marker for e.r. Concerning the fate of injected choline (T. Hallinan, *answering* D.J. Morré), at 30-40 min only about 1% of the counts are in sphingomyelin, although it constitutes say 3-4% of the microsomal phospholipids; slow labelling of sphingomyelin has been shown previously with both phosphate and choline. At 48 h it may contain about 10% of the counts.

H. Glaumann *summarizes as follows the evidence for the presence of cholesterol in endoplasmic reticulum membranes* (cf. Arts. 17 & 21).— From examination of sub-fractions from digitonin-treated microsomes, the Louvain group (including Drs. Amar-Costesec and Thinès-Sempoux) suggest that cholesterol is confined to a type of membrane identical with or related to plasma membrane. Our own observations [1, 2] are not in accord with this view or with the possibility of adsorption of cytoplasmic cholesterol on to microsomal membranes.-
(1) Little cholesterol is present in the cytoplasm in starved animals.
(2) Cholesterol was not preferentially removed in comparison with phospho-lipid by washing or by DOC treatment.
(3) Microsomal sub-fractions differed in acetate incorporation into cholesterol.
(4) Microsomal phospholipid and cholesterol increased concomitantly when proliferation of e.r. membranes was induced by phenobarbital treatment.
(5) Plasma membrane, if the sole locus of cholesterol, would seem by calcula-tion to comprise an unrealistically large proportion of the cell.
(6) Sphingomyelin is high in plasma membrane but low in microsomal membranes.
(7) Cholesterol was present in all sub-fractions obtained from smooth [2] or rough microsomes by zone centrifugation.

1. Glaumann, H. & Dallner, G., *J. Lipid Res. 9* (1968) 720.

2. Glaumann, H. & Ericsson, J.L.E., *J. Cell Biol. 47* (1970 555-567.

Remarks by O. Touster concerning 5'-nucleotidase as a p.m. marker (cf. Arts. 21 & 23).- The lysosomal nucleotidase reported by us in 1968 would hardly in-terfere at pH >7, nor would the literature point to trouble from e.r. and Golgi activity. We have found 1.44 u/mg of 5'-nucleotidase activity in our p.m. fraction P_2, and 0.245 u/mg in our Golgi apparatus preparation.

ELEMENTS FROM CELLS AND NON-HEPATIC TISSUES (Articles 25-35):
SUPPLEMENTARY MATERIAL

Comments by C.R. Krishna Murti relating to marker enzymes.- Working with a supposedly tightly bound enzyme, the stroma membrane-bound acid phosphatase of rabbit reticulocytes, Raja Baleu in our laboratories (Lucknow) has made the unexpected finding that the enzyme leaches out on merely keeping the stroma in water in the refrigerator overnight. If the evidence for the pres-ence of lysosomes in erythrocytes is based merely on assays for supposed lysosomal markers, the poor activity of erythrocyte acid phosphatases towards natural substrates calls for caution.

Comment by G. Siebert to G.B. Cline (cf. Art. 28).- Since the ^{67}Ga distribu-tion is essentially the same *in vivo* and in different *in vitro* systems, it would be worth studying the distribution before homogenization, e.g. by the electron probe technique or by anhydrous fractionation procedures.

Answering P. Rowe, G. Sachs (cf. Art. 29) said that alcohol precipitation was the means used to remove the detergent from the proteins. 'Brij' compares well with DOC in effectiveness, and marker enzymes survive (*in reply to* N. Crawford).

Answers to questions on the paper by D.J. Taylor & N. Crawford (Art. 30).- Is there a Mg^{2+} ATPase associated with platelet plasma membranes (G. Sachs)? - Yes [reported by Harris & Crawford in *Biochim. Biophys. Acta* 219 (1973) 720] but we feel that a membrane-associated actomyosin-like protein can account for it, the activity varying with the amount of myosin complexed with the membrane actin. We have not so far been able to unequivocally demonstrate a $Na^+K^+Mg^{2+}$ ATPase which shows consistent ouabain inhibition.
Why use pig rather than bovine platelets (J.R. Harris)? - (a) We believe that analytically and in their functional activity pig platelets more closely resemble human platelets, although they do have a slightly higher 5-HT content. (b) In the Birmingham abattoir, pigs are slaughtered earlier in the morning than cattle, and the important steps in platelet isolation can be completed on the same day.
Difference in pH optimum between the two phosphodiesterase (PDE) activities (T.D. Prospero)? - In platelet homogenates the optimum is 5.5 for *bis*(*p*-nitrophenyl)phosphate PDE and 7.9 for 5'-thymidine-*p*-nitrophenyl phosphate PDE, so that they are quite distinct enzyme activities. Further, 10 mM fluoride inhibits the latter completely and the former only slightly (<10%).

Points arising from paper by P. Tulkens on cultured cells (Art. 31).- Besides streptomycin, rubidomycin (a potential anti-leukaemic agent?) has been tested (*answer to* G. Siebert): it gets into nuclei, and its complex with DNA is pinocytosed into lysosomes.
Answering I. Friedberg, P. Tulkens reckoned that there were two possible explanations of streptomycin entry into lysosomes: there could be membrane transport due to a pH differences between the lysosomes (pH $\sim$5?) and the medium (pH 7), the streptomycin being positively charged, or there could be pinocytotic entry of a streptomycin-albumin complex, insofar as equilibrium dialysis shows that 4 moles of streptomycin bind to 1 mole of albumin.
Comments by C.R. Krishna Murti.- Early work by Bernard Davis in Boston has shown that streptomycin binding to *E. coli* plasma membrane leads to leakage of cytoplasmic constituents such as K^+, nucleotides and amino acids. One wonders if there might be similar leakage effects, possibly leading to cell death, when streptomycin accumulates in fibroblast lysosomes.

D. Lloyd (*answering* D.B. Roodyn; cf. Art. 32) agreed that the observed mitochondrial heterogeneity in yeast could be due to heterogeneity of cells in the culture, because (a) it was exponential when fractionated, and not all the cells may be undergoing glucose derepression at the same rate, and (b) the cells are at different stages of the life cycle.— There is in fact preliminary evidence (R.K. Poole) for age-dependent heterogeneity in *Schizosaccharomyces pombe*. Whilst we cannot rule out mitochondrial damage, it is likely that there is indeed physiological heterogeneity as enzyme heterogeneity has also been seen in rate-separations on shallow sucrose gradients, and also in Ficoll.

Questions to R.K. Poole (Art. 33) by P. Sheeler.— With the 'synchrony by selection' technique, what precautions are taken to avoid contamination by other microorganisms ? *Ans.:* none; the simple defined medium and low pH disfavour bacterial growth, as has been verified. Does the cell band continue to sediment during dynamic unloading in rate runs ? - Only slightly (likewise during loading), with some broadening which is viewable and allowed for.

EFFECT OF FREEZING IN LIQUID NITROGEN ON ENERGETIC PARAMETERS IN ISOLATED CHLOROPLASTS

Karl Kaminski and Betty Rorive

Institut de Botanique, Université de Liège, Belgium

As was previously reported [1], suspensions of tobacco leaf chloroplasts can retain their cyclic and non-cyclic phosphorylation rates without loss for at least 8 months, when frozen in liquid nitrogen. The fine structure of such chloroplasts, as seen with an electron microscope, is the same as in freshly extracted ones. Subsequently it was observed that the preservation can be sustained even for longer times, exceeding one year. It was checked too that the time-and-space-saving devices commonly used for the preservation of blood in liquid nitrogen such as 'straws', 'vidotubes', 'cannisters' and cryobiological dewars [2] can be used without modification for the chloroplast suspensions. Thus a stock of homogeneous chloroplasts can conveniently be kept for the study of phosphorylation.

It was also investigated whether or not such frozen chloroplasts retain their capacity for photosynthetic control (P.C.), i.e. the stimulation of the non-cyclic electron flow by a limited quantity of ADP (state 3) and its subsequent decrease (state 4) after the phosphorylation of the added ADP, as defined earlier [3]. The electron transport in the presence of ferricyanide (Hill reaction) was measured as the O_2 evolution with a Rank O_2 electrode. The actinic red light was saturating. In such conditions ADP had no effect on the Hill reaction of frozen tobacco chloroplasts but did affect spinach chloroplasts. However, a measurable P.C. is found with freshly extracted tobacco chloroplasts only after a second addition of ADP; the first addition of ADP gives no stimulation, but rather an inhibition, a few minutes after the addition. The addition of bovine serum albumin and ascorbate to the extraction mixture [4] restores this situation in frozen tobacco chloroplasts, again with a similar consequence of the first addition of ADP. Altogether, P.C. evidently could be a more sensitive paramater than phosphorylation to test the effects of freezing on energetic parameters in tobacco chloroplasts.

1. Kaminski, K. & Bronchart, R., in *Abstr. 6th Internat. Congr. Photobiology* [Bochum] (Schenck, G.O., ed.) (1972), p. 301.

2. Bouillenne, J.C., André, A., Brocteur, J. & Otto-Servais, M., in *Proc. 10th Congr. Eur. Soc. Haemat.*, Pt. 2, Karger, Basel (1967) pp. 1497-1501.

3. West, K.R. 7 Wiskich, J.T., *Biochem. J.* 109 (1968) 527-532.

4. Hall, D.O., Reeves, S.G. & Baltscheffsky, H., *Biochem. Biophys. Res. Commun. 43* 91971) 359-366.

ISOLATION OF LYSOSOMES FROM *TETRAHYMENA PYRIFORMIS*
USING RATE-ZONAL CENTRIFUGATION

Rosemary A. Cooper   and M. Dobrota
Dept. of Biological & Chemical Sciences   *Wolfson Bioanalytical Centre*
Llandaff College of Technology   *University of Surrey*
Western Ave., Cardiff CF5 2YB, U.K.   *Guildford GU2 5XH, U.K.*

This trial of the isolation of lysosomes illustrates the relative ease with
which eukaryotic microorganisms lend themselves to subcellular fractionation.
Due to rapid growth rates and high cell yields, some microorganisms serve as
a cheap, readily available source of material for fractionation experiments.
Growth under defined conditions provides a reproducible supply of starting
material, although the disadvantage of harvesting organisms from either a batch
or a continuous culture is that individuals in all stages of the cell cycle
will be present. This problem may be overcome by the synchronization of cultures, so that all the organisms divide simultaneously (*see* Art. 34, D. Lloyd
& R.K. Poole). Routine methods have now been established for synchronizing
the yeast *Schizosaccharomyces pombé* [1], and the ciliate *Tetrahymena pyriformis*, the flagellate *Crithidia fasiculata* and the colourless alga *Polytomella
caeca* (personal communications from D. Lloyd, C. Edwards and M. Cantor, respectively). In the present trial, a cell-free homogenate of *Tetrahymena pyriformis* was layered over a shallow gradient in a zonal rotor and subjected to
rate-zonal centrifugation in an attempt to isolate lysosomes on the basis of
their size.

MATERIALS AND METHODS

Growth of the organism. - Culture medium containing 2% (w/v) proteose peptone
(Difco), 0.1% liver digest (Oxoid) and 0.5% silicone MS antifoam RD was steri-
lized at 15 lb/in^2 for 20 min before inoculation with a 24 h starter culture.
A 10 1 culture was incubated overnight at 29° under gentle forced aeration.

Harvesting and preparation of cell-free extracts. - Half of the culture was
harvested by centrifugation at 1,000 *g* for 10 min in a MSE 6L Mistral centrifuge. Cells were washed in 0.2 M K_2HPO_4, then re-suspended in 3-5 vol of
disruption buffer (0.32 M sucrose, 24 mM Tris, 0.5 mM $MgCl_2$, pH 7.2). Organisms were disrupted in a hand-operated homogenizer (Kontes Glass Co.) at 4°.
About 10 complete cycles were sufficient to give 90% breakage. Unbroken
cells and pellicles were removed by centrifugation at 1,000 *g* for 5 min. The
remainder of the culture was harvested for the second zonal run (4 h later).

Fractionation by rate-zonal centrifugation. - An HS rotor (MSE) running in an
MSE '18' centrifuge was used. A gradient of complex shape, consisting of a
linear region between densities 1.04 to 1.14 g/ml, was followed by an exponential section to density 1.19 g/ml [2] and then a cushion of density 1.255.
After loading the sample and overlay, the' rotor was accelerated to the opera-
tional speed (9,000 rev/min). Finally the rotor was decelerated to 1,500 rev/
min (the loading/unloading speed) and unloaded by displacement of the gradient

with dense sucrose (2 M). The effluent was monitored for refractive index and for extinction at either 400 nm (Fig. 1a) or 280 nm (Fig. 1b). Fractions of 20 ml were collected.

Enzyme analysis. - Acid *p*-nitrophenyl phosphatase (EC 3.1.3.2) was assayed by an automated procedure [3].

RESULTS

The profile of light scattering in Fig. 1a suggests that the major portion of subcellular particles have sedimented from the sample zone to the cushion. The main peak of acid *p*-nitrophenylphosphatase corresponds with that of light scattering, although a smaller zone of activity located between fractions 23 and 28 may represent a population of slowly sedimenting lysosomes.

With a shorter centrifugation time (17 min instead of 40 min) as shown in Fig. 1b, the distribution profile of enzyme activity is more heterogeneous. At least three peaks of activity are observed, one at the sample zone, one between fractions 10 and 16, and another in fraction 24.

DISCUSSION

The foregoing initial attempts to isolate lysosomes from *T. pyriformis*, without assaying a range of marker enzymes, were in fact made as a practical demonstration during a Workshop Course on Lysosomes. The results indicate that the separation of *T. pyriformis* lysosomes by rate sedimentation should indeed be possible, and also serve to illustrate how a zonal separation problem can be approached.

Since *Tetrahymena* and mammalian lysosomes appear to be similar in size, we felt that the conditions described by Burge & Hinton [2] for liver lysosomes could be used for the initial attempt to separate *Tetrahymena* lysosomes by rate-sedimentation. However, from Fig. 1a it is obvious that after 40 min at 9,000 rev/min all the acid phosphatase-containing material has been banded hard against the cushion (where mitochondria are normally found, although the appropriate enzyme marker was not assayed). Since the lysosomes evidently sedimented faster than mammalian lysosomes, the centrifugation time in the next experiment was reduced to 17 min. The acid phosphatase distribution (Fig. 1b) now shows a quite different pattern, with three distinct regions all well clear of the cushion. To prove that this connotes three distinct populations of different-sized lysosomes would need further enzymic analysis, demonstration of latency, and electron-microscopic examination of fractions. It is, however, noteworthy that conditions for fractionation of cell-free extracts of *T. pyriformis* by rate-zonal centrifugation were established after only 8 h; moreover, a number of experiments could be performed on a single batch of harvested organisms. Possibly this protozoan may provide an excellent system for studying the mechanism of lysosome development, particularly when grown in synchronous culture.

References

1. Poole, R.K. & Lloyd, D., *Biochem. J. 136* (1973) 195-207.

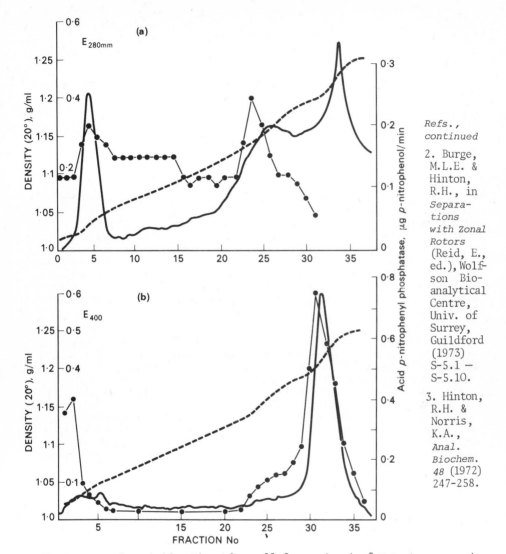

Refs., continued

2. Burge, M.L.E. & Hinton, R.H., in *Separations with Zonal Rotors* (Reid, E., ed.), Wolfson Bioanalytical Centre, Univ. of Surrey, Guildford (1973) S-5.1 – S-5.10.

3. Hinton, R.H. & Norris, K.A., *Anal. Biochem.* 48 (1972) 247-258.

Fig. 1. Rate-zonal centrifugation of a cell-free extract of *Tetrahymena pyriformis* in a sucrose gradient with an HS rotor operating at 9,000 rev/min, for (a) 40 min, (b) 17 min. ———, E_{400}(Fig. 1a) or E_{280} (Fig. 1b); —·—·—, acid *p*-nitrophenylphosphatase; ----, gradient. *Compare patterns, not absolute values.*

Re Art. 35 (Y. Shain; zymogen granules), G. Siebert wondered whether the action of trypsin related to the granule membrane or to the debranching enzyme itself.   C.R. Krishna Murti wondered whether, by analogy with the effect of water imbibition in releasing plant hormones, the observed proteolytic activity would cause such release.

MEMBRANE-ASSOCIATED PROTEINS (Articles 36-38): *SUPPLEMENTARY MATERIAL*

SEPARATION OF MEMBRANE GLYCOPROTEINS BY AFFINITY CHROMATOGRAPHY USING PHYTOHAEMAGGLUTININS

M.J. Crumpton and M.J. Hayman
National Institute for Medical Research
Mill Hill, London NW7 1AA, U.K.

Phytohaemagglutinins or lectins are proteins with an antibody-like specificity for carbohydrate residues [1].   Different phytohaemagglutinins possess different sugar specificities: for example, the phytohaemagglutinins of the jack bean (Concanavlin A, *abbreviated* Con A) and of the lentil (lentil phytohaemagglutinin, LcH) bind α-D-glucopyranosyl, α-D-mannopyranosyl and sterically related sugar residues, whereas soybean haemagglutinin binds D-*N*-acetylgalactosaminyl residues.   Various phytohaemagglutinins agglutinate animal cells and enveloped viruses and, in some cases, stimulate lymphocytes to transform and undergo mitosis.   These phenomena are most probably mediated by cell-surface glycoproteins.

Insolubilized lectins with a broad sugar specificity, such as Con A- and LcH-Sepharose, can be used to selectively purify membrane glycoproteins by affinity chromatography of membrane that had been solubilized in sodium deoxycholate.   Superior yields of glycoproteins were obtained using LcH-Sepharose as compared with Con A-Sepharose.   The application of LcH-Sepharose to the isolation of glycoproteins from the lymphocyte plasma membrane [2, 3] and various enveloped viruses [4] has been reported.   In each case, essentially a complete separation of the glycoproteins from the non-glycosylated membrane proteins was achieved.   The results suggest that the method is of general applicability.   On the other hand, it appears that membrane glycoproteins can be fractionated by using different phytohaemagglutinins with narrow sugar-specificities, such as soybean agglutinin.

1.   Sharon, N. & Lis, H., *Science 177* (1972) 949-959.

2.   Allan, D., Auger, J. & Crumpton, M.J., *Nature New Biol. 236* (1972) 23-25.

3.   Hayman, M.J. & Crumpton, M.J., *Biochem. Biophys. Res. Commun. 47* (1972) 923-930.

4.   Hayman, M.J., Skehel, J.J. & Crumpton, M.J., *FEBS Lett. 29* (1973) 185-188.

Answering P. Stahl, M.J.Crumpton mentioned evidence *(in press)* suggesting that Con A-Sepharose does specifically retain glycoprotein lysosomal enzymes.

INDEX of Subjects

The emphasis in the Index is on techniques and on types of bioconstituent examined, these being usually hepatic if not otherwise specified. There is no systematic indexing of enzymes examined or of conventional procedures applied, nor of contaminating elements or side-products. The indexing strategy was to facilitate any search for information, aided where appropriate by the Contents list (p. vi) or the Author list (p. x). The Index headings are largely similar to those in Vols. 1-3, where complementary information may be found.

Page entries such as 149- *signify that ensuing pages are also relevant, i.e. the* - *denotes a major entry.*

‡‡‡‡‡‡‡‡‡‡‡‡‡‡‡‡‡‡‡‡‡‡‡

CORRECTIONS TO PREVIOUS VOLUMES

Vol. 1: *see p. 273 of Vol. 3*

Vol. 2:
p. 6:   *Order of authors (as in text) should be* - O.P. SAMARINA...& G.P. GEORGIEV
p. 39:   Hjerten *should read* Hjertén *(as in* Contents list*)*
p. 124: adsorbent 4 *column heading in Table 2 should read* adsorbent 2
p. 216: *In line 4 of para. 2, the ref.* [1] *should read* [2].
p. 220: *Index entry* Zonal rotor separations: -63 *should read* 163-.

Vol. 3:
p. 175: *The concentration given halfway down the p. for Triton-X-100* (0.01%)
         should read 0.1%.
p. 212: *In axis legend to Figs. 5-8,* AT pase *should read* ATPase.
p. 268: *In line 1 of 1st para.,* diffe- *should read* different